Particles on Surfaces 2

Detection, Adhesion, and Removal

Particles on Surfaces 2

Detection, Adhesion, and Removal

Edited by
K. L. Mittal

IBM US Technical Education
Thornwood, New York

PLENUM PRESS • NEW YORK AND LONDON

Library of Congress Cataloging in Publication Data

Symposium on Particules on Surfaces: Detection, Adhesion, and Removal (1988:
  Santa Clara, Calif.)
  Particles on surfaces 2: detection, adhesion, and removal / edited by K.L. Mittal.
     p.   cm.
  "Proceedings of the Second Symposium on Particles on Surfaces: Detection, Ad-
hesion, and Removal, held in conjunction with the nineteenth annual meeting of the
Fine Particle Society, held July 18-22, 1988, in Santa Clara, California"—T.p. verso.
  Includes bibliographical references.
  ISBN-13:978-1-4612-7852-8   e-ISBN-13:978-1-4613-0531-6
  DOI: 10.1007/978-1-4613-0531-6

  1. Particles—Congresses. 2. Surfaces (Technology)—Congresses. I. Mittal, K. L.,
1945-   II. Title.
TA418.78.S96   1988   89-22943
620.'.44—dc20   CIP

Proceedings of the Second Symposium on Particles on Surfaces: Detection, Adhesion,
and Removal, held in conjunction with the Nineteenth Annual
Meeting of the Fine Particle Society, held July 18-22, 1988,
in Santa Clara, California
© 1989 Plenum Press, New York
Softcover reprint of the hardcover 1st edition 1989

A Division of Plenum Publishing Corporation
233 Spring Street, New York, N.Y. 10013

PREFACE

This volume documents the proceedings of the Second Symposium on Particles on Surfaces: Detection, Adhesion and Removal held as part of the 19th Annual Meeting of the Fine Particle Society in Santa Clara, California, July 20-25, 1988. The premier symposium on this topic was organized in 1986 and has been properly chronicled[1]. Based on the success of these two events and the high interest evinced by the technical community, we plan to regularly hold symposia on this topic on a biennial basis and the next one is slated for August 20-24, 1990 in San Diego, California.

As pointed out in the Preface to the first volume[1], the topic of particles on surfaces is of paramount importance in legion of technological areas. Particularly in the semiconductor device fabrication area, all signals indicate that the understanding of the behavior of particles on surfaces and their removal will attain heightened importance in the times to come. As the device dimensions are shrinking at an accelerated pace, so the benign particles of today will become the killer defects in the not too distant future. The tempo of research and development activity in the field of particles on surfaces is very high, and better and novel ways are continuously being devised to remove smaller and smaller particles. This Second Symposium had the same general objectives as the premier one and the intent here was to provide an updated account of activity and information in the realm of particles on surfaces. The program contained a total of 43 papers covering a variety of subtopics dealing with the various aspects of particles on surfaces. The interest level in the symposium was very high (as judged by the attendance) and there were brisk and illuminating discussions, both formally and informally, throughout the duration of the symposium.

This proceedings volume contains a total of 25 papers. It must be recorded here that all manuscripts were rigorously peer reviewed and most of them were substantially revised before inclusion in this book. In other words, this book is not simply a compilation of a bunch of unreviewed papers, but represents the collective wisdom and thinking of a large number of active practitioners which has passed the peer scrutiny.

The book is divided into four parts as follows: Part I. Particle-Substrate Interaction, Deposition and Adhesion; Part II. Particle Detection, Analysis and Characterization; Part III. Particle Prevention and Implications; and Part IV. Particle Removal. The topics covered include: factors affecting adhesion of small particles to surfaces; adhesion induced deformation between particles and substrates; detachable submicrometer particles from cleanroom garments; particle identification and characterization on surfaces using various techniques including infra-red, Raman and Auger electron spectroscopies; coatings to reduce particulate contamination; implications of particulate contamination in thin film growth and E-beam lithography; and various techniques (ultrasonic, megasonic, hydrodynamic, use of fluorocarbon surfactant solution and water) for removing particles from surfaces.

Yours truly sincerely hopes that the present and its predecessor volume[1] will serve as a useful source of contemporary R&D activity and trends dealing with particles on surfaces.

Acknowledgements: This section is always the fun part of writing the preface. First, I am thankful to the Fine Particle Society for sponsoring this event. My thanks are due to the appropriate management of IBM Corporation for allowing me to organize this Symposium and to edit this book. I would like to acknowledge the support and help of my wife, Usha, during the tenure of editing this Volume. Special thanks are extended to those behind the scenes (reviewers) who devoted their precious time in providing valuable comments. I would be remiss if I fail to acknowledge the co-operation, enthusiasm and contribution of the authors.

K.L. Mittal
IBM U.S. Technical Education
500 Columbus Ave.
Thornwood, NY 10594

1. K.L. Mittal, editor, Particles on Surfaces 1: Detection, Adhesion and Removal, Plenum Press, New York, 1988.

CONTENTS

PART III. PARTICLE PREVENTION AND IMPLICATIONS

PART IV. PARTICLE REMOVAL

PART I.   PARTICLE-SUBSTRATE INTERACTION,
           DEPOSITION AND ADHESION

THE ADHESION OF SMALL PARTICLES TO A SURFACE

J. Reed

Central Electricity Generating Board,
Berkeley Nuclear Laboratories,
Berkeley, Gloucestershire GL13 9PB,U.K.

In this paper the role of the force of adhesion in processes that govern deposition is discussed. The movement or transport of particles in gaseous systems is controlled by three basic processes:

* arrival to a surface;
* whether the particles bounce or stick upon impact; and
* whether the particles are resuspended.

In this paper the aim is to concentrate on particle and surface interactions. Developments are discussed in the behaviour of adhesion upon impact and the resuspension of small particles by a turbulent flow.

This is followed by a series of experiments designed to measure the force of adhesion of particles onto a surface. The forces of adhesion for 10um particles were found using a centrifuge to apply centripetal forces. It was necessary to take account of the size distribution of the particles in determining the true forces.

It was found that the particles were more strongly adhered to a copper substrate compared to that of steel. Various reasons for this are discussed, but one strong possibility is the increased contact area due to plastic deformation.

INTRODUCTION

The interaction of particles with a surface is of interest to a diverse range of topics relating to powder transport, air pollution and surface contamination. The suspension or resuspension of particles exposed to a moving fluid is a common occurrence in nature. Familiar as the origin of dust storms, it is influential in processes such as the erosion of soil and river beds as well as the 'wave of darkening' observed during the Martian springtime[1].

The movement or transport of particles in gaseous systems is controlled by three basic processes:

*   arrival to a surface;
*   whether the particles bounce or stick upon impact; and
*   whether the particles are resuspended.

In this paper the aim is to concentrate on particle and surface interactions. Therefore only the second two processes are considered in detail here and attention is drawn to a review of particle arrival rate by Hidy and Heisler[2].

The impact of particles with a flat surface has received considerable attention for both elastic and plastic impacts especially since the introduction of hardness tests[3],[4]. The adhesion of small particles to a surface was also well studied[5],[6]. However, theories that study the impact of particles with a surface and their subsequent adhesion or rebound are scarce.

Dahneke developed a theory to study aerosol particles impacting on a surface[7]. The possibility of plastic deformation in these experiments was discarded because it was felt that the impact velocities were insufficient to cause plastic deformation. It has been shown[8] that plastic deformation did occur for those particles and surfaces and therefore this limits the usefulness of Dahneke's model.

A second theory included the capacity for plastic deformation[9]. This theory suffers from a limitation of assuming that the surface is rigid and therefore cannot store elastic energy during the impact. Additionally, the energy of adhesion uses the projected area of contact and is not applicable to the general case of elastic-plastic impacts.

Described below is an impact-adhesion model that overcomes some of the above restrictions. Highlighted is the importance of energy of adhesion in determining the energy losses during the impact.

Once a particle has deposited onto a surface, there exists the possibility that the particle is resuspended. Resuspension is associated with the aerodynamic detachment of particles less than $100\mu m$ for which the principal force holding them onto the surface is the intermolecular force of adhesion.

In classic studies on aerodynamic removal[10],[11] it was observed that resuspension was initiated by flows whose velocity exceeded some threshold value. More specifically this threshold value decreases with increasing particle size. It has become traditional to explain this trend in behaviour in terms of a balance of the aerodynamic and adhesion forces; both forces increasing with particle size but the aerodynamic force has a greater size dependence[12].

However, this force balance takes no account of the timescales over which the particles are resuspended. For instance many measurements of resuspension in the environment indicate that removal is not instantaneous but persists for a period of time[13]. Several authors[14],[15] have suggested that this behaviour is evidence of a statistical origin of resuspension intimately associated with the random motion of the fluid close to the surface.

Several models for resuspension[16] all contain the implicit assumption that the resuspension rate is controlled by the frequency of the occurrence of aerodynamic lift force which exceeds the force of adhesion: in essence a force balance model.

Below a new model for resuspension is outlined[17] where it is recognised that the influence of the transfer of energy from the fluid to the particle is required to overcome the potential energy barrier formed by the force of adhesion. This energy balance approach is statistical in nature and is successful in explaining many features observed in resuspension.

THE IMPACT-ADHESION MODEL

The model presented here is based upon consideration of the energy balance during a normal impact between a sphere and a massive, plane body. It is only applicable where one body can be considered infinite.

If a sphere has an initial kinetic energy Q_I (the plane surface being stationary) then as it approaches the surface it will gain energy as a result of the attractive forces between the bodies. The total initial energy involved in the impact is then $(Q_I + Q_A)$ where Q_A is the energy due to the attractive forces. When the particle velocity is reduced to zero, part of this initial energy is converted into stored elastic energy, Q_E, while the remainder, Q_L, is dissipated. This is then followed by the recovery of the stored elastic energy which is converted into kinetic energy of the particle.

For a particle, with mass m, moving at velocity V we have

$$\frac{1}{2} mV^2 + Q_A = Q_E + Q_L \tag{1}$$

If the stored elastic energy, Q_E, is greater than the adhesion energy required to separate the particle from the surface, Q_A', then the particle will rebound. (Strictly speaking, the equation of motion for the particle ought to be solved. However, in this case, equation (2) below is sufficient.[18]

If the stored elastic energy, Q_E, is less than Q_A', then the particle will remain adhered to the surface. We define a critical impact velocity V_c when $Q_E = Q_A'$. Impact velocities above this value will result in the particle bouncing upon impact. Impact velocities below this value will result in particles adhering upon impact.

Both Q_E and Q_L will depend upon m, V, and Q_A. As shown later Q_A and Q_A' are functions of particle radius, R, mass, elastic constants of both the particle and the surface, and adhesion energy per unit area, γ. If the impact involves purely elastic deformations then Q_A and Q_A' are independent of velocity. However, if there is plastic deformation then Q_A and Q_A' will also depend upon the impact velocity and the elastic yield limit, y.

The rebound velocity is given by:

$$u = \left(\frac{2}{m} [Q_E - Q_A'] \right)^{1/2} \tag{2}$$

5

Energy Losses, Q_L

This section is divided into two parts. First, elastic impacts are considered. It is assumed that the only energy loss is due to elastic wave propagation. In the second part, elastic-plastic impacts are considered. It is assumed that the only energy loss mechanisms are elastic wave propagation and plastic deformation.

Any impact between a sphere and a flat surface will involve only elastic deformation as long as the maximum pressure between the bodies never reaches the value of the elastic yield limit, y, of the softer of the two bodies. For a sphere impacting onto a flat surface, it can readily be shown[19] that the elastic yield limit is only attained for impact velocities greater than a limiting velocity ϕ where

$$\phi = \left(\frac{2\pi}{3K}\right)^2 \left(\frac{2}{5\rho_1}\right)^{1/2} y^{5/2} \tag{3}$$

and

$$K = \frac{4}{3}\left[\frac{1-\nu_1{}^2}{E_1} + \frac{1-\nu_2{}^2}{E}\right]^{-1}$$

E_i, ν_i are Young's modulus and Poisson's ratio, respectively ($i=1$ particle, $i=2$ surface) and ρ_1 is the density of the particle. (In the present model it is assumed that only one of the bodies deforms plastically during the impact.)

If V is the velocity of approach of a particle beyond the range of the surface forces then an effective impact velocity including the effects due to surface forces, V*, can be defined by

$$V* = \left[V^2 + \frac{2Q_A}{m}\right]^{1/2} \tag{4}$$

where Q_A is evaluated for a sphere on a flat surface.

The impact will be elastic if

$$V* \leqslant \phi$$

which can also be expressed using equation (4) as

$$V \leqslant \left[\phi^2 - \frac{2Q_A}{m}\right]^{1/2}. \tag{5}$$

Clearly, if $\frac{2Q_A}{m} > \phi^2$ then the impact will involve plastic deformation.

Elastic Impacts. As already mentioned, the only energy loss mechanism considered for elastic impacts will be the propagation of elastic waves. In an impact between a small particle and a massive substrate the elastic energy stored in the bodies can be released before the elastic waves propagating in the massive body have returned to the impact zone. (We assume all the stored elastic energy is available for rebound.) This means the energy contained in these waves is unavailable to contribute to the kinetic energy of the particle and is, therefore, to be considered lost.

A model has already been presented to calculate the energy loss due to elastic waves propagating away from the impact zone during an elastic impact between a particle and a massive substrate[20].

The effects due to the adhesion forces were not considered in that analysis (i.e., the Hertz equations were used). I do not propose to re-analyse the problem but shall assume that it is sufficient to replace the impact velocity, V, with an effective impact velocity, V*, where it is assumed that the particle gains all the energy Q_A on approaching the surface (equation 4).

The fractional loss of initial energy, λ_W, is [20] as

$$\lambda_W (V*) = \Lambda_W V*^{3/5} \tag{6}$$

where Λ_W is a constant given by

$$\Lambda_W = \frac{6.58}{\rho_2 C_0^3} \quad \rho_1^{-1/5} K^{6/5} \tag{7}$$

with $C_0 = \left(\frac{E_2}{\rho_2}\right)^{1/2}$, and ρ_2 is the density of the substrate.

(Note: Strictly speaking, Λ_W is a function of the Poisson's ratio of

the surface. However, as shown by Hutchings[21] the variation within the typical range of Poisson's ratio of 1/4 to 1/3 is less than 1%. Therefore, it is assumed that Λ_W is a constant.)

The energy loss is, therefore, given by

$$Q_L = (Q_I + Q_A) \lambda_W(V*). \tag{8}$$

The critical impact velocity is given using equations 1, 2, 4, 7 and 8 when the rebound velocity, u, equals zero (equation 2), as

$$v_C = \left[\Lambda_W \left(v_C^2 + \frac{2Q_A}{m} \right)^{13/10} - \frac{2}{m} (Q_A - Q_A') \right]^{1/2} \tag{9}$$

which does not have an explicit solution.

Elastic-Plastic Impacts. A model for the adhesion upon impact of particles where the energy loss is due to plastic deformation alone has been presented elsewhere by Rogers and Reed[22]. The following two assumptions in this model, i.e.,

(i)   that the only significant energy loss mechanism is plastic deformation;
(ii)   that the energy gained by the particle due to the adhesion potential is negligible;

will be removed here. This means that, in conjunction with the model for elastic impacts, there now exists a consistent model with no restrictions upon impact velocity (except if the impact velocity $\gg \phi$, the impact equations are incorrect).

For elastic plastic impacts

$$Q_L = Q_p + Q_W \tag{10}$$

where Q_p is the energy lost in plastic deformation and Q_W is the energy loss due to elastic waves propagating from the impact zone in elastic-plastic impacts.

Using equation (10), equation (1) can be rewritten as

$$Q_I + Q_A(Q_I) = Q_E(Q_I,Q_A) + Q_p(Q_I,Q_A) + Q_W(Q_I,Q_A) \qquad (11)$$

From Rogers and Reed

$$Q_E = Q_e + Q_{pe} \qquad (12)$$

where Q_e is the elastic energy stored in the elastic region and Q_{pe} is the elastic energy stored in the plastic region with

$$Q_p = \frac{4}{15} \frac{Q_{pe}^2}{Q_e} \qquad (13)$$

$$Q_e = \frac{1}{2} m \phi^2. \qquad (14)$$

Also the radius of elastic deformation is

$$r_e = \left(\frac{5}{4} m \phi^2 \frac{R^2}{K} \right)^{1/5}, \qquad (15)$$

with the radius of plastic deformation

$$r_p = \left(\frac{2Q_{pe}}{\pi R y} \right)^{1/2} \left(\frac{3K}{2\pi y} \right). \qquad (16)$$

(These two quantities will be used to calculate the adhesion energy.)

The energy lost due to the elastic waves propagating away from the impact zone for an elastic-plastic impact has been evaluated[23]. We have

$$Q_W = 0.05291 \, \Lambda_W \phi^{3/5} \frac{m}{2} \left(v^2 + \frac{2Q_A}{m} \right) \frac{I(v^+)}{v^{+2}} \qquad (17)$$

where
$$v^+ = \frac{\left(v^2 + \frac{2Q_A}{m} \right)^{1/2}}{\phi} = \frac{v^*}{\phi}$$

and to a good approximation

$$\frac{I(v^+)}{v^{+2}} \approx 18.9 + 0.0439 \, (v^+-1) + 3.21.10^{-5}(v^+-1)^2$$
$$+ 4.07.10^{-8} \, (v^+-1)^3 + 2.10^{-10}(v^+-1)^4$$

Using equations (12-14) and (17), equation (11) can be rewritten as

$$\frac{1}{2} mv^2 + Q_A(Q_I) = Q_e + Q_{pe}(v^*) + \frac{4}{15} \frac{Q_{pe}(v^*)^2}{Q_e}$$
$$+ 0.05291 \, \Lambda_W \, \phi^{3/5} \left(\frac{mv^2}{2} + Q_A(Q_I) \right) \frac{I(v^+)}{v^{+2}} \qquad (18)$$

The critical impact-adhesion velocity can be obtained numerically using equations (2), (12-14), and (18) for u=0.

Adhesion Energy, Q_A and Q_A'

The equations for adhesion energy are similar to those given by Rogers and Reed[22]. They used a theoretical description of the adhesion forces acting between elastically deformable bodies presented by Johnson et al[24]. The total adhesion energy in the rebound, Q_A', is

8

shown to be the sum of the mechanical energy, Q_m, and the surface energy, Q_s, where

$$Q_A' = Q_m + Q_s \tag{19}$$

$$Q_m = \frac{P_0(P_1^{2/3} + 2P_0P_1^{-1/3})}{3K^{2/3}R_c^{1/3}} \tag{20}$$

and

$$Q_s = \pi\Delta\gamma\left(\frac{R_cP_1}{K}\right)^{2/3} \tag{21}$$

where P_0 is the external force (e.g., mass x acceleration due to gravity if the particle is resting on a horizontal surface; in practice for the cases considered here $P_0 \ll 3\pi\Delta\gamma R_c$) and $P_1 = P_0 + 3\pi\Delta\gamma R_c + (6\pi\Delta\gamma R_c P_0 + (3\pi\Delta\gamma R_c)^2)^{1/2}$. R_c is the contact geometry parameter and $\Delta\gamma$ is the adhesion energy per unit area. In the contact of two bodies of local radii of curvature R_1 and R_2,

$$R_c = \frac{R_1R_2}{(R_1 + R_2)} .$$

For an elastic impact between a sphere, radius R, and a flat surface the value of R_c is given by

$$R_c = R. \tag{22}$$

However, the contact geometry parameter, R_c, will alter when there has been plastic deformation[22].

Since it is assumed that the rebound process involves only elastic deformations the Johnson, Kendall and Roberts adhesion model can be used to give the adhesion energy, providing the relevant contact geometry parameter, R_c, is used. The contact geometry parameter, R_c, is given by[22]

$$R_c = K\left[\frac{(r_e^2 + r_p^2)^{3/2}}{\pi y r_p^2 + KR^{-1}r_e^3}\right] \tag{23}$$

where r_p and r_e are the projected radii of plastic and elastic deformation and y is the elastic yield limit of the softer body.

Let us turn attention to the difference between the energy of adhesion upon approach, Q_A, and the energy of adhesion upon separation, Q_A'. This can be illustrated if one considers the force, P_0, as a function of displacement, α.

The force-displacement curve is shown in Figure 1[25]. What one should realize here is that at the point of separation ((d) in figure 1) the surfaces have stretched slightly such that the displacement is negative.

During an impact, the particle and surface come into initial contact when $\alpha = 0$ ((a) in figure 1). The contact area then grows by virtue of the adhesion forces with α remaining zero ((a) to (b)). Subsequently the interaction follows the force-displacement up to (c), where the kinetic energy has been converted into stored elastic energy. During the recovery of the stored elastic energy, the force-displacement curve is followed exactly (c)-(b)-(d) up to separation.

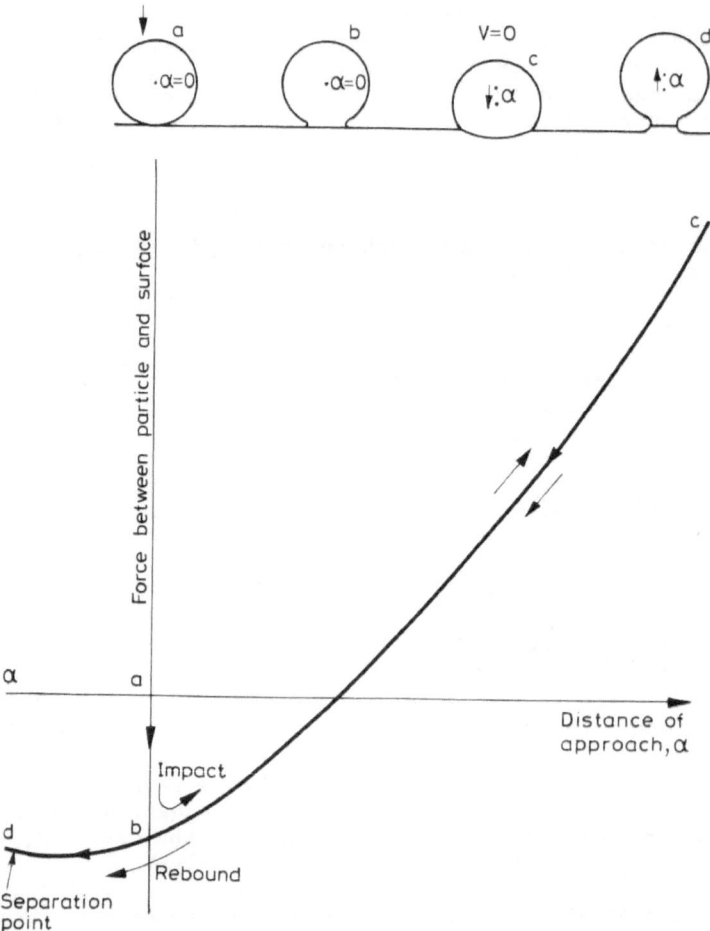

Figure 1. Schematic Representation of the Impact of a Particle
Including the Effects of Adhesion (from Ref 25)

Consequently, the path taken upon approach is not the same as that upon recovery. This is the reason for the difference between Q_A and Q_A'.

We have

$$Q_A' - Q_A = \int_0^{\alpha_c} P_0 \, d\alpha$$

Using Johnson, Kendall and Roberts[24], it can be shown that

$$Q_A = Q_A' - 0.629 \, (\pi\Delta\gamma)^{5/3} K^{-2/3} R_c^{4/3} \tag{24}$$

PARTICLE RESUSPENSION

Recently, a model for the resuspension of small particles by a turbulent flow was developed[26]. This model is based upon the view that an adhering particle and substrate are deformed elastically by their surface adhesion forces. In static equilibrium (no external forces or flow) there is a balance between the adhesion force and the elastic restoring force. However, when a particle is exposed to a turbulent flow, there is a continuous transfer of turbulent energy to the particle which causes the particle and surface to deform continuously about their static equilibrium configuration, continually changing the area of contact. Particles in contact with the surface are confined to motion within a surface adhesion potential well derived from the dependence of the adhesion and elastic restoring forces upon the area of contact (i.e. no rolling or slip).

In this approach, a particle is released from a surface (resuspended) when it receives enough energy to escape from the adhesion potential well. The transfer of energy takes place through the agency of aerodynamic lift forces whose:

(i)   average component modifies the shape and height of the adhesion potential well,
(ii) random fluctuating component causes the particle and surface to deform in a random oscillatory fashion about their static equilibrium deformation.

The motion is represented by that of a very stiff lightly damped harmonic oscillator with a forcing term provided by the random fluctuations in the aerodynamic lift force.

The probability per unit time of a particle being resuspended from a surface (resuspension rate constant), ρ is of the form

$$\rho = \frac{\omega_0}{2\pi} \exp \frac{-Q}{2\langle PE \rangle} \tag{25}$$

where Q is the height of the potential well,
$\langle PE \rangle$ is the average potential energy of a particle in the well, and ω_0 is the typical frequency of particle motion within well.

Significantly, this equation is similar in form to that for the desorption rate constant of molecules from a surface.

Consideration of the resuspension rate of particles from a surface where there is a spread of adhesion forces (due to surface roughness) shows that resuspension can be subdivided into two regimes:

(1)   Initial resuspension (occurring in times typically less than a second) in which a large fraction of the particles that eventually resuspend actually do resuspend. This effectively reduces the net deposition rate to a surface.
(2)   Longer term resuspension (resuspension occurring after a second) which controls the asymptotic decay of gas-borne concentration in a recirculating flow. The fractional resuspension rate, $\Lambda(t)$,

varied almost inversely with the time, t, of exposure for $1<5<10^5$ s., i.e.

$$\Lambda(t) = \xi t^{-\varepsilon}, \tag{26}$$

where $\varepsilon \simeq 1$. Although the resuspension constant, ξ, is influenced by particle size, surface roughness, and local flow, the inverse time relationship for Λ remains intact. In fact, it is extremely robust to a wide variation in flow and surface roughness. Such behaviour has been confirmed by experiment[27].

FORCES OF ADHESION

The adhesion between a particle and a surface depends upon three main factors:

i)   surface forces, i.e., the type of bond formed at an inter-
       face;
ii)   mechanical properties, which determine the mode of contact
       and separation, e.g., plastic deformation;
iii)   surface roughness, which determines the area of contact over
       which the adhesion occurs.

In most applications, the only type of surface force will be due to van der Waals interactions. Similarly, only in exceptional circum-stances will the mode of separation not be brittle. Therefore, the only two major facets that we can investigate are the effects of sur-face roughness on adhesion together with the plastic deformation upon contact which will alter the surface roughness.

The classical analysis of the contact between a sphere and a flat surface was first presented by Hertz in 1881. However, it had been noted (Derjaguin[28], Johnson[29]) that the adhesion forces pull the two bodies together thus increasing the contact area over that given by Hertz. Two distinct methods of treating the elastic adhesion developed. On one hand Derjaguin, Muller and Toporov (DMT)[30] suggested that the adhesional interaction would come from outside the contact region, while (JKR)[24] showed that within the contact region the surface forces modify the stress field. We will not discuss the two models in detail (see Tabor[31], for an excellent comparison of the two models) but the end result from both is that the force of adhesion is given by

$$F_a = b \, \pi \, \Delta\gamma \, R_c \tag{27}$$

where $\Delta\gamma$ is the adhesion energy per unit area, and R_c is the contact geometry parameter. The constant b is 2 from the DMT theory and 3/2 from the JKR theory. The predicted force of adhesion differs slightly in magnitude between the two theories, but it retains a similar form in both. Both these theories were developed for the contact of perfectly smooth bodies. This idealised situation can only be achieved in special circumstances, and in general all surfaces will have some measure of roughness associated with their profile, whether it be on a macroscopic or microscopic scale.

When real surfaces are pressed together, they touch at a large number of high spots, asperities, which deform elastically or plasti-cally forming micro-contact areas. The sum of the micro-contact areas is ordinarily a small fraction of the nominal or apparent area over which the two bodies are likely to be in contact. Due to the random

12

nature of the surface roughness, not all the particles will have the same contact area; consequently, there will be a distribution of adhesion forces. It is this spread that interests us here.

CENTRIFUGE ANALYSIS

An ultracentrifuge was used to determine, accurately, the removal forces that were applied to particles which had been deposited onto a surface. In many experiments where the adhesion is measured, improper allowance is made for the effect of particle size. Most removal forces are a function of particle size (e.g., blowing off) but not the same function of size as the adhesion. For example in the centrifuge experiments here, the removal force at any given speed is proportional to the cube of the radius, while the adhesion varies as the radius itself. Consequently, larger particles are subject to relatively larger forces than smaller forces. Therefore, in order to obtain the spread of adhesion forces, we need to subtract the effect of the size distribution from the results.

Let us assume that the force is reduced by a constant ψ, due to surface roughness, i.e.,

$$F_a = \frac{3}{2} \pi \Delta\gamma R\psi \tag{28}$$

In the centrifuge the removal force is

$$F_c = \frac{4}{3} \pi R^3 \rho L \omega_c^2 \tag{29}$$

where L is the distance of the surface from the centre of rotation, ρ the density of the particle, and ω_c is the angular velocity of the centrifuge rotor ($\omega_c = s_c \, 60/2\pi$ where s_c is the centrifuge speed in rpm).

Using these two equations we can define a critical size above which particles will be removed

$$R_s = \left(\frac{9 \, \Delta\gamma\psi}{8 \, L \, \rho \, \omega_c^2} \right)^{1/2} \tag{30}$$

and, therefore, assuming a distribution of adhesion reduction factor, the fraction of particles removed at any spin speed is given by

$$P(\omega_c) = \int_0^\infty \Phi(\psi) \int_{R_s(\omega_c, \psi)}^\infty \phi(R) \, dR \, d\psi \tag{31}$$

$\phi(R)$ is the size distribution.

Experimental Measurement

The centrifuge used in these experiments was a Measuring and Scientific Equipment (MSE) Superspeed 75 centrifuge with a titanium angle rotor capable of attaining accelerations of up to 4×10^5 g when used with the specimen holders described below. The titanium angle rotor is shown in figure 2 with the specimen holder in place. The specimen holder is shown in greater detail in the insert in figure 2.

The surfaces used in the experiment were a stainless steel (grade B) and copper of micro Vickers hardness 24.3×10^8 and 3.11×10^8 Nm-2 respectively. The surfaces were polished using standard techniques to give centre line average (c.l.a.) of less than 0.08 µm. The surfaces were cleaned by washing in acetone followed by a distilled water rinse.

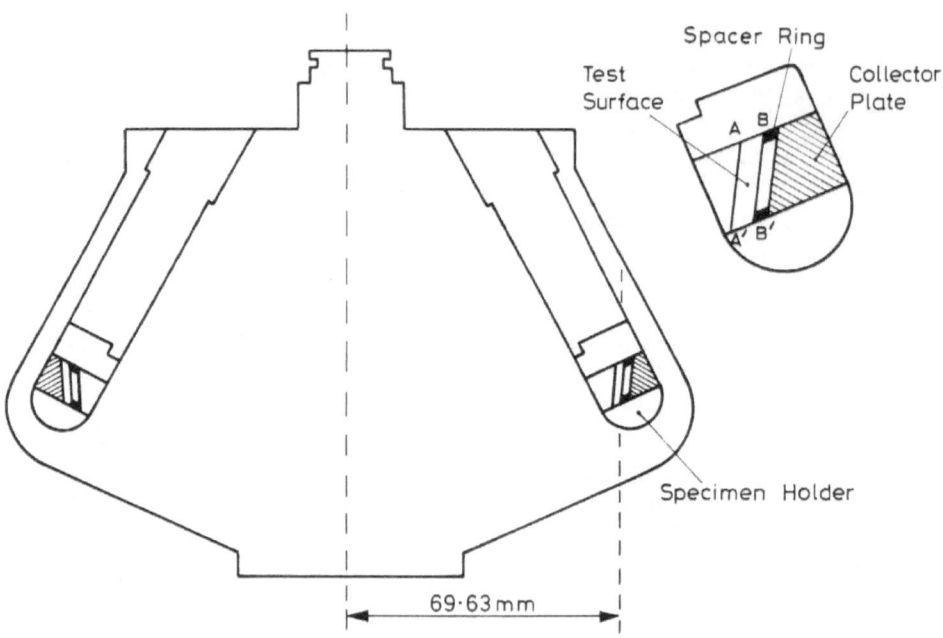

Figure 2.   Titanium Angle Rotor with Specimen Holder

The particles used in the experiment were Aluminium oxide Spherisorb A10Y. The particle size distribution was determined using a microscope and examining about 1000 particles (see figure 3). The particles were allowed to settle onto a surface after being tapped from a fine brush held 5 cm from the surface. The same technique was used in the size analysis and the adhesion experiment, and therefore it is assumed that this would not introduce any inconsistencies.

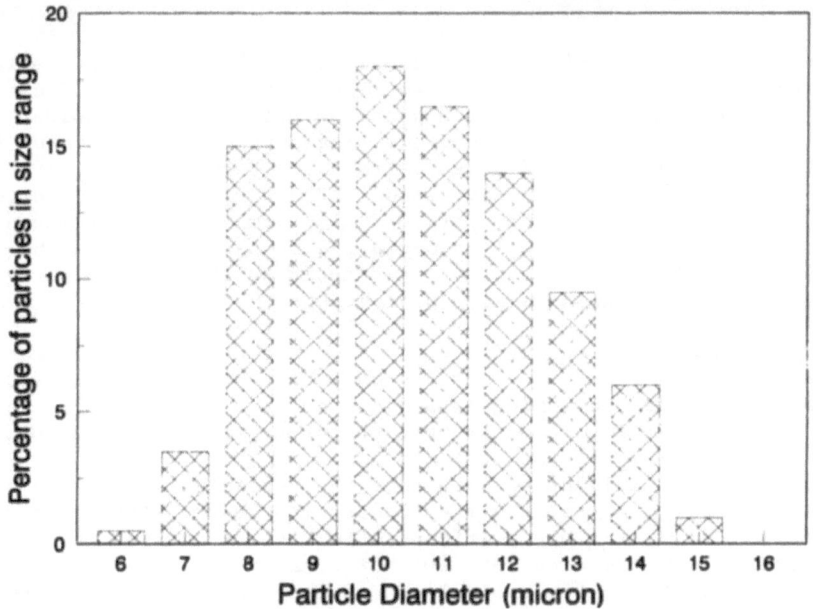

Figure 3. Particle Size Distribution of the Particles used in the
Centrifuge Experiments

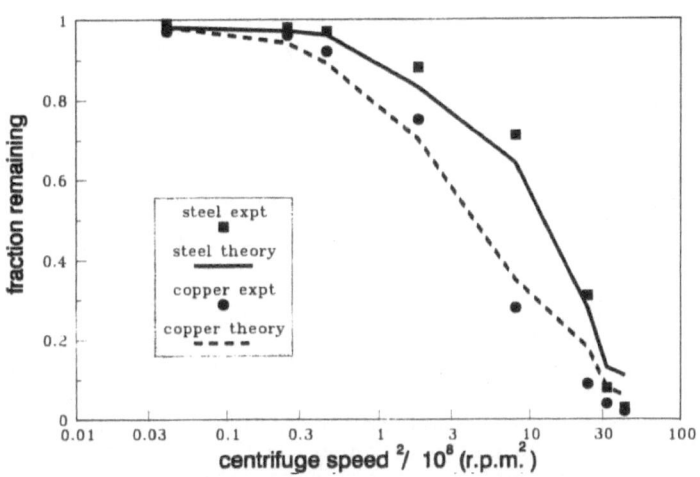

Figure 4. Fraction of Particles Removed as a Function of Centrifuge
Speed for both the Copper and Steel Substrates with the
~10 μm Aloxite Particles

When the particles were first deposited on the centrifuge surface, a photograph was taken of the whole surface using a high magnification close-up 35 mm camera. The surface was then put inside the holder which, in turn, was put into the rotor. The surfaces were rotated in the centrifuge at incrementally increased rotor speeds (5,000, 10,000, 20,000 rpm, etc) until the maximum speed (75,000 rpm) was reached (each desired speed being held steady for 10 min). After each incremental speed, the surface was carefully removed and re-photographed. The number of particles removed at any speed could be then easily determined by comparing successive photographs.

In order to reduce uncertainty, and check reproducibility, more than one test surface were used at the same time. The results of the fraction of particles remaining as a function of the centrifuge speed squared are shown in figure 4 for both the steel and copper surfaces.

It is very significant to note that the particles seem to be held on to the copper surface more strongly compared to the steel surface. We would not expect such a significant effect to be due to different surface energy per unit area, $\Delta\gamma$, for these materials. What is more likely is the lower elastic yield limit of the copper ensures a greater contact area because significantly larger asperities are deformed in the copper.[22] We do not have room for a detailed analysis here.

We have already said that our interest is in the spread in adhesion forces. We are now in a position to use equation 31 to eliminate the effect of particle size in the data presented in figure 4.

There are several functions that can be chosen for Φ. However, it seems sensible to take a log-normal distribution.[23] Therefore

$$\Phi(\psi) = \frac{1}{2\pi} \frac{1}{\psi \ln \sigma} \exp\left(-\frac{\{\ln (\psi/\mu\}^2}{2 \ln(\sigma^2)}\right) \qquad (32)$$

where μ is the mean adhesion coefficient and σ the spread in adhesion

Table 1. Fitted values for log normal distribution

	STEEL SURFACE	COPPER SURFACE
μ	31 ± 4	19 ± 3
σ	18 ± 4	12 ± 2

force and the fitted values are shown in figure 4. Any other choice of function for the distribution of adhesion forces Φ does not result in an acceptable fit to the theoretical data.

DISCUSSION AND CONCLUSIONS

The role of adhesion in deposition and resuspension has been high-lighted. It has been shown that the spread of adhesion forces on particles on a surface can be explained in terms of a log-normal distribution. The spread and reduction in adhesion forces on a copper surface is less than that for a steel surface. This is consistent with particles having a greater area of contact on the copper surface due to its lower elastic yield resulting in greater deformation of the asperities.

ACKNOWLEDGEMENT

This work was carried out at the Berkeley Nuclear Laboratories of the Technology Planning and Research Division and the paper is published with the permission of the Central Electricity Generating Board.

REFERENCES

1. C Sagan and J B Pollack, Nature, 223, 791-4 (1969)
2. G M Hidy and S L Heisler, in "Recent Developments in Aerosol Sci." Ch 7, Wiley, New York (1978)
3. J A Brinell, Second Cong. Int. Methods d'Essan, Paris (1900)
4. E Meyer, Zerits d. Vereines Deutch. Ingenieure, (52, 645(1908)
5. R S Bradley, Phil. Mag. 13, 853, (1932)
6. B V Derjaguin, Yu P Toporov and P Aleinikova, J. Colloid Interface Sci, 54, 59 (1976)
7. B Dahneke, J. Colloid Interface Sci, 32, 342, (1971)
8. J Reed, "Elastic/Plastic Impact-Adhesion Model and its Application to the Measurement of Adhesive Energy between Small Particles and a Flat Substrate", CEGB report No RD/B/0078/N82, (1982)
9. S S Brenner, H A Wriedt and R I Oriani, Wear, 68, 169 (1981)
10. R A Bagnold, "Physics of Wind Blown Sand and Desert Dunes" Methuen, London, (1941)
11. W S Chepil, Soil Sci, 60, 305-20 (1945)
12. M. Phillips, J Phys D, 13, 221-33 (1980)
13. G A Schmel, Environ. Int., 4, 167-27 (1980)
14. M Corn and F Stein, Am. Indl. Hyg. Assoc. J. 26, 325-36 (1965)
15. D E Aylor, "Plant Disease", Academic, New York, (1978)
16. J W Cleaver and B Yates, J. Colloid Interface Sci, 4, 464-74 (1973)
17. M W Reeks, J Reed and D Hall, J. Phys. D., 21, 574-89 (1988)
18. L N Rogers, "The Cleaning of Radioactively Contaminated Surfaces by Means of Particle Impact Processes", CEGB Report No. RD/B/N4784, 1980
19. R M Davis, Proc. Roy. Soc. A 197 417-432 (1949)
20. J Reed, J. Phys. D 18 2329-2337 (1985)
21. I M Hutchings, J. Phys. D 12 1819-1824 (1979)
22. L N Rogers and J Reed, J. Phys. D. 17 677-689 (1984)
23. J Reed, "Impact Adhesion Behaviour of U_3O_8 Particles Striking Steel Surfaces", CEGB Report No. TPRD/B/0724/N85, 1985
24. K L Johnson, K Kendall and A D Roberts, Proc. Roy. Soc. A 324, 301-313 (1971)
25. K L Johnson, in "Theoretical and Applied Mechanics", W T Koiter (ed), pp. 133-143, North Holland Publishers, Amsterdam, 1976
26. M W Reeks, J Reed and D Hall, J. Phys D., 21, 574-89 (1988)
27. D Hall and J Reed, J. Phys. D., 21, 1481-5 (1988)
28. B V Derjaguin, Kolloid. Z., 69, 155 (1934)
29. K L Johnson, Brit. J. Appl. Phys. 9, 199 (1958)
30. B V Derjaguin, V M Muller and Yu P Toporov, J. Colloid Interface Sci. 40, 1 (1975)
31. D Tabor, J. Colloid Interface Sci., 58, 2 (1977)
32. A D Zimon, "Adhesion of Dust and Powder", Consultants Bureau, New York (1982)

APPLICATION OF IMPACT ADHESION THEORY TO PARTICLE

KINETIC ENERGY LOSS MEASUREMENTS

Stephen Wall and Walter John

Air and Industrial Hygiene Laboratory
California Department of Health Services
2151 Berkeley Way, Berkeley, CA. 94704

Simon L. Goren

Department of Chemical Engineering
University of California, Berkeley CA. 94720

Although a number of models have been proposed to describe the
adhesion of particles to surfaces, most theories are restricted
to the case of the removal of precollected particles from a
surface. This provides an estimate only of adhesion surface
energy between the particle and substrate. In the dynamic case
of an impact, it is necessary to consider the adhesion surface
energy and the loss of impact kinetic energy as well as the
interaction between the two mechanisms. Data from laser-Doppler
measurements of kinetic energy loss were used to evaluate the
few available impact-adhesion models from low impact velocities,
where the major influence of adhesion surface energy is
expected, up to high velocities, where plastic deformation
dominates the impact.

PREVIOUS WORK

Impact Adhesion Models

The adhesion energy required to dislodge a small sphere from a surface
has been extensively investigated[1-4], but few theories model the adhesion
of a particle on impact with a surface. Dahneke[5] proposed an energy
balance model to explain the energy loss observed when particles impact
smooth surfaces. Plastic deformation was not considered to be as an energy
loss mechanism since the impact velocities were considered too low to
reach the plastic limit without stress concentration. Dahneke rejected
plastic damage due to stress concentration since he believed the contact
area was comparable to the scale of the surface roughness. However, as
discussed by Wall et al.[6], the polysytrene particles used by Dahneke in
impact velocity measurements would have been expected to suffer plastic
damage even at the critical velocity for bounce.

Plastic deformation was included in the adhesion theory of Brenner et
al.[7] in order to explain the energy loss in impacts of large metal
particles with a hard surface. Metal particles are known to undergo
plastic deformation at velocities less than the critical velocity for
bounce. Brenner et al. evaluated the Hertz equations for elastic
deformation at the yield limit stress to determine the elastic energy

available on rebound to overcome the surface forces. This neglects the additional elastic energy that is stored after the plastic limit is exceeded and underestimates the contact area used to determine the adhesion energy.

Rogers and Reed[8] first considered both elastic and plastic deformations occurring during large particle-surface impacts. During the compression phase of the impact, the elastic-plastic deformation theory of Bitter was utilized to determine the elastic energy available at the start of rebound to overcome the adhesion surface energy. The adhesion surface energy is calculated from the Johnson, Kendall and Roberts[3] theory (JKR) by modeling the rebound using a Hertz equivalent diameter particle. This is a particle with a diameter equivalent to a perfectly elastic particle which, when compressed to develop the same elastic energy that is available for rebound in the actual plastically deformed particle, yields the impact contact area. The model assumes no effect of adhesion forces during the compression phase of the impact and the absence of plastic deformation during the rebound process.

Fichman and Pnueli[9] used an approach similar to Rogers and Reed but were the first to introduce an additional energy loss through adhesion induced plastic deformation at the perimeter of the contact area, where the particle is under large tensile forces. Unfortunately, as in the impact model of Dahneke, the elastic energy stored in the region of plastic deformation was neglected.

Until recently, experimental data available for micrometer size particles were inadequate to test the theories discussed above. In real systems, complications such as surface coatings, adsorbed layers, surface roughness and the angle of impact can have a large influence on the impact energy loss and the development of adhesion forces during the time of contact. To understand the energy loss mechanism for aerosol sized particles, measurements are needed with accurately determined impact and rebound velocities for well-known bounce geometries, well-characterized solid particles and a variety of impact surfaces. Wall et al.[6] recently undertook such measurements. Their results are summarized below and are used to test the theoretical models.

EXPERIMENTAL

System and Approach

Measurements of particle bounce normal to the surface were made to determine the particle energy loss on impact with a solid surface. Incoming and rebounding velocities were determined by the laser-Doppler technique. A detailed discussion of the experimental system and procedures has already been given elsewhere (Wall et al.[6]), and only a brief description is given here for completeness.

The particles were freshly generated, solid, monodisperse spheres of ammonium fluorescein from 3 to 7 μm diameter. Particles were spray dried from solution drops (vibrating orifice aerosol generator) to form monodisperse solid, non-hygroscopic spheres, which were smooth down to near-molecular scale (3 nm). The impaction surfaces were molybdenum and silicon, hard materials with surface roughness determined by a high mechanical polish; muscovite mica, cleaved to obtain a molecularly smooth surface; and Tedlar, a deformable fluorocarbon polymer cast as a thick film.

Both molybdenum and Tedlar were relatively smooth surfaces with asperities less than 0.3 μm as determined by scanning electron microscopy.

The silicon surface was as featureless as mica within the 0.1 μm resolution of the instrument. The microscale roughness of the molybdenum target was less than 5.0 nm as compared with 0.5 nm for the best polished metal standards.

A cross sectional view of the velocity measurement region of the bounce cell including the acceleration nozzle, impaction surface and the active measuring volume portion of the fringe region is shown to scale in Figure 1. Particles were accelerated in a modified nozzle from the TSI Aerodynamic Particle Sizer. Incoming and rebounding particle velocity measurements were made for impact velocities from 1 to 100 m/s, with a dual beam laser Doppler system. Particles crossing the interference fringes, in the 126 μm diameter measuring volume, produced a Doppler signal with a frequency proportional to velocity. Since the particles decelerate under the aerodynamic drag force, velocity measurements were made as close as 10 μm from the surface to minimize extrapolation to the values at impact.

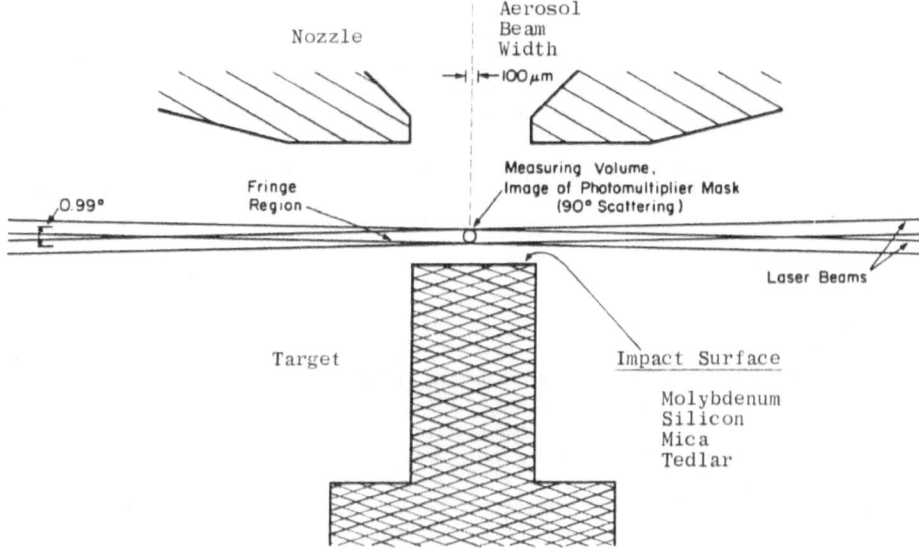

Figure 1. Cross section of the particle velocity measurement region showing the particle acceleration nozzle, impaction target and laser beams.

Impact Velocity

The theory for large Reynolds number flow (potential flow with a boundary layer correction based on Hiemenz flow) past a flat disc was used to develop an expression for the gas velocity along the centerline. The flowfield was established by fitting the trajectories of 0.516 μm tracer particles, which follow the gas velocity closely due to their low inertia. The calculated trajectory for the tracer particles is in good agreement with the experimental points.

Measurements of mean incoming and rebounding particle velocities were made for 0.516, 2.58, 3.44, 4.90 and 6.89 μm geometric diameter particles as a function of distance above the impaction surface along the particle beam centerline. To extrapolate these velocities to impact, a non-dimensional form of the equation of particle motion was used (Wall et al.[6]).

THEORETICAL BACKGROUND

Energy Balance

The particle bounce process can be treated as a simple energy balance as proposed by Dahneke[5]. The fraction of the impact energy recovered by the particle can be expressed in terms of the incoming velocity, V_i and the rebounding velocity, V_r as:

$$V_r^2/V_i^2 = e^2 + 2\Delta E/MV_i^2, \qquad (1)$$

where:
$\Delta E = E_i - E_r$,
$e = (1 - K_L/K_i)^{1/2}$ is the coefficient of restitution,
K_i is the kinetic energy of the incoming particle,
E_i is the adhesion energy of interaction at impact,
E_r is the adhesion energy of interaction at rebound,
K_L is the energy loss in the impact,
M is the mass of the particle.
Here the classical definition of e is used in contrast to the form used by Dahneke, $e = (1 - K_L/(K_i+E_i))^{1/2}$, which does not reduce to the classical form even when $\Delta E = 0$. At sufficiently low impact energy, the elastic energy at rebound just equals the adhesion surface energy. This is the critical velocity for particle capture defined as:

$$V_i^* = (-2\Delta E/(Me^2))^{1/2}. \qquad (2)$$

The classical expression for adhesion surface energy is

$$E = 4 \pi z_o R \gamma \qquad (3)$$

where R is the particle radius, z_o is the distance of separation between the surfaces of contact and γ is the adhesion surface energy parameter. This yields an inverse dependence of critical velocity on particle diameter for a point contact between the particle and the surface. The adhesion energy due to elastic flattening of the particle to form a finite contact area has been shown to be equivalent to the point contact case without flattening (Derjaguin et al.[2]), and the dependence of critical velocity on particle size is expected to be the same. Ammonium fluorescein particles impacting the four different target materials exhibited a stronger inverse dependence of critical velocity on particle size (powers from -1.2 to -1.8) than can be explained by the classical adhesion model.

The adhesion energy which must be overcome for the particle to rebound from the surface is the difference between the incoming and rebounding adhesion energy. Accordingly if the impact is symmetric $E_i = E_r$, $\Delta E = 0$ and the adhesion energy has no effect on the impact energy loss. Both Brenner et al.[7], and Rogers and Reed[8] have proposed asymmetric impact adhesion models in which energy loss occurs due to plastic deformation on impact only and the particle rebound is purely elastic.

Plastic Deformation

In the simple energy balance, plastic deformation is not considered as a function of the mechanical properties of the materials involved. The Hertz equations for purely elastic impacts and the elastic yield limit, Y, of the material can be utilized to determine the limiting velocity, V_y, above which plastic deformation begins to occur:

$$V_y = (2\pi/3K)^2 (2/5\rho)^{1/2} Y^{5/2}, \qquad (4)$$

where ρ is the density of the particle, the mechanical constant is

22

$K = 4/3(\kappa_p + \kappa_s)$ with $\kappa = (1 - \nu^2)/\xi$, ν is the Poisson ratio, ξ is Young's modulus and the subscripts p and s refer to the particle and the surface. The limiting velocity is determined by the bulk material properties and is independent of particle size. Using equation 4, lines representing constant V_y are shown in Figure 2 as a function of the yield limit stress Y_p and the bulk mechanical constant κ_p for variety of particle materials. The plastic deformation is assumed to occur only in the particle materials such that $\kappa_p \gg \kappa_s$. For all these materials, except for glass, plastic damage occurs for impact velocities above 1 m/s. Critical velocities for the size range of ammonium fluorescein particles used in this study were greater than 1 m/s (Wall et al.[6]).

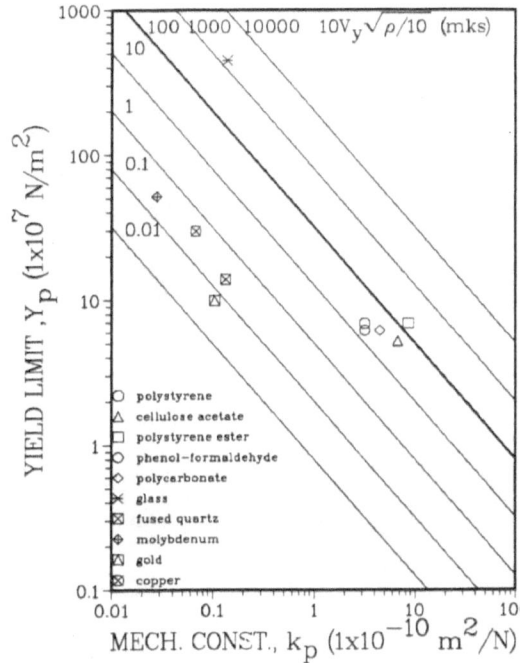

Figure 2.   Limiting elastic velocity as a function of mechanical constants for a variety of organic and inorganic materials.

Elastic-Plastic Adhesion Model

The Rogers and Reed impact model includes both surface energy acting over the contact area and mechanical energy loss through plastic deformation which applies at impact velocities greater than a fixed yield limit velocity, V_y. The model combines an impact compression phase, in which both elastic and plastic deformations are considered to obtain the area of contact at full compression, and an elastic rebound phase, in which surface forces acting over the contact area must be overcome for the particle to leave the surface. Plastic deformation energy loss is derived from a model due to Bitter[10] and no surface forces are considered to act during the compression. The compressive phase of the impact includes both purely elastic and elastic-plastic deformation. During the elastic phase shown in Figure 3a, the pressure distribution over the contact area is hemispherical and reaches a maximum stress at the center of contact which is equal to the yield limit stress. The radius of the circular elastic contact area is given by:

$$r_e = (5 \, K_y \, R^2 \, / \, 2 \, K)^{1/5} \qquad\qquad (5)$$

where $K_y = M \, V_y^2 / 2$ is the elastic energy at maximum elastic compression.

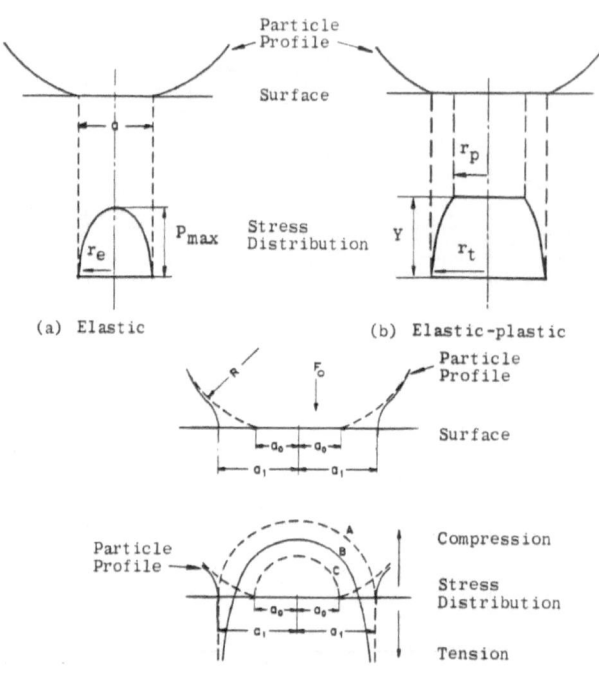

Particle Profile

Surface

P_{max} Stress Distribution

r_e

(a) Elastic

Particle Profile

r_p

Y

r_t

(b) Elastic-plastic

Particle Profile

F_o

Surface

a_o a_o

a_1 a_1

Compression

Particle Profile

A
B
C

Stress Distribution

a_o a_o

a_1 a_1

Tension

(c) Surface forces

Figure 3. Schematic representations of the stress distribution across the contact area for (a) the elastic and (b) the elastic-plastic compression phase of the impact according to the model of Bitter[10] and (c) the adhesion model of JKR[3], which were used to derive the elastic-plastic impact adhesion model of Rogers and Reed[8].

After the yield limit is reached at the center of contact, further compression produces both elastic and plastic deformation. As indicated in Figure 3b, the contact area consists of a plastic circle surrounded by an elastic ring. The pressure over the plastic circle is constant and equal to the yield limit stress, Y. A Hertzian pressure distribution remains over the elastic ring, which has an area equal to the maximum contact circle during the purely elastic deformation phase. The radius of the plastic contact circle is given by:

$$r_p = 3 K [(15/16) K_y K_p]^{1/4} / [(\pi Y)^{3/2} R^{1/2}] \qquad (6)$$

and the total contact area at maximum compression over which the surface forces are applied is $r_t = (r_o^2 + r_p^2)^{1/2}$. In the plastic deformation region, the elastic energy is a function of the yield limit stress and is given by: $K_{pe} = h_e \pi r_p^2 Y / 2$, where h_e is the Hertz elastic deformation. The permanent energy loss to plastic deformation can be expressed as a function of the yield limit kinetic energy as:

$$K_p = [(K_i - K_y/16)^{1/2} - (15 K_y/16)^{1/2}]^2. \qquad (7)$$

In the rebound phase, elastic energy from both the elastic and elastic-plastic deformations is available to overcome the adhesion surface energy. The critical energy condition for particle rebound is:

$$K_i - K_p = K_y + K_{pe} = E_a + E_m. \qquad (8)$$

An expression for the adhesion energy, E_a, is derived from the JKR theory for equilibrium contact adhesion in the form:

$$E_a = \Delta\gamma \pi r_t^2 (F_t/F_m)^{2/3} \qquad (9)$$

which depends on the adhesion surface energy, $\Delta\gamma$, the contact area at maximum compression, πr_t^2, and a factor representing the augmentation of the contact area due to the surface forces, $(F_t/F_m)^{2/3}$. The mechanical energy term E_m which is the force x distance contribution to the energy change:

$$E_m = F_o F_m^{1/3} [(F_t^{2/3} + 2 F_o F_z^{-1/3})/3K r_t] \text{ is usually}$$

insignificant for aerosol size particles.

Without surface forces, at equilibrium the only externally applied load, F_o, acting to retain the particle on the surface would be the gravitational force and the contact area is a_o as shown in Figure 3c. The surface force, F_s, acting over the contact area yields a total applied load, F_t, which is larger than F_o and results in an enhanced contact denoted by a_1 in Figure 3c.

The total adhesion force due to the surface energy of interaction and an external load can be written in terms of three components of force, as has been shown (Wall et al.[6]):

$$F_t = F_o + 2 F_s + 2 [F_s F_o + F_s^2]^{1/2} \qquad (10)$$

where:
$F_s = 3/2 \Delta\gamma \pi r_t^3 (K / F_m)$ is the absolute value of the equilibrium pull-off force,
$F_m = F_e + F_p$ is the total elastic force,
$F_e = (R^{1/2} K)^{2/5} (5 K_y / 2)^{3/5}$ is the force in the elastic region,
$F_p = \pi r_p^2 Y$ is the elastic force in the plastic region,
$\Delta\gamma = 2 (\Delta\gamma_p \Delta\gamma_s)^{1/2}$ is the adhesion surface energy which can be closely approximated by the geometric mean of the particle and target individual surface energies (Fowkes[11]).

In the Rogers and Reed model, plastic deformation occurs in the central region of the circle of contact and is surrounded by an annular region of elastic deformation. A similar model for elastic-plastic impacts was subsequently proposed by Fichman and Pnueli[9], who added a second region of plastic deformation at the circumference of the contact circle where the material is under tension. This is an additional energy loss mechanism and removes a deficiency in the JKR model which requires the deforming material to be under infinite tension at the perimeter of contact as indicated in Figure 3c. This adhesion-induced plastic deformation has been suggested by Fichman and Pnueli to provide the only energy loss criterion necessary to establish critical velocity. However, this approach was found to greatly underestimate the magnitude and particle size dependence of critical velocity (Wall et al.[6]).

RESULTS

Kinetic Energy Loss

The fraction of kinetic energy recovered in the impact of particles with surfaces is simply V_r^2/V_i^2, where the impact and rebound velocities are determined from the extrapolation of incoming and rebounding velocity measurements to determine the values at the surface. Ratios of rebound to impact velocities measured for 4.90 μm ammonium fluorescein particles over a wide range of impact velocities are shown in Figure 4 for four different targets.

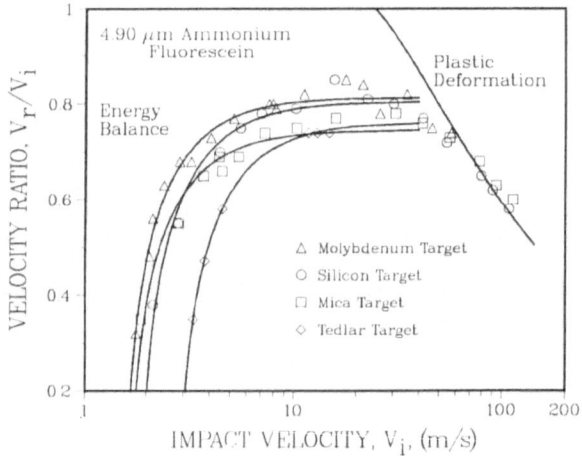

Figure 4. Velocity ratio (V_r/V_i) measurements over a full range of impact velocities (V_i) showing a dependence on target materials at low velocity, which was fit with the energy balance equation to obtain the critical velocity, and a lack of dependence at high velocity, which was fit with the classical plastic deformation equation.

The energy balance (equation 1) fitted to the rebound velocity data is shown for each target material. Application of the simple energy balance utilizes the fitting parameters ΔE, the adhesion energy and e, the coefficient of restitution. The model provides an adequate representation of the energy loss as a function of impact velocity at low velocity, where adhesion energy ΔE is the most sensitive fitting parameter. At sufficiently high velocity the impact energy, K_i, far exceeds the adhesion energy of the surface, ΔE, and the velocity ratio is primarily a function of the mechanical energy loss represented by $1 - e^2$. As applied by Dahneke, the model assumes e^2 to be a constant, which underestimates the energy loss at high velocity.

As indicated in Figure 4, the high velocity data can be fit using the plastic deformation expression for the velocity ratio:

$$V_r/V_i = (1 - K_p/K_i)^{1/2} \cong [1.94 \ V_y/V_i - 0.88 \ V_y^2/V_i^2]^{1/2}. \qquad (11)$$

Accordingly, the velocity ratio is a function only of the elastic yield limit velocity and is independent of particle size. The small difference in velocity ratio observed between the different targets at high velocity indicates that the elastic yield limit velocity was the same in each case. This is consistent with mechanical damage occurring only in the particle, which is believed to have a lower elastic yield limit than the target materials molybdenum, silicon, and mica. However, the fitted value for the limit yield velocity of 20 m/s appears to be an order of magnitude too high when compared with the organic materials as shown in Figure 2.

The elastic-plastic impact model of Rogers and Reed was developed to predict the critical velocity from a knowledge of the materials properties: surface energy, yield limit and mechanical constant. This model has previously been applied to the laser-Doppler data to fit the critical velocity particle size dependence for all four target materials (Wall et al.[6]). The critical velocity fitting parameters are included in Table I.

Table I. Fitting Parameters (equation 8) for the Critical Velocity Particle Size Dependence of Ammonium Fluorescein Particles Impacting Various Targets.

Target	Mechanical Constant 10^{-10} m^2/N k_p	k_s	Elastic Yield Limit 10^7 N/m^2 Y	Yielding Material	Surface Energy J/m^2 $\Delta\gamma$
Mica	7.6	0.100	6.8	particle	0.28
Molybdenum	7.6	0.028	7.4	particle	0.34
Silicon	7.6	0.050	7.4	particle	0.38
Tedlar (PVF)	7.6	4.5	5.8	target	0.47

From equation (8),

$$V_r/V_i = [1 - (K_p + E_{adh})/K_i]^{1/2}. \qquad (12)$$

This assumes no plastic damage occurs during the rebound phase. A comparison of the energy balance model (for constant ΔE and e) and the elastic-plastic model with the velocity ratio data for different diameter ammonium fluorescein particles impacting the silicon target is given in Figure 5. The elastic-plastic model of Rogers and Reed was applied by using the parameters (surface energy, yield limit stress, and mechanical constants of Table I.) established by fitting the particle size dependence of the critical velocity.

As the impact velocity is increased above critical velocity, velocity ratios calculated from the elastic-plastic model initially increase and are in close agreement with the energy balance equation fitted to the data, except for the 2.58 μm particles. However, at higher velocity the decrease in the velocity ratio (indicating a higher degree of energy loss) with increasing impact velocity, which is characteristic of plastic deformation, begins to occur at a much lower velocity than the experimental data. At the highest impact velocities, the velocity ratios for the different size particles calculated from the elastic-plastic model converge as would be expected when the primary energy loss is through plastic deformation. However, the predicted energy loss is almost a factor of forty greater than that derived from the experimental measurements. The model assumes that the yield limit stress for plastic deformation is a constant, independent of impact velocity. Experimental measurements have shown that the yield limit stress of a material is not always constant, but can depend on the rate of deformation , and is referred to as the dynamic yield limit (Taylor[15]; Bitter[10]).

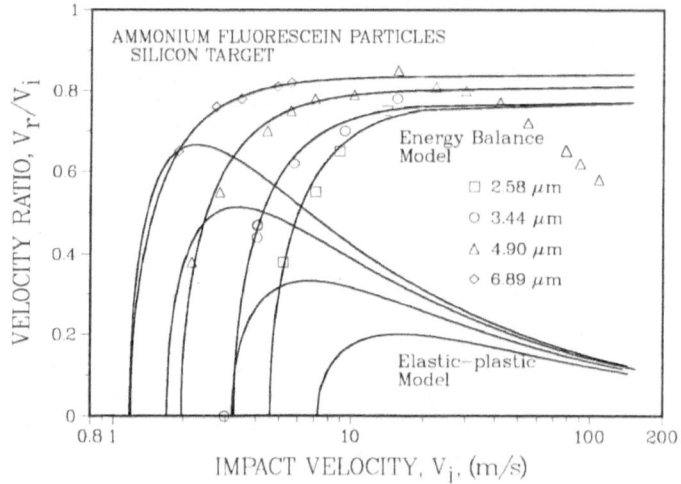

Figure 5.   A comparison of the energy balance model and the elastic-plastic model with the velocity ratio data for different diameter ammonium fluorescein particles impacting the silicon target.

Dynamic Yield Limit

A dynamic yield limit, which is significantly greater than the static yield limit usually reported for materials can be explained by the strain rate effect (Malvern[12]). According to Malv ?>, a material is brought to the point of initiation of plastic flow aft r a given amount of elastic strain independent of the elastic strain rate. However, plastic flow requires time in which to develop, so that the additional strain beyond the static yield velocity is at first mainly elastic. The strain rate effect, which is a function of the deformation <u>rate</u>, should not be confused with work hardening, which is a function of deformation independent of the rate. Plastic materials are known not to work harden (Bitter[10]), but do undergo strain rate effects.

Dynamic yield limits which are greater than the static yield limit by more than a factor of 3 have been reported for perfectly plastic elemental metals (Tabor[13]) and for polymers (Bitter[10]). The yield limits for both macro- and micro-crystalline waxes have also been reported to increase with the rate of deformation (Mozes[14]).

The expression for energy loss to plastic deformation (equation 11) can be solved for the yield limit velocity as a function of the measured impact and rebound velocities as:

$$V_y = 1.11 \ V_i \ [\ 1 - (\ 1 - 0.933 \ V_r^2/V_i^2 \ )^{1/2} \ ]. \quad (13)$$

Using equation 13, yield velocities calculated from the laser-Doppler data for impacts on silicon and mica targets are plotted as a function of

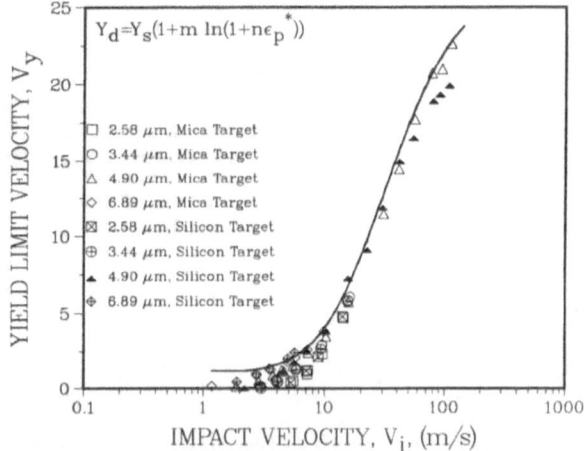

Figure 6. The points are the yield limit velocities calculated from the experimental data, using equation (13) which neglects the adhesion energy. The line is calculated from the velocity ratio data for 4.90 μm particles impacting the molybdenum and silicon targets using a dynamic yield limit expression derived from Malvern[12].

impact velocity in Figure 6. The yield velocity increases from the static value V_y with a dependence which becomes steep above 10 m/s. At the highest velocities, the yield limit increases more slowly. This is consistent with previous results for soft metals (Taylor[15]) which show a static yield limit at small strain rates and a transition region at intermediate strain rates which approaches an almost constant dynamic yield limit at sufficiently high deformation rates. A dynamic yield limit can explain the apparent inconsistency between the yield limit velocity of 20 m/s, obtained from fitting the classical mechanical deformation equation to the high velocity data (see Figure 4), and the static yield limit velocity, expected to be <1 m/s (see Figure 2) for the physical properties of organic materials similar to ammonium fluorescein.

Malvern[12] has presented a theory of the strain rate effect where the apparent increase in the yield limit is a function of the plastic strain rate:

$$Y_d = Y_s [1 + m \ln(1 + n \epsilon_p^*)] \qquad (14)$$

where m and n are empirical fitting constants, ϵ_p^* is the plastic strain rate, and Y_s, Y_d are static and dynamic yield limits respectively. The plastic strain rate can be determined from the maximum deformation, α_{max}, and the duration of the plastic portion of the impact, t_2. Using the Goldsmith plastic deformation model, Wang and John[16] have developed expressions for the deformation and the time of contact. For impact velocities well above the yield limit velocity:

$$\alpha_{max} = V_1 (M/\pi Y_s R)^{1/2} \qquad (15)$$
$$t_2 = (\pi/2) (M/\pi Y_s R)^{1/2} \qquad (16)$$

$$\text{where} \qquad V_1 = (V_i^2 - V_y^2)^{1/2}.$$

Accordingly, the plastic strain rate normalized to the yield limit velocity,

$$\alpha_{max}/(t_2 V_y) = \epsilon_p^* \simeq (2/\pi) (1/V_y) (V_i^2 - V_y^2)^{1/2}. \qquad (17)$$

A best fit of the experimental data (shown as the solid line in Figure 6) was obtained by expanding the logarithm in the equation of Malvern:

$$Y_d = Y_s [1 + 2m/(1 + ((2/n) (\pi/2) V_y / (V_i^2 - V_y^2)^{1/2}))^p]. \quad (18)$$

The exponent was added to allow adjustment of the slope of the yield curve. The physical significance of the fitting parameters can be identified. Here m=1.20 is the increase in the yield limit stress above the static value at the midpoint between the static and the maximum dynamic yield limits, 2/n=10 is the ratio of the impact velocity to the static yield limit at which the midpoint of the increase in yield limit is reached and p=1.30 is the slope of the yield limit curve. The above numerical values are for the deformation of the ammonium fluorescein particles impacting silicon and mica surfaces.

The apparent increase in the yield limit velocity for ammonium fluorescein (a factor of 25) corresponds to an increase of a factor of 3.6 in the yield limit stress. This is in close agreement with the maximum increase in the yield limit stress reported for impact deformation of a variety of plastic surfaces. An increase in the yield limit for polyvinyl chloride, PMMA and polyethylene of a factor of 3.4 - 3.6 has been reported by Bitter[10] and Tabor[13] has reported a factor of 3.2 increase for perfectly plastic metals (copper, aluminum, tellurium-lead). However, the corresponding increases for non-polymeric organic materials such as ammonium fluorescein have not been previously reported.

The stress rate effect has been applied to the Rogers and Reed model by substituting the dynamic yield expression of equation 18 for the static yield limit. The elastic-plastic model, modified by the introduction of the strain rate effect, is compared with the rebound velocity ratio data for the silicon target in Figure 7a and the mica target in Figure 7b. For both targets, close agreement between the modified model and the velocity measurements is achieved for impacts with the 4.90 μm particles over the entire velocity range. Agreement with the other particle sizes was within experimental error (10%) except for the largest particle impacting the silicon target and the smallest particle size impacting the mica target. In these cases the particle recovered about 20% more impact energy than the theory predicts in the velocity range less than 10 m/s.

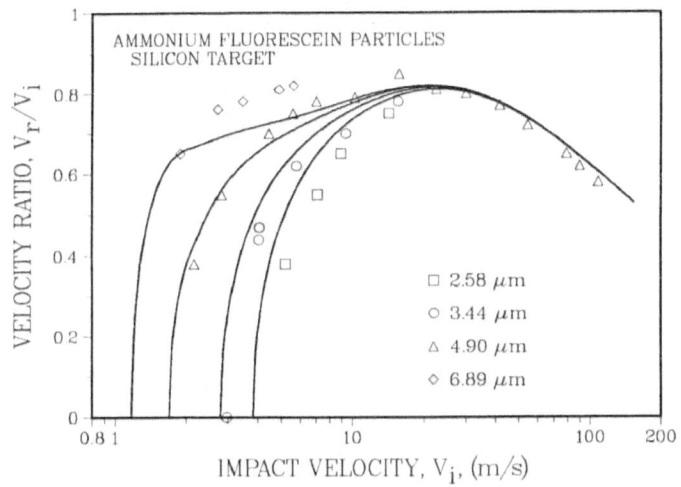

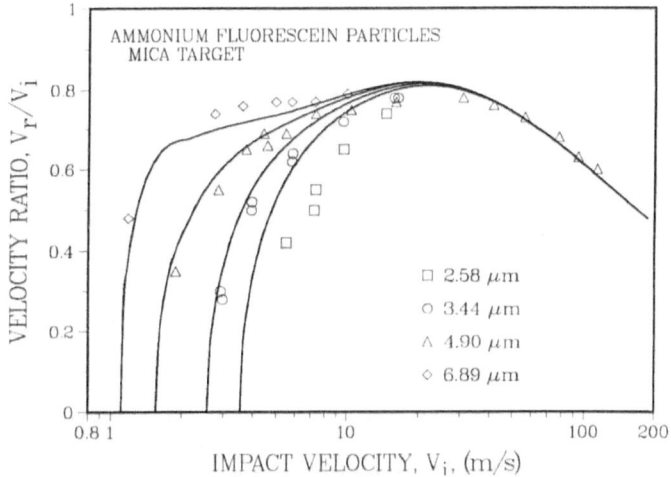

Figure 7. Modified Rogers and Reed model including the dynamic yield limit function applied to the velocity ratio data for the (a) silicon target and (b) mica target over the full range of impact velocities.

The critical velocity particle size dependence for ammonium fluorescein particles from 2.58 μm to 6.89 μm impacting the molybdenum, mica, silicon and Tedlar targets has previously been fitted (Wall et al.[6]) using the Rogers and Reed model and the fitting parameters given in Table I. Using the same parameters, the modified model, which includes the dynamic yield limit expression, is compared with the model using a static (constant) value for the yield limit stress in Figure 8. The effect of

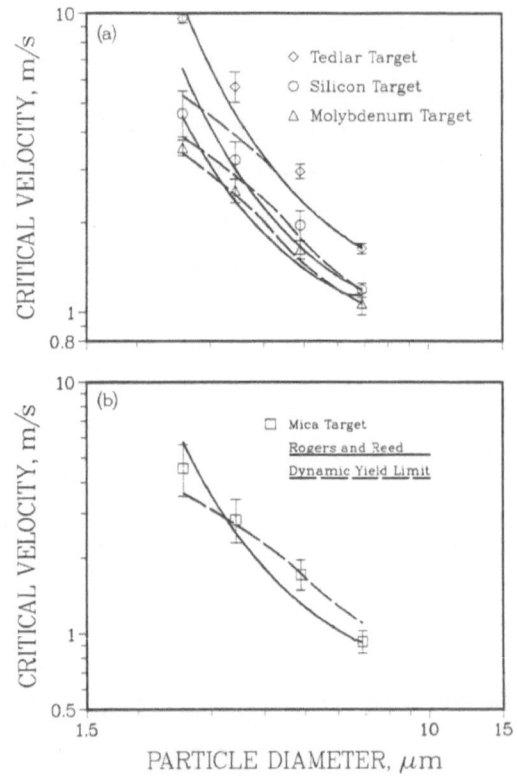

Figure 8. Comparison of the critical velocity particle size dependence, for the elastic-plastic model of Rogers and Reed (static yield limit) and the modified model incorporating a dynamic yield limit, with the experimental data.

including the dynamic yield limit, which is a function of impact velocity, is to decrease the degree of plastic deformation and to reduce the dependence of critical velocity on particle size.

The modified model agrees with the experimental points within the experimental error except for the Tedlar target where the theory underestimates the critical velocity at the smaller particle sizes. This is not unexpected since the dynamic yield limit function used is valid for deformation in the ammonium fluorescein particles only. Evidence has already been given that plastic deformation probably occurred in the Tedlar target which may well have a different yield limit function than ammonium fluorescein.

DISCUSSION

The simple energy balance as proposed by Dahneke[5], with the adhesion energy estimated from the classical theory for a point contact and with the assumption of a constant degree of energy loss to plastic deformation, underestimates the critical velocity particle size dependence and overestimates the energy loss at high velocity.

In the adhesion model for elastic-plastic impacts presented by Rogers and Reed, the adhesion energy is neglected in the compressive (incoming) phase of the impact and the plastic deformation energy loss is calculated from the physical properties of the materials of contact. At the start of rebound the Johnson, Kendall and Roberts (JKR) theory for equilibrium is applied to determine the adhesion surface energy and the contact area. These assumptions applied to a dynamic impact underestimate the contact area, requiring an apparent overestimate of adhesion surface energy to fit the critical velocity data.

The Rogers and Reed theory also assumes a constant yield limit, which produces increasingly larger overestimates of the energy loss as impacts occur further above the critical velocity. A dynamic yield limit function, derived from the equations of Malvern was fit to the laser-Doppler data and used to modify the Rogers and Reed model. The modified model agreed well with the experimental measurements over the entire impact velocity range from 1 to 100 m/s. However, the required surface energy parameters remain unrealistically large.

Fichman and Pnueli calculated critical velocity from the energy loss which is considered to occur in the ring of plastic deformation at the edge of contact area, where the material is in tension. Even when numerous errors in the equations are corrected, the model of Fichman and Pnueli greatly underestimates the critical velocity for particle bounce and the associated dependence on particle size.

CONCLUSION

There is currently no comprehensive theory for particle surface impacts which fully accounts for the components of energy loss. The currently available models include assumptions which are introduced to simplify the interaction between the mechanical and adhesion forces which occur throughout the impact.

An improved theory would include the surface forces dynamically throughout the impact process. Additional energy losses which should be considered are the plastic damage at the edge of the contact area where the material is under tension and the enhancement of the plastic damage at the center of contact due to the action of the adhesion force during compression. A dynamic yield limit should be used in calculating the plastic damage.

REFERENCES

1. R.S. Bradley, Phil. Mag., $\underline{13}$, 853 (1932).

2. B.V. Derjaguin, V.M. Muller and Yu.P. Toporov, The effect of contact deformations on the adhesion of particles, J. Colloid Interface Sci., $\underline{53}$, 314-326 (1975).

3. K.L. Johnson, K. Kendall and A.D. Roberts, Surface energy and the contact of elastic solids, Proceedings of the Royal Society of London, $\underline{324}$, 301-313 (1971).

4. D. Tabor, Surface forces and surface interactions, J. Colloid Interface Sci., $\underline{58}$, 2-13 (1977).

5. B. Dahneke, The influence of flattening on the adhesion of particles, J. Colloid Interface Sci., $\underline{40}$, 1-13 (1972).

6. S.M. Wall, W. John, H.C. Wang and S.L. Goren, Measurements of kinetic energy loss for particles impacting surfaces, Submitted to J. Aerosol Sci. and Technol. (1988).

7. S.S Brenner, H.A. Wrient and R.A Oriani, Impact adhesion of iron at elevated temperatures, Wear $\underline{68}$, 169-190 (1981).

8. L.N. Rogers and J. Reed, The adhesion of particles undergoing an elastic-plastic impact with a surface, J. Phys. D: Appl. Phys. $\underline{17}$, 677-689 (1984).

9. M. Fichman and D. Pnueli, Sufficient conditions for small particles to hold together because of adhesion forces, J. Appl. Mechanics $\underline{52}$, 105-108 (1985).

10. J.G.A. Bitter, A study of erosion phenomena - Part I, Wear $\underline{6}$ 5-21 (1963).

11. F.M. Fowkes, Attractive forces at interfaces, Industrial and Engineering Chemistry $\underline{56}$, 40-52 (1964).

12. L.E. Malvern, The propagation of longitudinal waves of plastic deformation in a bar of material exhibiting a strain-rate effect, J. Appl. Mechanics $\underline{18}$, 203-208 (1951).

13. D. Tabor, Hardness of solids, Endeavour $\underline{13}$, 27-32 (1954).

14. G.Y. Mozes, "Paraffin Products: Properties, Technologies, Applications", Elsevier Scientific Publishing Co. New York (1982).

15. F.R.S. Taylor, The testing of materials at high rates of loading, James Forrest Lecture, Cambridge University (1946).

16. H.C. Wang and W. John, Dynamic contact charge transfer considering plastic deformation, J. Aerosol Sci. $\underline{16}$,399-411 (1988).

ADHESION FORCE DUE TO A MENISCUS IN A CROSSED-FIBER SYSTEM

Thomas M. Wentzel and William S. Bickel

Department of Physics
University of Arizona
Tucson, Arizona 85721

In certain circumstances, the adhesion force which holds
two particles together can be due chiefly to a water meniscus
that forms in the contact zone. The source of this meniscus
is water vapor, which condenses from the surrounding
atmosphere. We have measured the adhesion force due to a
water meniscus between a pair of circularly cylindrical
quartz fibers held in a crossed cylinder geometry. The
experiment permitted many repeated measurements with the same
fiber geometry and environmental conditions, as well as
control of the geometry and environment. We report
measurements of the adhesion force as a function of relative
humidity and fiber size, compare our measurements to theory,
comment on hysteresis observed on several time scales, and
discuss statistics of our measurements. We also discuss the
application of light-scattering techniques to the study of
adsorbed surface layers.

INTRODUCTION

When two smooth, hard, curved surfaces are brought into contact,
many different forces act upon the surfaces and many different phenomena
occur to cause adhesion.[1] Electrostatic and van der Waals forces may
help to bring the surfaces into contact, and once contact takes place,
they will hold the surfaces together.[2] Small surface irregularities
called asperities may be present and hold the bulk surfaces apart from
each other. Both the asperities and the bulk surfaces may deform with
time. The ratio of force from each contributing source, as well as the
total force, will depend critically on the geometry of the two surfaces.
Once the surfaces have come into contact, a new force can develop to hold
them together; a capillary force. Unless the surfaces have been severely
deformed where they touch each other, the two surfaces will meet at a
very small angle. This highly acute angle between the surfaces acts as a
nucleation site for water vapor and any other volatile materials that are
in the atmosphere around the two surfaces.[3]

A pure vapor will condense into a stable, spherical liquid drop only
when its partial pressure (p/p_s) exceeds 100%. However, if the mean
curvature of the liquid surface is negative, as in a concave meniscus,

instead of positive, as in a spherical droplet, the liquid can be in equilibrium with its vapor at a partial pressure below 100%. The relationship between the mean curvature and the vapor pressure is given by the Kelvin equation.[3,4] So, if the two surfaces curve smoothly away from each other, and if the angle between the surfaces is highly acute, a vapor present in the surrounding atmosphere will condense at the nucleation site to form an annular meniscus with a highly negative mean curvature. This effect is known as capillary condensation. The meniscus will grow, and the surface's mean curvature will increase (but stay negative) until it is in equilibrium with the vapor in the atmosphere. When such a meniscus is present, it will likely be the dominant source of adhesion force between the two surfaces.[2]

When the surfaces are separated, the meniscus remains temporarily as a liquid bridge between them. It changes shape and, if there is time, changes volume through condensation or evaporation in order to retain a negative curvature in equilibrium with the vapor. When the separation distance becomes great enough, the meniscus becomes unstable and splits. Now the drops of liquid left on each surface have a liquid-vapor interface with positive mean curvature, and they evaporate in the unsaturated atmosphere. The force between the separating surfaces varies as a function of distance due to the variation in meniscus shape and will show a velocity dependence due to the viscosity of the liquid.

In our experiments, we measured the force between crossed fibers due to the presence of a meniscus of condensed liquid, typically water. We studied how this force depended on the partial vapor pressure and on the fiber radii of curvature and surface composition.

EXPERIMENTAL APPARATUS

The smooth, hard, curved surfaces that we have used in our experiments are small, circularly cylindrical quartz fibers of various radii. Such fibers can be easily mounted and micromanipulated to establish various contact geometries. Deryagin[5] has shown in his thermodynamic theory of adhesion at "convex" contact that the shape of the contact region between two equal-sized cylinders at right angles to each other is geometrically equivalent to that of sphere-sphere or sphere-plane contact. (In addition, if the cylinders are of different radii, or are not at a right angle, their contact region resembles that of ellipsoids in contact.) Geometry-dependent forces between the cylindrical surfaces will thus also be equivalent, as long as the size of the region of consideration (in our case, the meniscus) remains much smaller than the radii of curvature of the surfaces. Thus, we can use our quartz fibers in a crossed cylinder geometry, as shown in Fig. 1, to gather data which can be extended to systems of spheres (or ellipsoids) in contact. The area A in which the separation of the two cylinders is less than or equal to some distance d (where $d \ll R_1, R_2$) is given by the equation

$$A = 2\pi d \sqrt{R_1 R_2} / \sin\theta \quad .$$

The shape of this area is circular when the cylinders have equal radii and are crossed with $\theta = 90°$; otherwise it is elliptical. The term $\sqrt{R_1 R_2}$ is called the "effective radius" of the cylinder pair.

One of the goals of our research is to study the statistics of the adhesion force between small particles.[6] We designed an apparatus that performs thousands of near-identical contact and separation events between the same two quartz fibers under controlled environmental conditions. Our results are strengthened by the statistics of many experimental trials, which we use as a probe of the adhesion process.

Our experimental apparatus contains a pair of crossed quartz fibers (Fig. 2). One of the fibers is suspended vertically and the other is held horizontally on an arm which cyclically moves it toward, touches it against, and pulls it away from the vertical fiber at a known rate. After the two fibers make contact, if they adhere to each other, the vertical fiber is pulled sideways by the crossed horizontal fiber until the restoring force in the vertical fiber becomes greater than the adhesion force between the two fibers. At this time, contact is broken and the vertical fiber swings back and returns to its equilibrium

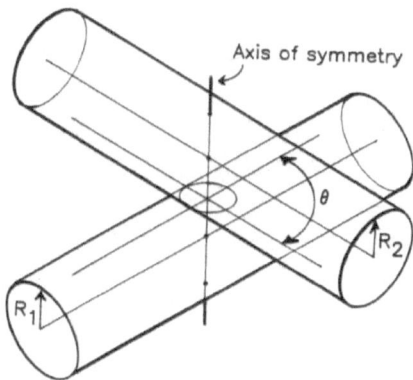

Figure 1. Two circular cylinders of radii R_1 and R_2 held in a crossed cylinder geometry, touching at a point. The area A in which the cylinder separation is less than or equal to some distance d is shown by the oval around the contact point.

position. The time between pulses of scattered light from a properly positioned laser beam is monotonically related to the time during which the vertical fiber was pulled sideways by the crossed fiber. We calibrate the device by analyzing the geometry of the crossed-fiber motion and transforming a measured contact time into a vertical fiber deflection distance. Then, by using equations for the deflection of a circularly cylindrical beam, we can calculate the force associated with this deflection. In this way, the vertical fiber acts as its own force gauge.

We can operate our crossed-fiber apparatus under a variety of experimental conditions. We control the gas mixture used in the experimental chamber, as well as its humidity. The humidity is monitored with a Vaisala HUMICAP capacitive thin-film sensor, and can be held constant in the chamber, varied slowly over many hours, or changed from one value to another in a matter of seconds. We take data with different fiber geometries, relative humidity changes, vapor types, temperature changes, and repetition rates.

The data we gather with our crossed-fiber apparatus consist of a set of the contact times just described. We store these data on a computer

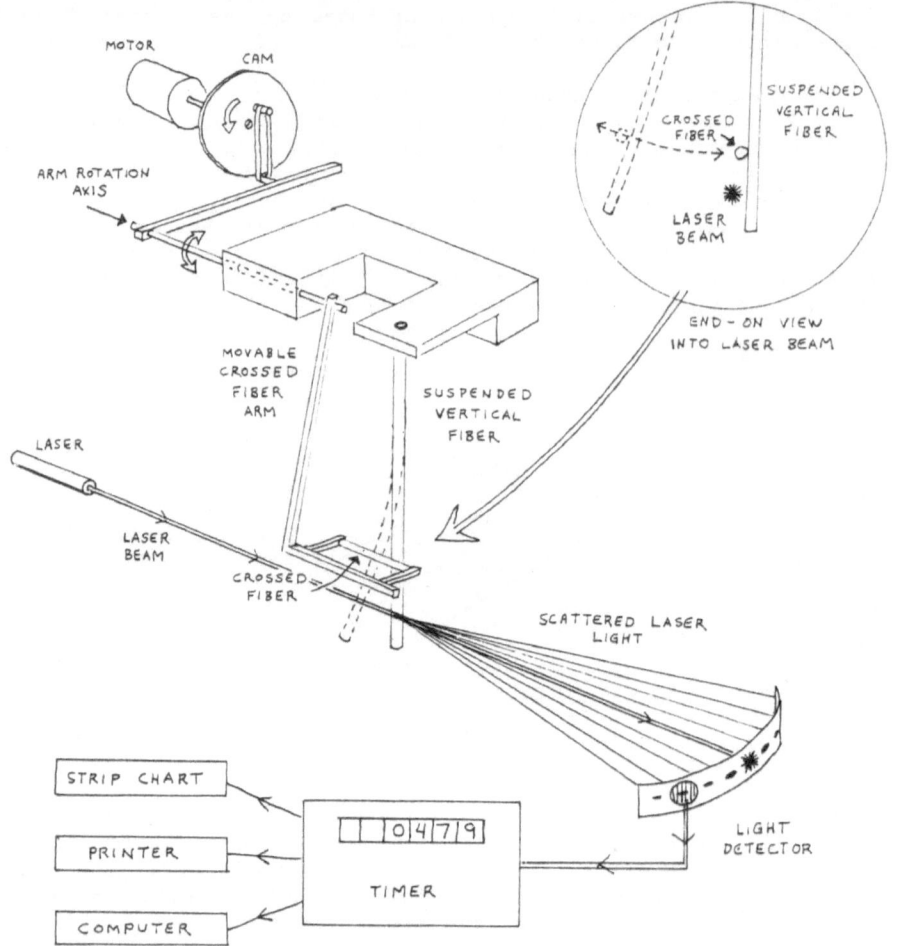

Figure 2. Schematic diagram of the apparatus to measure adhesion force
between crossed fibers.

and analyze them. Figure 3 shows a set of 550 contact times, followed by
a histogram of the data. A curve-fitting routine is then applied to the
histogram to compare the data to a Gaussian distribution. Changes in the
distribution of contact times can be easily analyzed in this way. It is
important to note that the width of the distribution of contact times is
not noise, but is due primarily to actual variations in the adhesion
force between the two crossed quartz fibers from one trial to another.
However, by taking hundreds or thousands of trials, the average adhesion
force can be determined with great accuracy.

EXPERIMENTAL RESULTS

     For all the results reported in this paper, data were taken at the
rate of one contact and separation event every 6 seconds, which allowed
enough time after fiber separation for the broken water meniscus to
evaporate from both fibers and for the vertical fiber to return to its
equilibrium position. The fibers were cleaned with ethyl alcohol

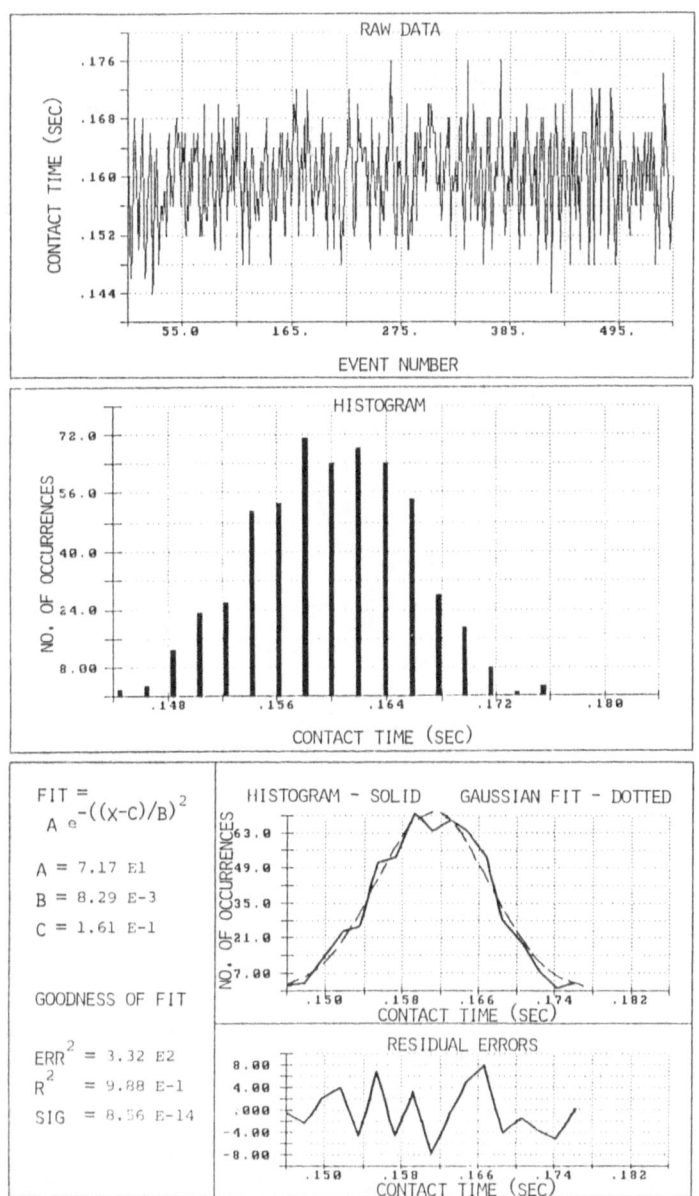

Figure 3. Top: A series of measured contact times, monotonically related to the adhesion force. Center: Frequency histogram of the contact times. Bottom: Curve-fit of a gaussian to the histogram.

followed by acetone, placed in a nitrogen atmosphere humidified with
water vapor (unless otherwise stated), and held at a temperature of 21°C.
In each trial, the relative humidity (partial pressure of water vapor,
abbreviated "RH") was cycled slowly from 0% to 98% and back to 0% again
over approximately two hours' time. RH measurements have an uncertainty
of 1%. Figure 4 shows a plot of the contact times measured during a
typical data trial. For this trial, the two crossed fibers used were
somewhat large -- 335 μm and 2150 μm radii, giving an effective radius of
424 μm. The graph shows two curves, one for increasing RH and one for
decreasing RH. Error bars on the data points indicate the extrema of the
data. Both curves show that contact times are low at low RH, have a
maximum around 65% RH, and drop to an intermediate value at high RH.
They coincide, except for the range from 40% to 70% RH, where there is a
sizable hysteresis. Of special interest is the sharp drop at 44% RH on

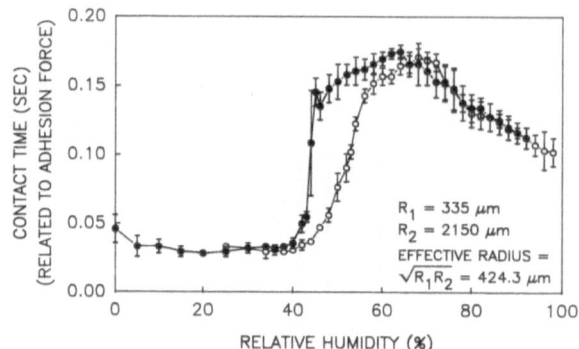

Figure 4. Contact times measured over a full cycle of RH, showing
          hysteresis in the adhesion force. (○) Increasing RH; (●)
          decreasing RH. Error bars show extrema in data.

the decreasing RH curve, which occurred between two consecutive contact
and separation events. The phenomenon causing the hysteresis is
initiated by high RH and remains as long as the RH is kept high. The
phenomenon is not extinguished by separating the fibers at high RH, since
several hundred contact and separation events took place as the RH was
lowered from saturation to 40%.

     A further example of long-term hysteresis is shown in Figs. 5 and 6.
For these trials, both fibers had a radius of 85 μm. Prior to cycling
the RH for the trial of Fig. 5, the experimental chamber had remained at
20% RH for several hours. Each trial took approximately three hours, and
the trial of Fig. 6 immediately followed that of Fig. 5. Extrema in the
data, indicated by error bars, cover a wider range in these figures,
especially at lower RH where the contact time often alternated between
two levels. The increasing RH curve in Fig. 5 shows that the contact
times gradually increase with increasing RH, whereas the decreasing RH
curve shows that contact time has a marked hysteresis in the form of a
large peak and sharp drop at 48% RH. When the RH increases for the trial
of Fig. 6, the contact time rises to a peak earlier than for the previous
trial; but as RH decreases, the contact time behaves much the same as
before. Subsequent trials in which RH was cycled produced curves
resembling Fig. 6. Only after drying the system for several hours could
a curve resembling Fig. 5 again be recorded. These curves show evidence

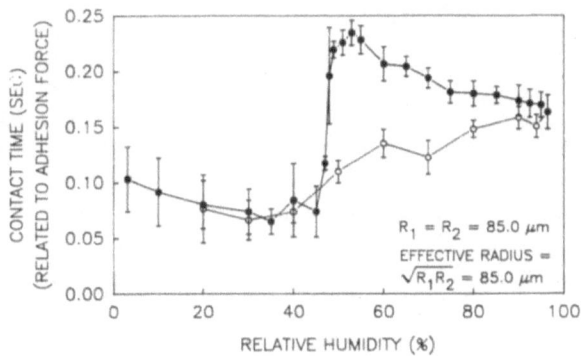

Figure 5. Contact times measured over the first of two full cycles of RH from 0% to 96% and back. (○) Increasing RH; (●) decreasing RH. Error bars show extrema in data.

of a phenomenon which is initiated by high RH, and which can survive for many minutes at low RH to influence the adhesion behavior once RH increases again. Since the curves are reproducible, the observed hysteresis is likely not due to any surface contaminants or the leaching of silica from the fibers by the water meniscus.

Figures 7 and 8 show data from a trial using the same fibers as for the trial of Fig. 4, but using ethyl alcohol as the vapor which condenses to form a meniscus, instead of water. (Figure 8 is an enlargement of the right side of the graph in Fig. 7.) Ethyl alcohol has a viscosity, density, dipole moment, and saturation vapor pressure similar to that of water, but its surface tension is less than a third of that of water. These curves have a very different shape than the prior curves for water

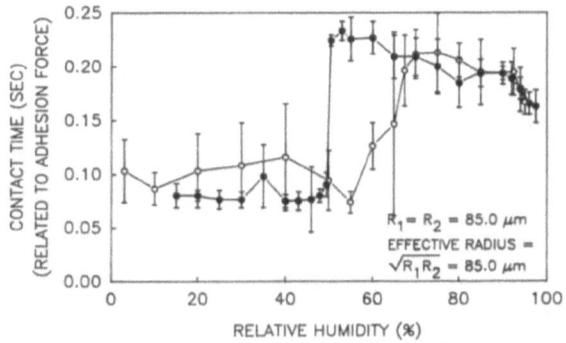

Figure 6. Contact times measured over the second of two full cycles of RH from 0% to 96% and back. (○) Increasing RH; (○) decreasing RH.   Error bars show extrema in data.

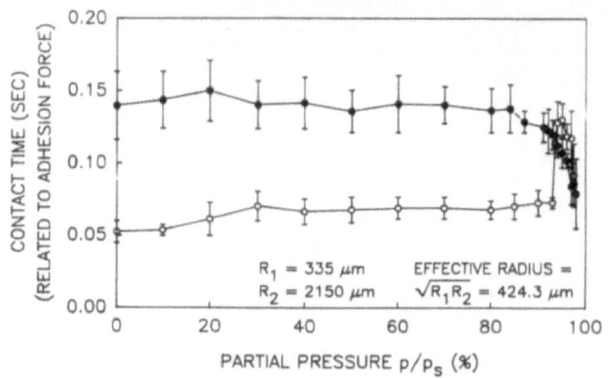

Figure 7. Contact times measured over a full cycle of partial pressure (p/p_s) of ethyl alcohol vapor. (○) Increasing p/p_s; (●) decreasing p/p_s. Error bars show extrema in data.

vapor. The curve for increasing p/p_s shows that contact times remain essentially constant and then have an abrupt peak at a high p/p_s of 94%. As p/p_s is lowered from near saturation, the contact times are first low, then rise to a maximum and stay high even as p/p_s is reduced to zero. This is a very pronounced hysteresis effect. Contact times remained high for several hours after finishing this data run.

Both the shape and magnitude of the curve for adhesion force vs. RH changes dramatically as a function of effective fiber radius. This is shown in Fig. 9. Measured values of the adhesion force for seven fiber pairs are plotted as a function of increasing RH. The curves for the smallest fiber pairs have peaks at or near saturation, while curves for larger fiber pairs have peaks at lower RH, extending down to 50%. The adhesion force's dependence on RH obviously varies with fiber radius. It

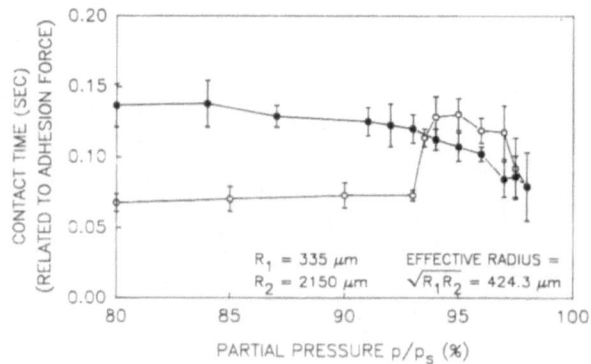

Figure 8. Enlargement of right side of Fig. 7. Error bars show extrema in data.

is important to discover how the meniscus between the fibers controls
this behavior, and if there are other phenomena responsible for this
behavior.

DISCUSSION

Fisher and Israelachvili[7] have an excellent presentation of meniscus
force predictions made by macroscopic thermodynamics. This theory is
valid for surfaces with large radii of curvature and for menisci much
smaller in extent than the surfaces' radii of curvature. We can compare
this theory to our experimental results if we keep in mind several
important reservations. First, our fibers have small radii of curvature.
At high RH, both the meniscus and the fibers are on the same size scale,
so a small meniscus approximation is not valid. Second, at low RH the

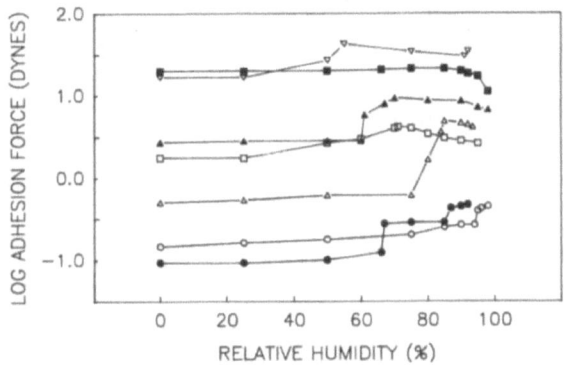

Figure 9. Adhesion force vs. increasing RH for various effective fiber
radii: (○) 17.7 μm; (●) 39.5 μm; (△) 68.9 μm; (▲) 85.0 μm;
(□) 97.5 μm; (■) 145 μm; (▽) 424 μm.

meniscus is very sharply curved, and macroscopic thermodynamics no longer
applies.[7] The liquid surface tension, normally considered a constant,
becomes a function of the liquid surface curvature. Lastly, the theory
predicts no dependence of adhesion force on RH, whereas a dependence of
adhesion force on RH is commonly witnessed.[7-9]

Figure 10 shows adhesion force values from the previous graph
plotted as a function of effective fiber radius. (Here, error bars
indicate one standard deviation in the data.) The straight line
indicates the prediction of macroscopic thermodynamics, for a meniscus
contact angle $\alpha = 0°$. The data for high RH lie on the theory curve when
radii are larger than 100 μm, and fall below it when fiber radii are
smaller than that value. The data for low RH lie consistently below
those for high RH, with the exception of the point for effective fiber
radius = 145 μm. This graph demonstrates in two ways the breakdown of
macroscopic thermodynamics' accuracy in predicting meniscus forces
between small particles. Fisher and Israelachvili[7] have already reported
that the adhesion force due to a meniscus between large (~20 mm) radius
surfaces drops below the theoretical value at RH below 90%. Our results
show that this is also true for a meniscus between surfaces of less than
100 μm effective radius, independent of RH.

One of the reasons for the theory breakdown is that as RH is held constant and particle size decreases, the size of the meniscus increases relative to the particle, until a small meniscus approximation no longer holds. The small meniscus approximation also assumes the meridional profile of the meniscus to be an arc of a circle, a poor assumption for menisci that are large relative to the particle. These drawbacks can be overcome by using "exact meniscus theory," i.e., equations for menisci based on solutions to the Laplace-Young equation,[10] combined with the Kelvin equation. This theory has the disadvantage of being computationally intensive.

In spite of its greater accuracy in modeling meniscus shape, exact meniscus theory does not predict values of adhesion forces which match experimental data. Force vs. separation curves for constant volume menisci between two spheres, measured by Mason and Clark,[11] show a maximum at small separations rather than at zero. Erle et al.[12] have shown that

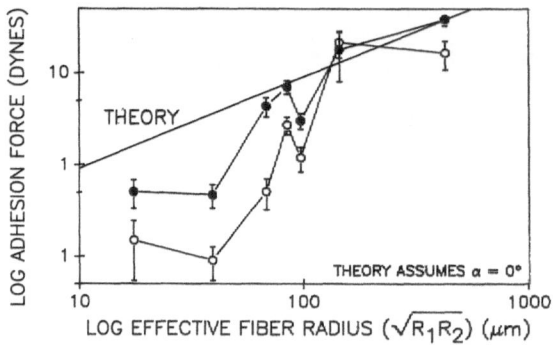

Figure 10. Adhesion force vs. effective fiber radius, for 0% (○) and 96% (●) RH, and a comparison to predictions of macroscopic thermodynamics (theory).

the theory predicts a maximum at zero separation, in disagreement with this data; but they believed the difference to be due to experimental difficulties. We have written a computer code incorporating the equations of exact meniscus theory,[10] and used this to show that the theory predicts that the maximum force occurs at zero separation for a meniscus of constant mean curvature as well (Fig. 11). Thus, whether surfaces are separated quickly (constant volume meniscus) or slowly (constant mean curvature meniscus), theory predicts that the force necessary to begin separation will be sufficient to complete the separation and split the meniscus. The computer code has been used to generate curves of adhesion force vs. RH for various sizes of spheres in contact with a plane (Fig. 12). These show no similarity to the data of Fig. 9, with the possible exception of the curve for 145 μm effective fiber radius. The reasons for the mismatch result from insufficient modeling of the contact region. First, the liquid surface tension was treated as a constant rather than as a function of radius of curvature, which would affect the shape of the curve below 90% RH. Second, surface deformation, though slight for quartz, can become appreciable as the size of the fibers and their contact region decreases. Deformation would affect the curve at lower RH, when the meniscus is small. The computer code was written with the assumption that the surfaces were undeformed.

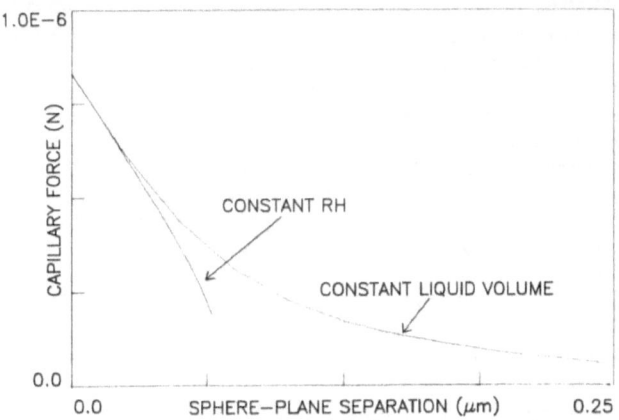

Figure 11. Exact meniscus theory calculation of capillary force between a
sphere (R = 1.0 μm) and a plane as a function of separation
distance, plotted for constant liquid volume = 0.021 μm^3 and
for constant RH = 99.0%. Contact angle = 0°.

Third, the computer code does not account for the films of water adsorbed
on the surfaces that could affect meniscus shape at high RH.

Various researchers have reported the presence of an adsorbed film
of liquid on surfaces exposed to an unsaturated vapor of the liquid,[13-17]
including the case of water adsorbed on quartz. We have observed the
growth of such an adsorbed layer on a sub-micron quartz fiber using

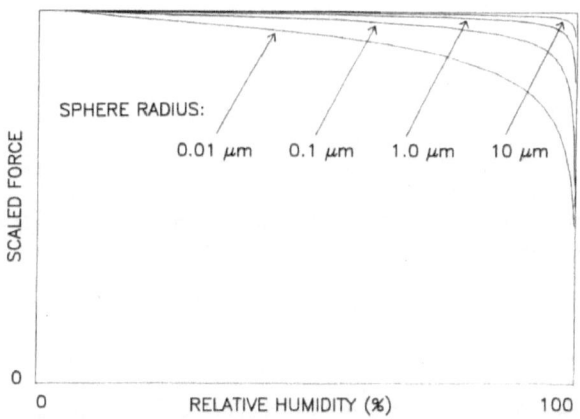

Figure 12. Exact meniscus theory calculation of capillary force between a
sphere and a plane as a function of RH, plotted for four
sphere radii. Sphere-plane separation = 0.0 μm. Contact
angle = 0°.

light-scattering techniques. The four unique, polarized light-scattering Mueller matrix elements[18] were measured for a circularly cylindrical quartz fiber in a dry nitrogen atmosphere using a light-scattering nephelometer, and the fiber size was determined by matching the matrix elements with those of exact light scattering theory for a cylinder. For this particular fiber radius, the matrix element S_{12} was most sensitive to changes in fiber radius. This element is related to the percentage of horizontally polarized light scattered by a sample illuminated by unpolarized light, as a function of scattering angle. Positive 100% polarization corresponds to all-horizontal polarization, and negative 100% to all-vertical polarization. This matrix-element curve was measured for various RH values as the RH around the fiber was increased. Figure 13 shows the S_{12} matrix element curve for a cylindrical quartz fiber in dry nitrogen and a second curve taken with the same fiber in humid nitrogen. Changes in this matrix element match predictions of light scattering theory for a cylindrical fiber clad with a concentric second layer. By matching the theory to the data, the water film thickness can be determined with an uncertainty of less than a nanometer.

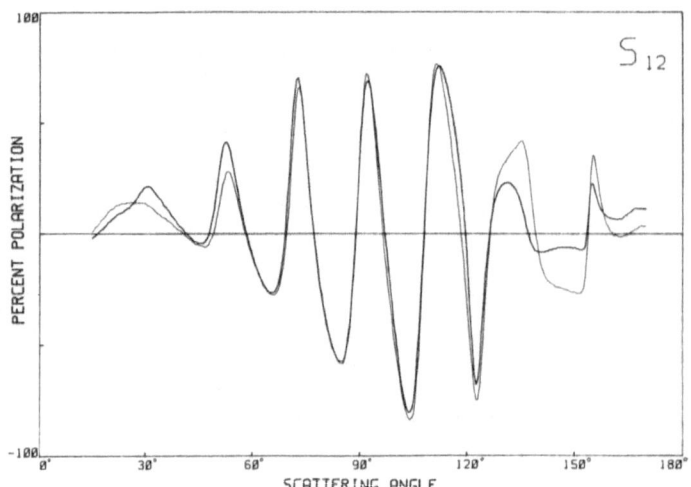

Figure 13. Experimental light-scattering Mueller matrix element S_{12}, the percentage of horizontally polarized scattered light, for a circularly cylindrical quartz fiber illuminated by unpolarized light in dry nitrogen gas (dark curve) and in nitrogen gas at 98% RH (light curve), showing change in the light-scattering signal due to build-up of a surface layer of condensed water.

Evidently, an absorbed layer of water coats the somewhat larger fibers used in our adhesion measurements. The presence of such a wetting film will alter the geometry of the meniscus formed between the fibers, as well as the adhesion force, though the theoretical effect of this alteration is not known. However, researchers are still studying the influence of adsorbed layers on menisci in simpler geometries, such as planar gaps.[16]

CONCLUSIONS

It is likely that the presence of adsorbed layers on quartz fibers causes some or all of the observed hysteresis phenomena. This hysteresis effect changes character and becomes very pronounced when water is replaced with liquid ethyl alcohol. Both the adsorbed layers and the meniscus which forms in the contact region influence the shape of the adhesion force vs. RH curves, though it is not yet clear how they do so. Surface deformation also likely influences the curve shape, especially at lower RH. The sharp discontinuities in mid-range on some force vs. RH curves may indicate the upper limit in RH of this influence. We are continuing to examine the effect of surface deformation in our experiments.

The adhesion force drops below the Kelvin equation prediction for effective fiber radii below 100 μm. This effect may be explained by the presence of the adsorbed layers and by the liquid meniscus' very small meridional and neck radii of curvature. The surface tension at the water-air interface may be different from that of bulk water for these small radii of curvature, and the exact functional dependence of surface tension on curvature is unknown.[17] Experiments similar in concept to this one, using repeated contacts between surfaces of known geometry, can be used to map out this dependence. Before this can be done, though, we need better control and prevention of any surface deformation, as well as a better understanding of the interaction between the meniscus and adsorbed surface layers.

ACKNOWLEDGMENTS

We wish to thank the University of Arizona's Center for Microcontamination Control for support of this research.

REFERENCES

1. J. N. Israelachvili, "Intermolecular and Surface Forces with Applications to Colloidal and Biological Systems," Academic Press, London, 1985.
2. R. A. Bowling, J. Electrochem. Soc., 132, 2208-2214 (1985).
3. L. R. Fisher and J. N. Israelachvili, J. Colloid Interface Sci., 80, 528-541 (1981).
4. J. C. Melrose, A.I.Ch.E.J., 12, 986-994 (1966).
5. B. V. Deryagin, N. A. Krotova, and V. P. Smilga (R. K. Johnston, Translator), "Adhesion of Solids," Consultants Bureau, New York, 1978.
6. W. S. Bickel and T. M. Wentzel, in "Particles on Surfaces 1: Detection, Adhesion, and Removal," pp. 225-236, K. L. Mittal, Editor, Plenum Press, New York, 1988.
7. L. R. Fisher and J. N. Israelachvili, Colloids Surfaces, 3, 303-319 (1981).
8. A. D. Zimon (R. K. Johnston, Translator), "Adhesion of Dust and Powder," 2nd Edition, Consultants Bureau, New York (1982).
9. W. J. Whitfield, in "Surface Contamination: Genesis, Detection, and Control," Vol. 1, pp. 73-82, K. L. Mittal, Editor, Plenum Press, New York (1979).
10. F. M. Orr, L. E. Scriven, and A. P. Rivas, J. Fluid Mech., 67, 723-742 (1975).
11. G. Mason and W. C. Clark, Chem. Eng. Sci., 20, 859-866 (1965).
12. M. A. Erle, D. C. Dyson, and N. R. Morrow, A.I.Ch.E.J., 17, 115-121 (1971).

13. N. L. Cross and R. G. Picknett, Trans. Faraday Soc., 59, 846-855 (1963).

14. R. M. Pashley and J. A. Kitchener, J. Colloid Interface Sci., 71, 491-500 (1979).

15. L. R. Fisher, Adv. Colloid Interface Sci., 16, 117-125 (1982).

16. Z. Zorin, D. Platikanov, and T. Kolarov, Colloids Surfaces, 22, 147-159 (1987).

17. H. K. Christenson, J. Colloid Interface Sci., 121, 170-178 (1988).

18. B. W. Bell and W. S. Bickel, Appl. Optics, 20, 3874-3879 (1981).

ADHESION INDUCED DEFORMATIONS BETWEEN PARTICLES AND SUBSTRATES

L. P. DeMejo, D. S. Rimai and R. C. Bowen

Copy Products R & D, Eastman Kodak Company, Rochester, New York 14650 and *Analytical Technology Division, Eastman Kodak Company Rochester, New York 14650

Deformations of planar substrates and contacting particles arising from adhesion forces had been postulated long ago. We report direct observations of such deformations between submicron particles of polyvinylidene fluoride (PVF_2) and a polyester–siloxane block copolymer or polished silicon substrates and between approximately 2–3 μm nickel particles and the same substrates. These observations were made using scanning electron microscopy (SEM). For the case of the silicon substrate, only the relatively soft particles of PVF_2 were observed to deform. On the other hand, substantial deformations of the polyester–siloxane block copolymer were observed when contacted by either type of particle. The magnitude of the deformation of the polyester–siloxane substrate in contact with the PVF_2 particles was calculated, assuming that the adhesion forces arise from van der Waals interactions and from the interfacial tension between the contacting surfaces. The cases of only Hertzian (elastic) and of both elastic and plastic deformation were considered. It was found that previously proposed models do not accurately predict the diameter of the contact area. However, the observed diameter was accurately estimated by combining the plastic response model of Krupp with the interfacial energy model proposed by Johnson, Kendall, and Roberts. The contributions of various experimental artifacts, such as joule heating, space charge effects resulting from the electron beam in the SEM, and the effects of conductive sputtered coatings on the observations, are also discussed.

1. INTRODUCTION

An understanding of particle–to–substrate adhesion is important both from a fundamental scientific viewpoint as well as in many technological areas. This is illustrated by the wealth of scientific literature discussing such topics as adhesion and van der Waals interactions.[1-7] The magnitude of the adhesion force between materials depends upon their contact area. However, the adhesion forces can cause the materials to deform and, thereby, increase their contact area. This would further increase the adhesion. This effect has long been recognized.[8,9] Derjaguin[8] first calculated the expected increase in contact area assuming that the deformations were elastic (Hertzian). This was later expanded upon by Krupp[9] to include both elastic and non–elastic (plastic) response of the materials to the adhesion induced stress.

Despite the scientific and technological significance of understanding surface force induced deformations, little has been done to directly measure and quantify such effects for particles of the order of a micrometer. Recently, observations of such deformations were made using nominally 3 μm diameter gold spheroids on an electrically conducting substrate.[10] However, since these materials are electrically conducting and the particles are much harder than the substrate which deformed, the results and experimental techniques could not be generalized to many common materials. This prompted additional studies of surface force deformations caused by soft, insulating, spherical particles on even softer, insulating substrates.[11] In this paper observations of adhesion induced deformations of a soft, electrically insulating, planar substrate (a polyester–siloxane block copolymer henceforth referred to as PSBC) by submicron particles of polyvinylidene fluoride (Kynar 301F™) and by 2–3 μm diameter particles of nickel are reported. These observations are contrasted to a lack of any observable deformation when much harder polished silicon wafers are used as the substrate. Finally, the deformations resulting from the PVF$_2$ particles on the PSBC substrate are analyzed in terms of the Lifshitz model of adhesion, based on van der Waals interactions, and the measured crater diameters are compared to predictions assuming elastic (Hertzian) response, plastic (non–elastic) response, and interfacial tension effects.

2. EXPERIMENTAL

In this study particles of PVF$_2$ (Kynar 301F™, manufactured by Pennwalt), having an average diameter of approximately 0.3 μm and a Young's[12] modulus of approximately 1×10^9 N/m^2 were deposited onto polished silicon wafers (Young's modulus approximately 1×10^{11} N/m^2) and onto the PSBC (Young's[13] modulus approximately 2×10^7 N/m^2). These materials were chosen because the particles were small, spherical, essentially monodisperse in size, and lacked asperities, which could complicate observations and analysis. The substrates were sufficiently smooth and had the proper hardnesses to facilitate the observations. Particles of nickel, with diameters typically of the order of 2 μm, were also sprinkled onto these substrates. While these particles were neither spherical nor monodisperse, they were sufficiently hard so as not to visibly deform. Moreover, the asperities on these particles allowed the thickness of the sputtered conductive coating to be directly estimated. The samples were imaged using the secondary electron emission from an SEM. To avoid a beam induced surface charge, which would distort the image pattern and introduce artifacts, the samples were first coated with a conductive layer inside a Polaron E5100 high resolution sputter coater. Typically, a 12 nm thick 60:40 gold/palladium (Au/Pd) coating was deposited onto the samples by sputtering with argon for 45 seconds at 2.5 kV and 20 mA. The temperature rise during the sputtering process, under these conditions, was found to be less than 20°C. While this temperature increase should not be significant, the samples were mounted on a cold stage during sputtering so as to further reduce any temperature rise.

The samples were mounted onto cross sectional stubs. These permitted viewing them at high tilt angles. The samples were then placed into either a Phillips 515 SEM or a JEOL 100 CX TEM (transmission electron microscope) with a side entry goniometer and an auxiliary secondary imaging detector, which allowed the TEM to be used as an SEM. Typically, a 30 kV accelerating voltage and a 10 nm beam size were used. Appropriate magnifications typically ranged from 20,000X to 50,000X. It was observed that, after viewing a region for several minutes, a broad deformation could be induced in some substrates by joule heating caused by the beam. This effect was readily distinguishable from the adhesion induced deformations reported herein. Moreover, viewing times were kept under one minute so as to avoid this artifact. The coating and viewing of the samples were done approximately 2 weeks after the particles were deposited on the substrates.

3. RESULTS

Examples of SEM micrographs of the adhesion induced deformations by the PVF$_2$ particles at 80° and 88° tilt angles are shown in Figs. 1 and 2, respectively. Micrographs of the particles on the silicon wafer are shown in part A of each figure while the micrographs of the particles on the PSBC substrate are shown in part B.

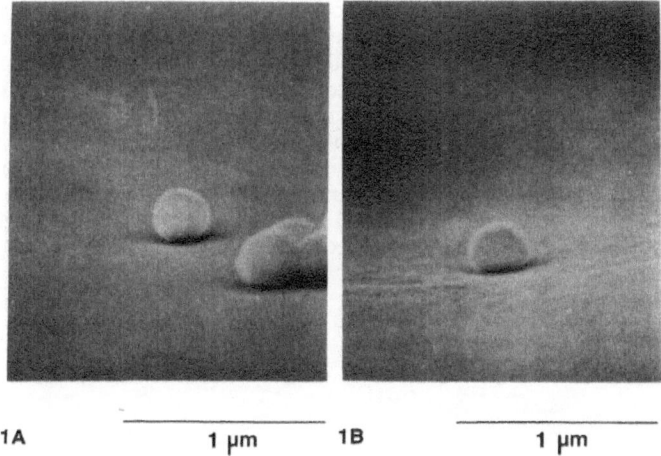

1A   1 µm   1B   1 µm

Fig. 1. SEM micrographs of particles of Kynar 301F™ on a silicon wafer (1A) and on a polyester–siloxane block copolymer (1B), as viewed with the normal to the substrate tilted 80° from the electron beam. The particles are observed to embed into the soft polymeric substrate but not into the silicon, although the Kynar 301F™ particles on the silicon wafer appear to flatten. The measured crater diameters are consistent with predictions based on models assuming the validity of van der Waals type interactions when plastic response is taken into account combined with contributions arising from interfacial tension.

      Several interesting features are apparent in these figures. As can be seen in Fig. 1, the particles appear to embed deeply into the PSBC. Some distortion also appears to occur when the particles are in contact with the silicon wafer. In the former case, where the particles are much harder than the substrate, the deformation is attributable to the substrate deforming. In the latter example the relative hardness of the silicon precludes any appreciable deformation of the substrate. Rather, the particles, themselves, are deforming. This is more clearly discernible from the electron micrographs shown in Fig. 2, which were taken with the sample viewed almost on edge. For a random sample of 10 PVF_2 particles on the polyester–siloxane block copolymer, the mean diameter of the crater and the standard deviation are listed in Table I.

Table I. Diameter of the Crater Formed by the Kynar 301F™ Particles on the Polyester–Siloxane Block Copolymer Substrate

Number of particles	10
Mean diameter of crater	0.23 µm
Standard deviation	0.02 µm

Space charge effects in scanning electron microscopy frequently give rise to artifacts which can distort images. To ensure that no such effects had occurred in these images, nickel particles were deposited onto a silicon wafer and this sample was examined by SEM. In this instance both materials are hard. Figure 3 shows an electron micrograph of a nickel particle on the silicon substrate. No apparent deformation of either the particles or the substrate is observed, despite the fact that the particle is contacting the substrate at only one "point" and, therefore, should be exerting considerable pressure on the substrate. Since the resolution of the SEM is sufficient to readily resolve "point" contacts such as the one shown in Fig. 3, it is most probable that the deformations reported in this study are real rather than experimental artifacts.

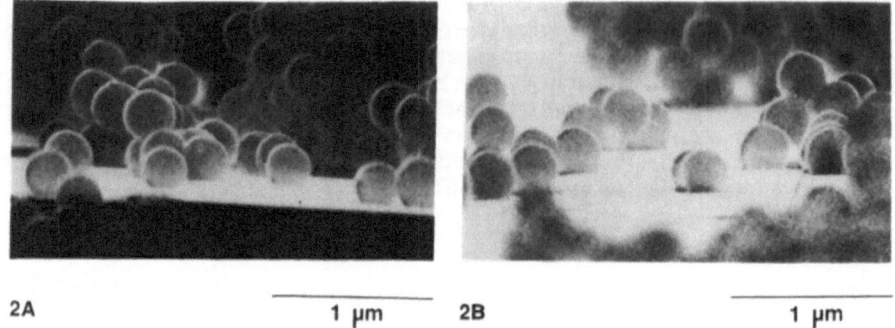

2A   1 μm   2B   1 μm

Fig. 2. SEM micrographs of Kynar 301F^{TM} particles on a silicon wafer (2A) and on a polyester–siloxane block copolymer substrate (2B) as viewed with the normal to the substrates by an 88° angle to the electron beam. The flattening of the particles against the silicon substrate is clearly observable.

Another factor which can give rise to artifacts which simulate embedding is the application of an electrically conductive layer on the samples to avoid a charge buildup by the electron beam. As previously discussed, a 60:40 gold/palladium alloy was sputter coated onto the samples after the particles were deposited onto the substrate. If the coating becomes sufficiently thick, it is conceivable that the particles would start to become buried. In this study, for the submicron PVF$_2$ particles, it was estimated that the particle diameter to coating thickness ratio was approximately 30:1. A coating of this thickness would not be expected to significantly bury the particles. This is supported by calculations which model the effect of coating thickness on the appearance of the particle–substrate system and by further experimental evidence, presented elsewhere.[10,11] It was found that the 12 nm thick gold/palladium coating, used in this study, was sufficient to eliminate most of the beam induced space charge. Residual charge may still be present at the particle–substrate interface and may have caused the shadows which are evident in this region in both Figs. 1 and 2. This suggests that it may be difficult to obtain a sufficiently conductive layer to avoid space charge effects without masking the indentations if the particle diameter is significantly smaller than those used in this study.

Further information can be obtained by comparing nickel particles on silicon wafers to those on the PSBC. For consistency, both samples were sputter coated with 60:40 gold/palladium, even though both the nickel and the silicon are sufficiently conductive electrically not to require it. Samples coated for 45, 180, and 720 seconds are shown in Figs. 4–6. In each instance silicon substrates are shown in part A, while micrographs of the PSBC substrate are shown in part B. After sputter coating for 45 seconds, under the same conditions as had been used previously in this study, the nickel particles on the

1 µm

Fig. 3. SEM micrographs of a nickel particle on a silicon substrate. No conductive layer was sputter coated in this example. Within the limits of resolution, no deformation of either the particle or the substrate is observable. This result demonstrates that the reported deformations, such as those shown in Figs. 1 and 2, are real rather than experimental artifacts.

silicon wafers still clearly show undeformed regions of contact, as can be seen in Fig. 4A. This is similar to observations using uncoated samples, such as that shown in Fig. 3, and supports the argument that the coatings are sufficiently thin so as not to introduce additional artifacts. On the other hand, the nickel particles appear to embed into the PSBC, as is evident from Fig. 4B. The estimated coating thickness is 12 nm. After sputter coating for 180 seconds (estimated thickness is approximately 48 nm), some of the asperities, which are of the order of approximately 0.2 µm, appear to now end at the interface with the silicon, with more planar contact (Fig. 5A). Embedding is still evident for the nickel particles on the PSBC substrate (Fig. 5B). This may be due to the coating now becoming sufficiently thick so as to bury the contact points. After coating for 720 seconds (Fig. 6A and 6B), the particles appear well coated with a fluffy material (which is, presumably, the

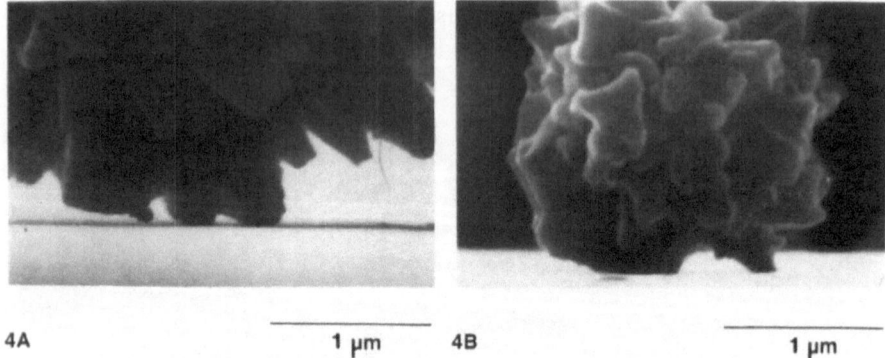

4A   1 µm   4B   1 µm

Fig. 4. SEM micrographs of nickel particles on a silicon wafer (4A) and on a polyester–
siloxane block copolymer (4B). A 60:40 gold/palladium alloy was sputter coated
in argon onto these materials at 2.5 kV and 20 mA for 45 seconds. The estimated
coating thickness was 12 nm. No apparent deformations are seen when the nickel
is on the silicon. The appearance of the particle–substrate interface in 4A is
similar to that of the uncoated sample in Fig. 3. The nickel particle appears to
embed into the PSBC substrate, in a qualitatively similar manner to that which
was observed for the PVF_2 samples in Figs. 1B and 2B.

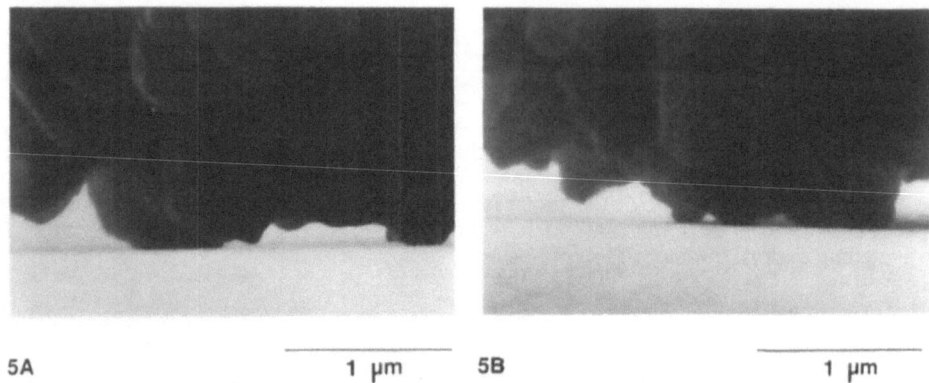

5A   1 µm   5B   1 µm

Fig. 5. SEM micrographs of nickel particles on silicon (5A) and on PSBC (5B) substrates
after sputter coating 60:40 gold/palladium for 180 seconds. The coating thickness
was estimated to be approximately 48 nm. The results are similar to those shown
in the previous figures, suggesting that the coating is still sufficiently thin so as
not to substantially mask effects.

gold/palladium) and to have embedded into both substrates. The latter observation is strictly an artifact due to the coating thickness, which is now estimated to be approximately 200 nm, or the same size as the asperities. These results support the argument that the deformations reported with the samples of PVF2 are real rather than experimental artifacts, as well as confirming the estimated coating thickness of the conductive layer.

4. DISCUSSION

The surface diameter of the crater caused by the adhesion forces between the substrate and the particle can be calculated using several different models. For simplicity only the case of hard, spherical particles in contact with a soft substrate will be considered. This occurs when the particles of PVF2 contact the PSBC substrate. Although it

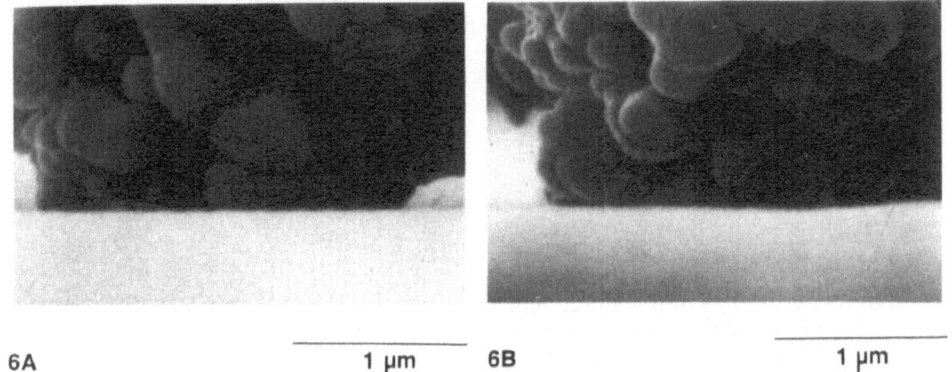

6A   1 μm   6B   1 μm

Fig. 6. SEM Micrographs of nickel particles on silicon (6A) and PSBC(6B) substrates after sputter coating 60:40 gold/palladium for 720 seconds. The coating thickness was estimated to be approximately 200 nm, which is of the same magnitude as the asperities. The particles appear to be coated with a fluffy material, which is, presumably, gold/palladium. The apparently flat interface, particularly between the nickel and the silicon, show that, for coatings of this thickness, the particles are buried up to the height of the asperities. This confirms the estimates of the thickness of the sputter coating conducting layer.

would be interesting to more quantitatively study the indentation caused by nickel particles on the two substrates, the irregular shape and the relatively large asperities of these particles make this overly complicated.

If the deformation is sufficiently small so that the elastic limit of the substrate is not exceeded, the diameter of the indentation arising from the adhesion forces can be calculated assuming Hertzian response. In this instance the radius of the indentation, ρ_0, is related to the force of adhesion between the particle and the undeformed substrate, F^0, the Young's modulus of the substrate, E, the Poisson ratio, ν, and the radius of the particle, R, by the equation[3]

$$\rho_0^3 = 3/4\ F^0\left(\frac{1-\nu^2}{E}\right)R \qquad\qquad [1]$$

F^0 can be calculated, assuming van der Waals type interactions, using the relationship

$$F^0 = \frac{h\omega}{8\pi z_0^2} R \qquad [2]$$

Using typical values reported by Krupp,[3] $z_0 = 4$ Å (which is comparable to the interatomic spacing of van der Waals bonded solids), $\hbar\omega = 2.5$ eV, and $R = 0.1$ μm, one finds that $F^0 = 10^{-8}$ N. This value is consistent with estimates made by extrapolating measured adhesion forces down to particles of this size.[15] If it is further assumed that $\nu = 1/2$, then it is found from Equation 1 that $2\rho_0 = 0.015$ μm. As stated in Table I, the average measured crater diameter is 0.23 μm. Reasonable errors in parameters such as F^0 or E would not significantly affect ρ_0. Therefore, the diameter of the crater cannot be calculated assuming only Hertzian deformations. This is not totally surprising because, as is evident from the micrographs, the distortions are relatively large and, therefore, likely to exceed the elastic limit of the PSBC.

Krupp[3] assumed that both elastic and non–elastic (plastic) deformations can occur resulting from particle–substrate adhesion. By assuming a time dependent hardness, $H(t)$, and dividing the crater into two regions—an inner, high stress area of plastic response and an outer, low stress area of elastic behavior, he was able to calculate the diameter of the surface deformation as well as relative sizes of the areas of elastic and non–elastic deformation. In his model ρ_0 is the radius of the entire

$$\pi\rho_0^2 = \frac{F^0}{H} + \frac{1}{3}\left[\frac{\pi^{3/2}(1-\nu^2)RH}{2E}\right]^2 \qquad [3]$$

and

$$\pi\rho_1^2 = \frac{F^0}{H} - \frac{2}{3}\left[\frac{(\pi^{3/2}(1-\nu^2)RH}{2E}\right]^2 \qquad [4]$$

crater and ρ_1 is the radius of the inner, high stress region. Using Krupp's[3] relationship between the hardness and the Young's modulus, $H = 2 \times 10^{-3}E$, it is found that $\rho_0 = \rho_1$. This implies that, according to this model, virtually the entire deformation is plastic. The small contribution to the crater size from elastic compression, as calculated from the Krupp model, is consistent with that obtained from Hertz. It was also found that $2\rho_0 = 0.11$ μm. While this is the correct magnitude, it is only half of the observed response. Again, reasonable errors in the parameters used in calculating the diameter according to this model should not significantly alter the results. Therefore, although the Krupp model gives more accurate predictions than does the Hertz model, it still is not fully satisfactory. The discrepancy between the measured crater diameter and those predicted by Krupp probably arise from two sources. First, the Krupp model only applies to adhesion induced compression. As will be discussed shortly, tensile effects arising from the forces of adhesion can significantly affect the crater diameter. Second, Hertz's model of compression assumes that the deformation is small compared to the diameter of the indenter. Any theory which builds on Hertz's theory, such as that proposed by Krupp, implicitly makes that assumption. This assumption is not valid here. However, a discussion of large deformations would involve solving the problem of large scale nonlinear stress–strain response of polymeric materials. This is beyond the scope of this paper.

Johnson, Kendall, and Roberts (hereafter referred to as JKR) proposed a model to explain an observed increase in contact area between spheres beyond that which would have been expected to be caused by the applied load. While the JKR model assumes elastic response, it differs from the classical Hertzian model in that, while Hertz assumed only compressive stresses, the JKR model assumes compressive stresses towards the center of the contact and tensile stresses near the edge. With no externally applied load, the JKR model predicts that two elastic spheres in contact will form a circular contact area having a radius, a, such that

$$a^3 = R(6w_A\pi R)/K \qquad [5]$$

56

where w_A is the work of adhesion, $R = R_1R_2/(R_1 + R_2)$, where R_i is the radius of the ith sphere and

$$K = 4/[3\pi(k_1 + k_2)] \tag{6}$$

The values of k_i are related to the Poisson ratio of the material, v_i, and the Young's modulus, E_i, by:

$$k_i = \frac{1 - v_i^2}{\pi E_i} \tag{7}$$

For the case of a relatively hard sphere contacting a compliant plane, and assuming that the work of adhesion, w_A, equals 0.030 J/m^2 (which is a typical value for these types of materials), it is found that $2a = 0.11$ μm. This is approximately equal to the diameter of the crater as predicted by the Krupp model and, therefore, approximately half of the observed diameter. As in the cases of the other two models, the discrepancy between the measured and observed crater surface diameters cannot be accounted for by reasonable errors in the choice of parameters. Therefore, the diameter of the crater is not accurately predicted by a simple application of the JKR model. The results of these calculations are summarized in Table II.

Table II. Comparison of predictions of various models with the observed crater diameter

	Hertz	Krupp	JKR	Combined Non-elastic Compression and Tensile Effects	Observed
Diameter(μm)	0.015	0.11	0.11	0.22	0.23

The JKR model, because it is based on the Hertz model, assumes both elastic response and small deformations compared to the particle diameter. Neither of these assumptions appear valid in this instance.

It should be noted that the JKR model calculates the contributions to adhesion induced crater diameters assuming that both elastic compression and tensile effects are significant. However, as previously shown using both the Krupp model and the Hertz model of compliance, the contribution to the crater diameter due solely to elastic compression is small. Therefore, it can be argued that the JKR radius is mainly due to tensile effects rather than compression. Thus, it appears that both tensile effects and nonelastic compression contribute significantly to the size of the adhesion induced crater. Indeed, if the tensile contribution is simply added to the non-elastic compressive term, it is seen that the calculated diameter equals 0.22 μm. This is in good agreement with the statistically measured crater diameter of 0.23 μm.

There has been much work in recent years to expand the JKR model to include non-elastic response, as suggested by Maugis and Bauquins.[16] Maugis related the force of adhesion and the radius of contact to the elastic yield point and hardness of metals.[17] This was further advanced by Maugis and Barquins,[18] who studied deformations resulting between punches and substrates. Further studies were reported by Pollock,[19,20] Pollock, Shufflebottom, and Skinner,[21] Pashley and Tabor,[22] Chowdhury and Pollock,[23] and Maugis and Pollock.[24] Derjaguin, Muller, and Toporov,[25] and Muller, Yushchenko, and Derjaguin[26] expanded upon elastic models.

Most of these papers deal with metallic contacts. The stress–strain relationships, hardness and yield strength characteristics, and rheological properties of metals differ significantly from those of polymers. Therefore, the comparison of the predictions of these models with the present observations would require the experimental determination of parameters which are beyond the scope of this manuscript.

5. CONCLUSIONS

Deformations resulting from particle–substrate adhesion have been observed by SEM. For the case of submicron spherical particles of PVF_2 in contact with a polyester–siloxane block copolymer the mean surface diameter of the craters formed was 0.23 μm with a standard deviation of 0.02 μm. Calculations assuming Hertzian compression of the substrate, arising from adhesion forces, predicted craters much smaller than those observed. However, nonelastic compression, such as predicted by Krupp and tensile effects such as those proposed by Johnson, Kendall, and Roberts both appear to contribute significantly to the size of the adhesion induced crater. Qualitatively similar effects were observed for nickel particles, approximately 3 μm in diameter, on the same substrate. When the particles were placed on polished silicon wafers, no deformation of the substrate was observed, although the particles of PVF_2 appeared to flatten.

ACKNOWLEDGEMENTS

The authors would like to thank Drs. R. Bucks, A. Chowdry, P. Hartley, J. May, J. Minter, and Mr. T. Poeth for their valuable comments and suggestions.

REFERENCES

1. K. L. Mittal, Editor, "Surface Contamination: Genesis, Detection, and Control," Vol. I, Plenum Press, New York (1979).
2. A. D. Zimon, "Adhesion of Dust and Powders," Consultants Bureau, New York (1982).
3. H. Krupp, Adv. Colloid Interface Sci. *1*, 111 (1967).
4. J. M. Georges, Editor, "Microscopic Aspects of Adhesion and Lubrication," Elsevier, Amsterdam (1982).
5. L. N. Rogers and J. Reed, J. Phys. D. Appl. Phys. *17*, 677 (1984).
6. K. L. Johnson, K. Kendall, and A. D. Roberts, Proc. R. Soc. Lond. A. *324*, 301 (1971).
7. J. Visser, in "Surface and Colloid Science," E. Matijevic, Editor, Vol. 8, pp. 3–84, Wiley, New York (1976).
8. B. V. Derjaguin, Kolloid Z. *69*, 155 (1934).
9. R. S. Bradley, Trans. Faraday Soc. *32*, 1088 (1936).
10. D. S. Rimai, L. P. DeMejo, and R. C. Bowen, J. Appl. Phys., *65*, 755 (1989).
11. L. P. DeMejo, D. S. Rimai, and R. C. Bowen, J. Adhesion Sci. Technol. *2*, 331 (1988).
12. Materials Engineering, Materials Selector (1986).
13. •Plastics Technology, Materials Buyer Guide (1983/1984).
14. J. W. May, unpublished results (1987).
15. See, for example, D. F. St. John and D. J. Montgomery, J. Appl. Phys. *42*, 663 (1971). Also, D. S. Rimai, unpublished results (1982).
16. D. Maugis and M. Barquins, J. Phys. D: Appl. Phys. *11*, 1989 (1978).
17. D. Maugis, in "Microscopic Aspects of Adhesion and Lubrication," J. M. Georges (Editor), pg. 221, Elsevier, Amsterdam (1982).
18. D. Maugis and M. Barquins, J. Phys. D: Appl. Phys. *16*, 1843 (1983).
19. H. M. Pollock, Vacuum, *31*, 609 (1981).
20. H. M. Pollock, J. Phys. D: Appl. Phys. *11*, 39 (1978).
21. H. M. Pollock, P. Shufflebottom, and J. Skinner, J. Phys. D: Appl. Phys. *10*, 127 (1977).
22. M. D. Pashley and D. Tabor, Vacuum, *31*, 619 (1981).
23. S. K. Roy Chowdhury and H. M. Pollock, Wear, *66*, 307 (1981).
24. D. Maugis and H. M. Pollock, Acta Metall. *32*, 1323 (1984).
25. B. V. Derjaguin, V. M. Muller, and Yu P. Toporov, J. Colloid Interface Sci., *53*, 314 (1975).
26. V. M. Muller, V. S. Yushchenko, and B. V. Derjaguin, J. Colloid Interface Sci. *77*, 91 (1980).

THE EFFECT OF EXTERNAL NOISES ON THE ATTACHMENT OF PARTICLES TO SOLID

SURFACES

J.P. Hsu

Department of Chemical Engineering
National Taiwan University
Taipei, Taiwan 10764
Republic of China

A stochastic analysis of the reversible attachment of
particulate material to solid surfaces in a batch system is
presented. The effect of external noises on the transient
behavior of the phenomenon is investigated through solving
the stochastic differential equation describing the system.
Two types of random sources have been considered: i) a random
initial condition, and ii) a random operating environment
which leads to a differential equation with random coefficients. The results of Monte Carlo simulation reveal that these
random factors can have significant effect on the transient
number of attached entities; their effect on its steady-state
value, however, is negligible.

INTRODUCTION

The attachment of particles to solid surfaces is one of the most
often observed phenomena in natural and physical processes. The analysis
of this phenomenon was pioneered by Langmuir[1], who investigated the adsorption of gas molecules onto a solid surface. The basic idea is that
the solid surface contains a number of active sites where the collision
of a gas molecule will result in an attachment. This mechanism has been
found to be successful in interpreting various interactions between particles and surfaces[2-8]. More often than not, the derivation of the
governing equation describing the dynamic behavior of the phenomenon
under consideration is based on deterministic principles. In studying
the transient behavior of bacterial adhesion to a substrate surface, a
stochastic analysis is presented[5,6,9]. The description of the evolution
of the phenomenon in the analysis is based on probabilistic laws. The
resultant model is capable of providing the information about its mean as
well as its fluctuating characteristic, and, therefore, is a generaliza-
tion of the corresponding deterministic model. It should be pointed out,
however, that in the stochastic approach, the random nature of the phenomenon is due to the microscopic characteristic of bacterial cells. In
other words, the random nature stems from the nonuniform behavior of individual entities (cells). Since it is induced by a source inside the
system, the term "internal noise" is usually used to represent this
fluctuation [10].

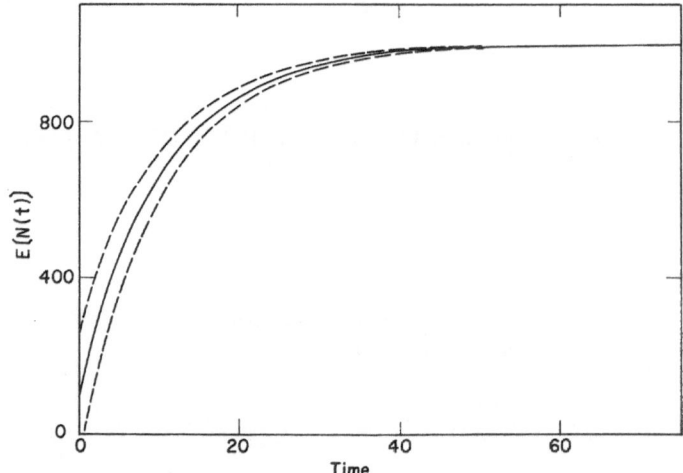

Figure 1. Result of simulation for the case of a random initial condition: $k_1=k_2=1.0 \times 10^{-5}$, $\lambda=0.01$. Solid line represents the variation of $E[N(t)]$, the dashed lines denote an approximate 95% confidence band.

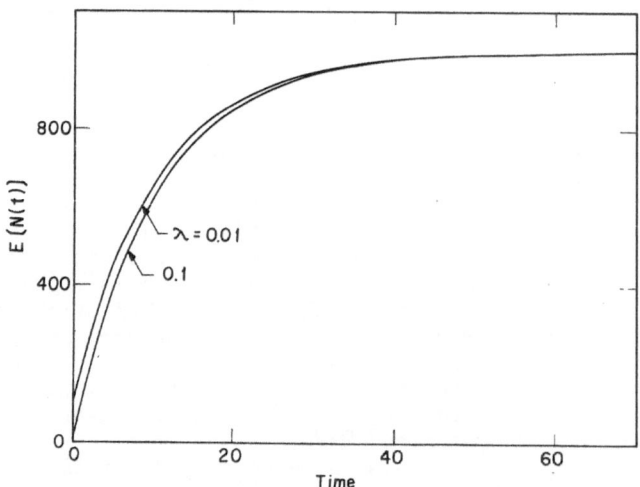

Figure 2. Variation of $E[N(t)]$ as a function of λ: $k_1=k_2=1.0 \times 10^{-5}$.

In reality, the fluctuating behavior of a dynamic system is also induced by other random sources. For instance, it is known that a variation in the medium temperature will affect cell adhesion[11]. Thus a fluctuation in the operating temperature around a certain set point will induce noises in the transient adhesion curve. Moreover, it has been found that a small error in the value of the initial conditions produces an enormous error in later predictions about the process[12]. This means that a finite uncertainty can make future predictions meaningless. These types of randomness are often due to the sources outside the system. To differentiate the randomness induced by an internal source, the term "external noise" is adopted[10]. The main objective of the present study is to examine the influence of external noises on the attachment of particles to solid surfaces in a batch system.

THEORETICAL CONSIDERATION

The analysis is begun by letting $N(t)$, M, and S be the number of attached particles at time t, total number of particles in the system, and the number of available sites on the solid surface, respectively. Here, we assume that the rate of attachment of particles to surface is proportional to the product of the number of unattached particles and the number of empty sites on the surface; the rate of detachment of the attached particles from the surface is proportional to the number of attached particles, i.e.,

$$\frac{dN(t)}{dt} = k_1(M-N)(S-N) - k_2N \qquad (1)$$

where k_1 and k_2 are rate constants. The solution of this equation subject to an appropriate initial condition yields the expression for the time dependence of the number of attached particles.

Random Initial Condition

Suppose that the initial value of the number of attached particles can not be determined exactly, due to, for example, instrumental limitations or other factors. The problem at hand may then be represented by a stochastic differential equation with a random or uncertain initial condition. The solution of Equation (1) subject to the initial condition $N(0)=N_0$ gives

$$\frac{N_0-r_2}{N_0-r_1} = \frac{N-r_2}{N-r_1} \exp[k_1t(r_1-r_2)] \qquad (2)$$

where N_0 is a random variable representing the number of particles on the surface initially, r_1 and r_2 satisfy the following quadratic equation of r

$$r^2 - (M+S+k)r + MS = 0 \qquad (3)$$

with $k=k_2/k_1$. Equation (2) becomes, after rearrangement

$$N_0 = \frac{(r_1X - r_2)}{(X - 1)} \qquad (4)$$

where $X=(N-r_2)\exp[k_1t(r_1-r_2)]/(N-r_1)$. Note that $N(t)$ is randomly distributed since N_0 is a random variable. If we denote $f_{N0}(n_0)$ as the probability function of N_0, it can be shown that the probability density function of $N(t)$ takes the following form[13]:

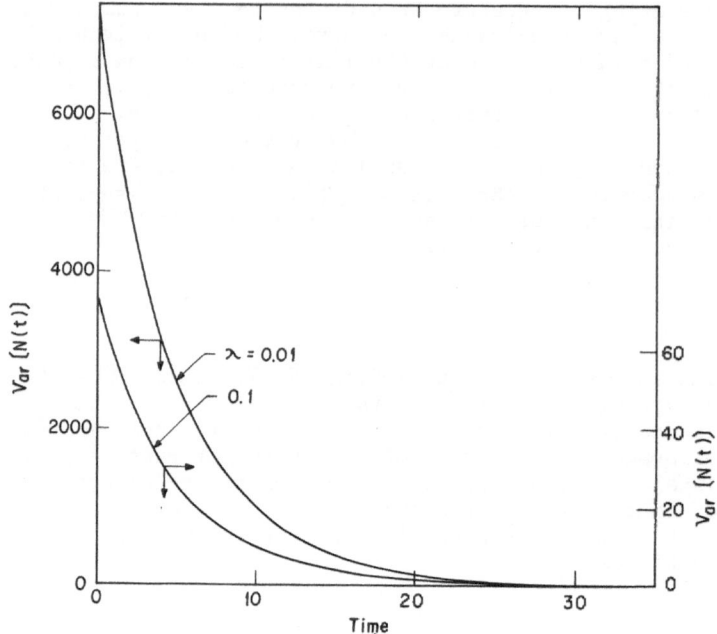

Figure 3. Variation of Var[N(t)] as a function of λ: $k_1=k_2=1.0\times10^{-5}$.

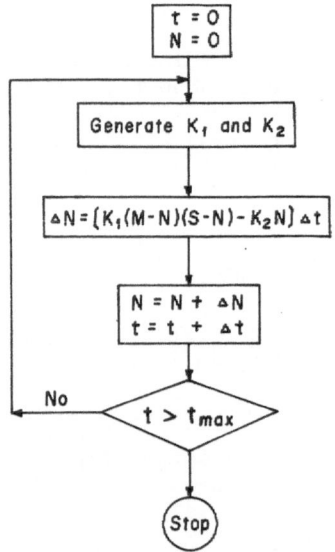

Figure 4. Simulation procedure for the case of random coefficients.

$$f_N(n) = f_{NO}[(r_1X-r_2)/(X-1)] \; |J| \qquad (5)$$

where the Jacobian J is evaluated by

$$J = \frac{\partial N_O}{\partial N}$$

$$= [\frac{r_2-r_1}{(N-r_1)(X-1)}]^2 \; \exp[k_1t(r_1-r_2)] \qquad (6)$$

The expected value of a random variable is a measure of its mean behavior; the variance of the random variable is commonly used as a measure of the significance of its random fluctuation. The expected value of the number of attached particles at time t, $E[N(t)]$, is

$$E[N(t)] = \int_{all \; n} n(t)f_N(n)dn \qquad (7)$$

The variance of the number of attached particles at time t, $Var[N(t)]$, is calculated by

$$Var[N(t)] = \int_{all \; n} \{n(t)-E[N(t)]\}^2 \; f_N(n)dn \qquad (8)$$

Random Rate Constants

When the system is subject to an external noise, for example, the temperature of the operating environment is fluctuating around a certain set point, the rate constants in Equation (1) can be random variables. Thus, we have a stochastic differential equation with random coefficients. Solving Equation (1) for $N(t)$ yields

$$N(t) = \frac{(r_2Y-r_1)}{(Y-1)}$$

$$= h(k_1,k_2,t) \qquad (9)$$

where $Y=r_1\exp[k_1t(r_1-r_2)]/r_2$. Therefore, given the joint probability density function of k_1 and k_2, $g_{K1,K2}(k_1,k_2)$, the characteristics of $N(t)$ can be calculated. For example, the mean of $N(t)$ is evaluated by

$$E[N(t)] = \int\int_{k1,k_2} h(k_1,k_2,t)g_{K1,K2}(k_1,k_2,t)dk_1dk_2 \qquad (10)$$

and the variance of $N(t)$ is estimated by

$$Var[N(t)] = \int\int_{k1,k_2} \{h(k_1,k_2,t)-E[N(t)]\}^2 g_{K1,K2}(k_1,k_2)dk_1dk_2 \quad (11)$$

Numerical Simulation

The effect of external noises on the transient behavior of the attachment of particles to a solid surface is examined through Monte Carlo simulation. For illustration, we assume that M=1000 and S=10000. Consider first the case when the initial number of attached particles is a random variable. It is expected that most of the particles are unattached initially, and, therefore, an exponential distribution for N_0 is assumed. An exponential probability density with parameter λ is defined by[14]

$$f_{NO}(n_0) = \lambda\exp(-\lambda n_0) \qquad (12)$$

In each run of the simulation, an exponential random variable is generated[15] to represent the value of N_0. Forty replications are conducted,

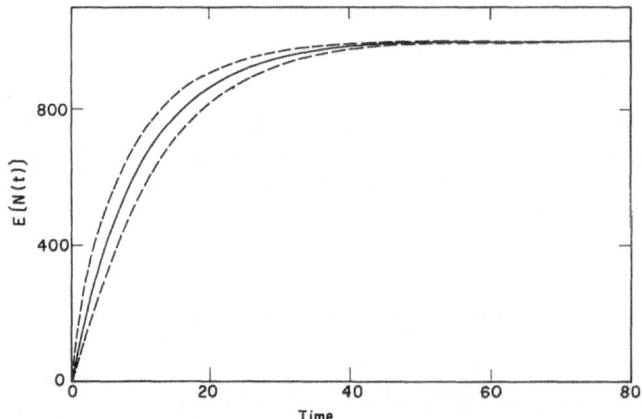

Figure 5. Results of simulation for the case of random coefficients: $E[K_1]=E[K_2]=1.0 \times 10^{-5}$, $Var[K_1]=Var[K_2]=1.0 \times 10^{-11}$. Solid line represents the variation of $E[N(t)]$, the dashed lines denote an approximate 95% confidence band.

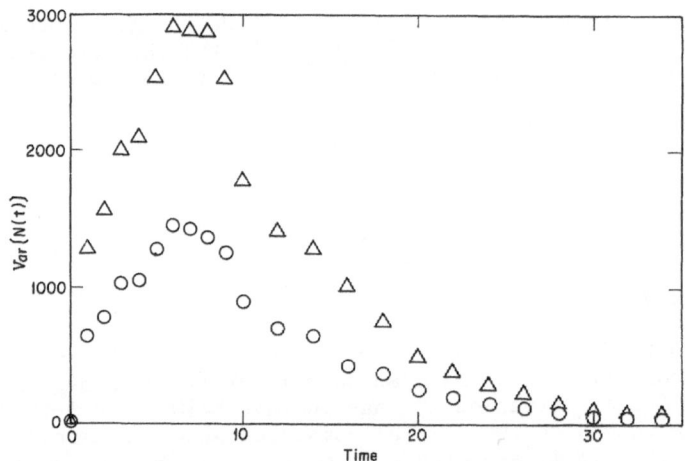

Figure 6. Variation of $Var[N(t)]$ as a function of $Var[K_1]$ and $Var[K_2]$: $E[K_1]=E[K_2]=1.0 \times 10^{-5}$. $\bigcirc$: $Var[K_1]=Var[K_2]=5.0 \times 10^{-12}$; $\triangle$: $Var[K_1]=Var[K_2]=1.0 \times 10^{-11}$.

and the mean and the variance of $N(t)$ are estimated based on these replications. Figure 1 shows the result of simulation for arbitrarily assumed values of adjustable parameters. An approximate 95% confidence band, as estimated by $E[N(t)] \pm Var[N(t)]^{1/2}$, is also presented in this figure. The variation of $E[N(t)]$ as a function of λ is illustrated in Figure 2, and the corresponding variances are illustrated in Figure 3.

In the case when k_1 and k_2 are random variables, we assume that they are independent of each other and each possesses a Gaussian density with means $E[K_1]$ and $E[K_2]$, respectively, and variances $Var[K_1]$ and $Var[K_2]$, respectively. The procedure for the Monte Carlo simulation in this case is summarized in Figure 4. Forty replications are performed to estimate the mean and the variance of $N(t)$. The results are presented in Figure 5. Figure 6 illustrates the variation of $Var[N(t)]$ as a function of $Var[K_1]$ and $Var[K_2]$; $E[N(t)]$ is found not to be sensitive to the variation of $Var[K_1]$ and $Var[K_2]$ in this case.

DISCUSSION

As is seen from Figure 1, a random initial condition has the effect of causing a wide confidence band for $E[N(t)]$ in the early stage of the attachment. The width of the confidence band decreases with time, and vanishes as time approaches infinity. Therefore, it is concluded that a random initial condition has no effect on the steady-state number of attached particles. Note that since the mean value of N_0 is $1/\lambda$[14], $E[N(t)]$ is functionally dependent on λ, as reflected by Figure 2. Figure 3 reveals that $Var[N(t)]$ is roughly proportional to $1/\lambda^2$. In the case when the rate equation has random coefficients, Figure 6 reveals that $Var[N(t)]$ has a maximum during the course of attachment; also, $Var[N(t)]$ is approximately proportional to $Var[K_1]$ and $Var[K_2]$. Again, as is in the case of a random initial condition, $Var[N(t)]$ vanishes as t goes to infinity, meaning that it has no effect on the steady-state number of attached particles.

SUMMARY

It has been shown that the mean and the variance of the number of attached particles are of the same order of magnitude. Thus the order of the random fluctuation caused by an internal noise is proportional to the square root of the mean number of attached particles. As the size of the system or the number of particles increases, the significance of this fluctuation decreases[5]. On the other hand, the fluctuation caused by an external noise is independent of system size, and thus, even for a large-sized system, the fluctuation caused by an external noise can still be significant.

REFERENCES

1. J. Langmuir, J. Am. Chem. Soc., 40, 1361-1403 (1918).
2. C. Peterson and T.K. Kwei, J. Phys. Chem., 65, 1330-1333 (1961).
3. L. Jankovics, J. Polymer Sci. Part A, 3, 3519-3522 (1965).
4. M.T. Boughey, R.M. Duckworth, A. Lips, and A.L. Smith, J. Chem. Soc. Faraday Trans. I 74, 2200-2209 (1978).
5. J.P. Hsu and H.H. Wang, J. Theor. Biol., 119, 435-444 (1986).
6. J.P. Hsu and H.H. Wang, J. Theor. Biol., 124, 405-435 (1987).
7. B. Vincent and C.A. Young, J. Chem. Soc. Faraday Trans. I 76, 665-673 (1980).
8. B. Vincent, M. Jafelicci, and P.F. Luckham, J. Chem. Soc. Faraday

Trans. I 76, 674–682 (1980).

9. J. Gani, Math. Biosci., 1, 545–554 (1967).

10. N.G. van Kampen, "Stochastic Processes in Physics and Chemistry," North-Holland, Amsterdam, 1981.

11. H.H. Wang, T.W. Chiu, and J.P. Hsu, Biotech. Bioeng., 29, 1122–1126 (1987).

12. J.P. Crutchfield, J.D. Farmer, N.H. Packard, and R. Shaw, Sci. Am., 235, 46–57 (1986).

13. T.T. Soong, "Random Differential Equations in Science and Engineering," Academic Press, New York, 1973.

14. A.M. Mood, F.A. Graybill, and D.C. Boes, "Introduction to Theory of Statistics," McGraw-Hill, New York, 1974.

15. A.C. Atkinson, and M.C. Pearce, J. Roy. Stat. Soc. A139, 431–461 (1976).

MEASUREMENT OF DETACHABLE SUBMICROMETER PARTICLES

AND SURFACE CLEANLINESS OF CLEAN ROOM GARMENTS

Mark B. Stutman

Clean Room Products Group
W. L. Gore & Associates, Inc.
2401 Singerly Road
Elkton, MD 21921

The quality and cleanliness of clean room garments is an important issue to the contamination control industry. Garment cleanliness may be evaluated with techniques such as the Helmke Drum tumble test, and surface examinations using Multiple Internal Reflectance (MIR) and Energy Dispersive Spectroscopy (EDS). The author summarizes four years of experience with tumble testing of various clean room fabrics and monitoring the effectiveness of clean room laundry processing. Common patterns of tumble test results and fabric comparisons are reviewed. Measurements of salts, metals and organic surface contaminants on clean processed garments are also presented. Together these methods may be used to diagnose and monitor the clean wash process. Critical factors in maintaining the quality of freshly laundered garments are identified, and a comparison of clean room laundries is presented.

HISTORICAL REVIEW

As a manufacturer of clean room garments, one element of our development effort has been to understand the contribution of laundering to the performance of a clean room garment. We have examined many in-house and most of the commercial clean room laundries in the United States and Western Europe. In the past five years, we have analyzed more than 150 freshly laundered garments for particulate characteristics, and surface residues.

Historically, clean room garments were intended to control visible human dusts, including clothing fibers, dirt, hair, skin flakes, body oil and spittle. Decades ago, cotton and staple fabrics were recognized as sources of lint and fibers. In response, uniform suppliers developed low-linting clean garments, constructed from fabrics woven totally from continuous synthetic fibers such as polyester or nylon. With this development came a major reduction in the particle burden of the clean room garment itself as a source of fiber contamination.

As semiconductor devices have increased in complexity and decreased in size, so have their requirements for contamination control and clean manufacturing technologies. For semiconductor processing, significant contamination includes submicrometer sized objects. Indeed, in some areas, cleanliness requirements have outpaced the ability to routinely monitor and measure particulate contamination.

As we understand it, the best clean room garment must function as a total "body filter", encasing the wearer in a suit possessing good filtration characteristics.[1] In order to fully exploit fabrics possessing superior filtration properties, garment construction must employ advanced features such as snug closures at the wrists, neck, and ankles, and sealed seams. Totally encased in a jumpsuit, headgear, mask, gloves, and boots, the wearer must remain cool and comfortable while working. Finally, the garment fabric must not shed particles to the environment, and laundered. garments must retain low residuals after cleaning.

Clean room laundries have developed various methods for clean processing of garments soiled with body oils, dirt, clothing lint, and chemical stains, while retaining only very low amounts of surface residuals. Many of these developments are proprietary, especially water (or solvent) polishing, wash chemistry and formulation. Other improvements were made by upgrading the environment of the laundry to that of a modern clean room. Features such as fully garmented workers, HEPA filtered room and drier air, low shedding room and work surfaces, barrier washers, bag sealers and protection of garments during handling and transport all can impact on clean room garment quality.[2]

METHODOLOGY

In our laboratories, garment cleanliness is evaluated with techniques such as the Helmke Drum tumble test, and surface examinations using Multiple Internal Reflectance (MIR) and Energy Dispersive Spectroscopy (EDS).

Detachable Particles

Detachable Particles measured on clean room garments are an indication of the inherent lint level of the fabric, and of the quality of the laundering process. ASTM F-51 (1968) describes a test method for evaluating clean garments based on the number of fibers counted per unit area of garment surface. The utility of this test method is largely restricted to older clean rooms and has diminished with advances in garment technology.

In 1982, George Helmke of Bell Laboratories published a description of an apparatus for measuring releasable submicrometer particles from clean room materials, called the tumble test.[3] The tumble test apparatus is comprised of an optical particle counter and a stainless steel drum, with internal vanes, which rotates at 10 rpm. In use, the drum must be located within a Class 10 space. The clean room garment to be tested is placed within the drum, and the air over the sample is monitored with the particle counter.

The test method recommended by Helmke, and adopted by IES RP-3[4], consists of 10 one-minute counts. Garments are classified according to the ten-minute average of the particles >0.5 μm/minute. The cleanest Class I coveralls (RP-3, Table 2) must release fewer than 1200 particles >0.5 μm/minute. The most common use of the test consists of quality assurance monitoring of individual laundry loads.

Surface Cleanliness

Surface cleanliness is an indication of the effectiveness of the wash process in decontaminating a dirty garment. Samples are obtained by carefully cutting a small (roughly 1 sq. cm.) piece of fabric from the test garment.

Multiple Internal Reflectance (MIR) infrared spectra will detect the presence of organic contaminants. Test spectra can be examined for specific chemical bond types to determine class of compound, or compared to library spectra for identification. The transmission spectra reported below were obtained on a Nicolet 10DX FTIR, using a Harrick Micro MIR accessory, with a 45 KRS-5 (KBr) crystal set at 60°. The depth of sample penetration is approximately one to five micrometers.

Energy Dispersive Spectroscopy (EDS) is useful in identifying elements heavier than fluorine, e.g., salts and metals, in major, minor or trace quantities. Our Princeton Gamma Tech System IIIA spectrometer provides a one to ten micrometer depth surface analysis of the composition of a sample in the SEM chamber.

CASE STUDIES

Patterns of Tumble Test Results

Figure 1 shows several characteristic patterns of tumble test results. A steady count with low variation (curves A & C) is the most common pattern observed when testing high quality clean laundered garments. A second, related pattern shows several minutes of high but decreasing counts, which settle into a lower steady state (curve B). A plausible explanation is that an initial shower of detachable particles is released when the garment is first tumbled, followed by lower counts indicative of the cleanliness of the garment.

A third pattern exhibits large random fluctuations suggestive of bursts (curve D), as if sources of particles are alternately smothered and uncovered by folds in the garment. We have observed such bursts even after an hour of tumble testing.

A fourth pattern (not shown) is of high and increasing counts, indicative of active shedding. This is more often characteristic of fabrics with some staple fibers (e.g., cotton blends), and some wipes and disposable face masks.

Comparison of Clean Room Fabrics

Several earlier discussions of Helmke Drum results on garments constructed from common clean room fabrics have been published. Differences between the common clean room fabrics were described by Swick and Vancho in 1985.[5] That work demonstrated clear differences between fabrics, with spunbonded polyolefin fabric contributing the highest numbers of releasable particles, followed by polyester fabrics of taffeta and herringbone weaves. Expanded polytetrafluoroethylene (PTFE) laminated fabric ranked lowest in detachable particles. No correlation between tumble test count and the number of wash/dry cycles experienced by the polyester or expanded PTFE garments was found. These new garments were only worn during tests conducted every ten wash cycles, and did not experience normal physical wear.

A related experiment measured the detachable particles contributed solely by the outside surface of the clean room garment.[6] Clean, woven polyester and spunbonded polyolefin garments showed relatively little decrease in detachable particles when their cuffs and closures were tied shut with non-linting thread. In contrast, tumble test counts for expanded PTFE garments decreased an average of 85% when tied shut. These results were attributed to a combination of the low shedding characteristics of the exterior PTFE fabric surfaces, and of the relative high filtration efficiency of the fabric.

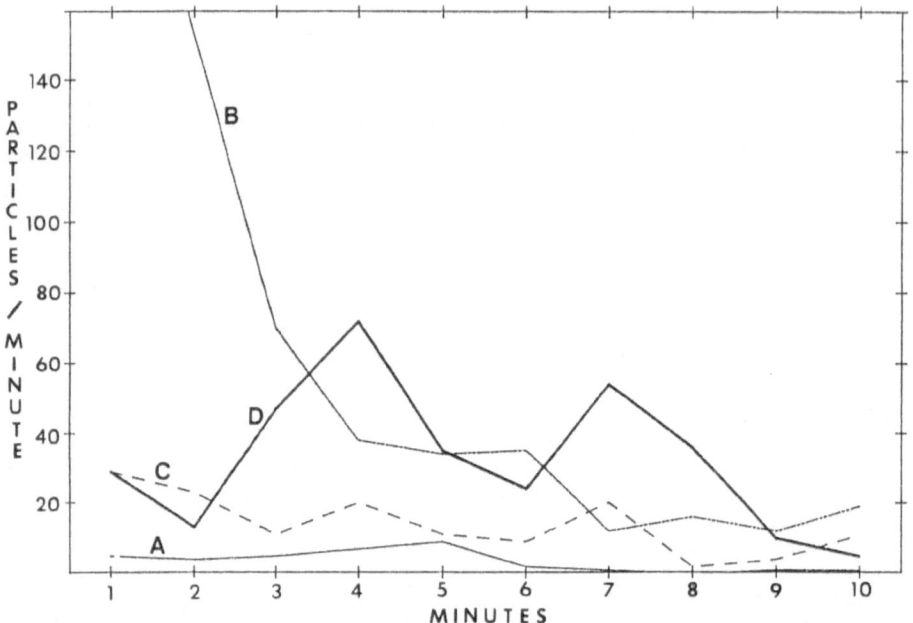

Figure 1. Patterns of Helmke Drum Tumble Test Results on Clean Room Garments [particles >0.3 μm/minute].

A,C. Steady Counts, Low Variation
B. Initial High Counts Decreasing to Lower Steady State
D. High Variation (Bursts)

Polyester fabric garments generally test higher than expanded PTFE garments when laundered identically. The best quality clean room laundries can produce polyester garments with counts less than 50 particles/minute (Table I, rows 1 and 2), and expanded PTFE garments with counts of only a few particles/minute (Table I, row 3).

Effect of Garment Wear and Laundering

In early experiments in our laboratory, water quality was shown to be a crucial parameter in obtaining high quality clean room garments with low tumble test counts. Under identical wash and drying conditions, garments subject to a deionized (DI) water rinse cycle between tap wash and drip dry under HEPA filtered air were frequently lower in counts than the garments simply washed in tap water. The effect of tap water washing was

Table I.   Selected Helmke Drum Tumble Test Results.

ID	particles >0.3 μm minute	particles >0.5 μm minute
A. EXEMPLARY CLEAN GARMENTS		
woven polyester suits		
1. Laundry A (n=15 suits)	37 ± 15	25 ± 11
2. Laundry B (n=15 suits)	34 ± 12	21 ± 9
two-layer expanded PTFE suits		
3. Laundry A (n=8 suits)	2.3 ± 1.9	
B. INADEQUATE RINSING OF CLEAN GARMENTS		
4.a. expanded PTFE	2700 ±1450	2190 ±1230
b. same gmt 4x DI rinsed	56 ± 46	39 ± 32
5.a. expanded PTFE	5210 ±1580	4170 ±1200
b. same gmt 4x DI rinsed	69 ± 15	55 ± 9
C. WORN GARMENTS		
6.a. woven polyester #202	3330 ± 570	2450 ± 430
b. same gmt 2x DI rinsed	2060 ± 380	1500 ± 270
7.a. woven polyester #203	2200 ± 470	1580 ± 380
b. same gmt. 2x DI rinsed	2080 ± 340	1500 ± 260
D. LAUNDRY EVALUATION		
8. expanded PTFE	310 ± 220	150 ± 170
9. woven polyester	3660 ± 540	1610 ± 220

NOTE: Numbers reported are the mean particle count ± the
standard deviation

highly variable; however, thorough DI water rinsing of garments resulted
in consistently low tumble test counts.

When a clean room garment exhibits high tumble test counts, the likely cause is either the garment, or the laundering process. A simple test
to distinguish between these possibilities is to thoroughly rinse and agi-
tate the test garment in multiple DI water wash cycles.

If tumble test counts are significantly lower after re-rinsing in DI
water, then the laundered garment was either insufficiently cleaned, or
subsequently re-contaminated. An example is shown in Table I, rows 4 and
5. The two garments reported here still produced visible sudsing in the
DI water rinse on the third rinse cycle. Overall tumble test counts were
reduced by 98% after the fourth rinsing.

If high counts are not substantially reduced by clean rinsing, then the garment itself is suspect. Two such garments are shown in Table I, rows 6 and 7. These two polyester garments were considerably frayed at the cuffs, and their counts remained high even after multiple DI rinses.

Laundry Evaluation

Improvements in the clean wash process can sometimes be made after evaluation of laundered garments using our methods. One such laundry sent us several garments, laundered and packaged in their facility. Surface residue analysis and tumble tests were performed on the polyester herringbone and expanded PTFE clean room garments, shown in rows 8 and 9 of Table I.

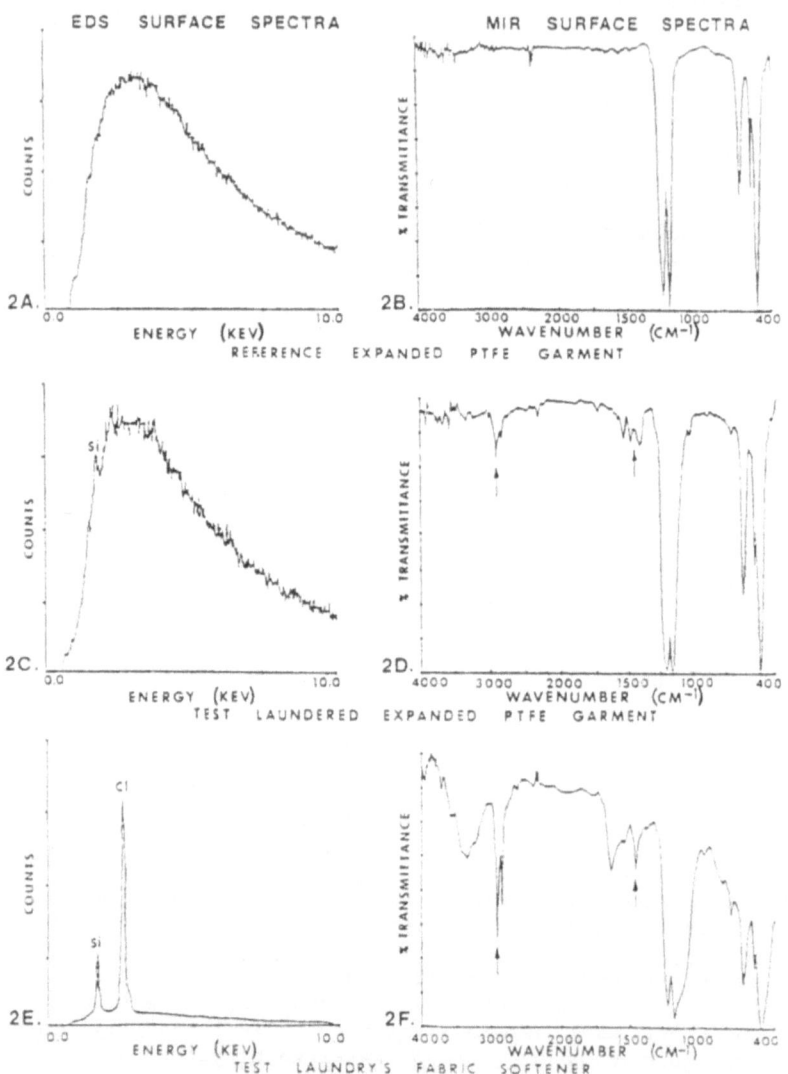

Figure 2. EDS and MIR Surface Spectra of Clean Laundered Expanded PTFE Garment.

A,B. Reference Spectra of Uncontaminated Garment Surface
C,D. Spectra of Laundered Garment Surface
E,F. Spectra of Laundry's Fabric Softener

Both counts were high, and the polyester garment was not a Class I garment by IES RP-3 standards. Examination of their wash process revealed the probable cause.

A brief tumble test of their fabric softener packet produced measured particle concentrations in excess of 150,000 particles >0.3 μm/minute. Figure 2 shows a series of EDS and MIR spectra from this laundry. The EDS spectra for the wash chemistry and water (not shown) indicated major amounts of sodium and chlorine, and lesser amounts of silicon, phosphorus, aluminum, calcium, titanium and iron. Only silicon was detected on the sample garment surfaces (Figure 2C). However, considerable residue was detected in the infrared scan (Figure 2D) of the surface of the laundered expanded PTFE garment. The signature of the surface residue closely matched that of the fabric softener (Figure 2F), which was added to the drier. Since some topical antistatic agent was required to reduce the static charge on the dried garments in the winter season, an alternate formula was recommended to reduce the particle burden to the clean garments.

Using these methods, we have determined several process improvements which result in cleaner garments. Two common recommendations involve reductions in concentration or elimination of certain wash chemicals, and improvements in the packaging, handling, and storage of the clean gar-

Table II. Evaluation of Eighteen Clean Room Laundries.

SAMPLE ID	HELMKE DRUM TUMBLE TEST pll>0.3um/min	ENERGY DISPERSIVE SPECTROSCOPY									INFRARED ALIPHATIC HYDROCARBON
		Na	Al	Si	P	S	Ca	Ti	Fe	Cu	
A	2										NONE
B	3	TR		TR							TRACE
C	5										TRACE
D	22			TR					TR	TR	TRACE
E	30	TR		TR						TR	TRACE
F	32			TR				MN	TR	TR	MINOR
G	40	TR		TR							TRACE
H	40			TR						TR	TRACE
I	49			TR							NONE
J	81		TR	TR		TR					TRACE
K	102										NONE
L	142			TR		TR					NONE
M	192										TRACE
N	240	MJ	TR	MN	TR	TR	MJ		TR		MINOR
O	331			TR							MINOR
P	341			MN					TR		TRACE
Q	844			TR							NONE
R	1263			TR		TR	TR		TR		TRACE
Median = 65		22%	11%	78%	6%	22%	11%	6%	28%	22%	61%

NOTE: MJ = Major   MN = Minor   TR = Trace

ments. A multiple rinse cycle using high quality wash solvent is a third critical parameter. For water, this might include filtering, softening, and final polishing with reverse osmosis or deionization.

Tabulation of Laundries

Table II is a compilation of our evaluation of eighteen clean room laundries. In all cases, the garment results reported were from the outside surface of a clean, unworn expanded PTFE garment. In general, we have found surface residuals to be the same for woven polyester garments. While tumble test counts are higher for worn or polyester woven

fabrics than for the expanded PTFE laminate, the ranking of the effectiveness of the cleaning process is generally the same.

Tumble test counts on these test garments varied nearly a thousand-
fold, from essentially background levels in the laboratory, to higher
than 1200 particles >0.3 µm/minute. For each laundry, the tumble test
counts shown here correspond to the reported surface residue analyses.
While they have been sorted by tumble test count, the reader should not
conclude that one laundry is necessarily better than another one below it
in the Table. Figure 3 shows the relative distribution of the tumble
test results. Median particle density was 65 particles >0.3 µm/minute,
but the distribution of detachable particles measured on these ostensibly
identical test garments appears positively skewed. This indicates that
most laundries are able to produce a Class I garment.

Most of the garments examined had detectable quantities of silicon
contamination on their surfaces (78%). Iron was the next most common element detected (28%), followed by sodium, sulfur and copper (22%). Calcium, aluminum, titanium and phosphorus were also detected at least once.
Aliphatic hydrocarbons, probably components of surfactants and other wash
chemicals, were detected on 61% of the garment surfaces.

DISCUSSION

The research methods and case studies described above were selected
to illustrate their power in understanding the nature and sources of lint
from clean room garments. By itself, the Helmke Drum can be used to

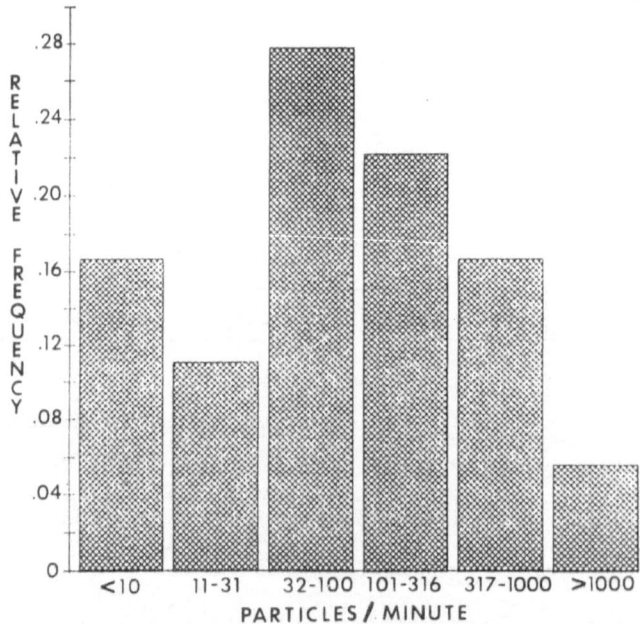

Figure 3. Helmke Drum Tumble Test Results of Eighteen Clean Room Laundries [particles >0.3 µm/minute].

monitor releasable particles from a clean garment. Using simple experimental designs, comparisons can be made between different garments laundered under identical conditions, and also between different laundry treatments of identical garments. The effects of new versus worn or tattered garments may also be investigated.

In a production setting, the Helmke Drum is a cost-effective tool for monitoring the quality of a freshly laundered load of clean room garments. Results are available in real time, and can be used to routinely monitor the cleaning process. When warranted, additional tumble testing, and more expensive surface analytical techniques such as MIR and EDS can be employed to investigate and diagnose out-of-control clean room processing. Currently, a four-stage impactor is being operated in tandem with the particle counter in order to identify the particles collected with the drum.

CONCLUSIONS

Most of the clean room laundries tested produce clean room garments with low levels of surface detachable particles and surface residuals. There are, however, significant variations in the quality of freshly laundered garments. Using the Helmke Drum, our research has identified the most critical factors in maintaining laundry cleanliness: partial loading of the washer, multiple rinses in filtered, deionized water, and careful packaging. We have shown that undamaged garments with high tumble test counts can be restored to their original low shedding levels by subjecting them to multiple rinsing and agitation in filtered, deionized water.

When properly constructed and processed, the clean room garment itself is not a significant source of contamination in the clean room. Rather, it is the lint and dust from the wearer's street clothes and skin, which work through the clean room fabric and into the air. Garments with sealed closures constructed of expanded PTFE laminate exhibit the highest degree of particle control and comfort, and the lowest level of releasable particles and surface residuals among the available clean room fabrics.

ACKNOWLEDGMENTS

The author would like to express his appreciation to all of those individuals in the industry who have welcomed our inquiries and examinations of their proprietary processes.

REFERENCES

1. M.B. Stutman, J. Parenteral Sci.Technol., in press.

2. D.L. Barclay, Microcontamination 4(7), pp. 28-32ff, (1986).

3. G.E. Helmke, 1982 Proceedings of the Institute of Environmental Sciences, pp. 218-220.

4. IES RP-CC-003-87-T, "Garments Required in Clean Rooms and Controlled Environmental Areas," Institute of Environmental Sciences, Mt. Prospect, Illinois, 12pp., October 1987.

5. R. Swick and V.Vancho, Microcontamination, <u>3(2)</u>, pp. 46-51, (1985).

6. J. Bowser, Paper presented at the 16th Nordic R^3-Symposium, Ronneby, Sweden, April 1985.

ACCUMULATION OF PARTICLE DERIVED IONIC CONTAMINANTS ON ELECTRONIC EQUIPMENT: AIRBORNE CONCENTRATIONS AND DEPOSITION VELOCITIES

J. D. Sinclair and L. A. Psota-Kelty

AT&T Bell Laboratories
Murray Hill, New Jersey 07974

C. J. Weschler and H. C. Shields

Bell Communications Research
Red Bank, New Jersey 07701

We previously reported deposition velocities for chloride, nitrate, sulfate, sodium, ammonium, potassium, magnesium, and calcium associated with fine and coarse particles at telephone company switching equipment locations in Wichita, Kansas and Lubbock, Texas. Preliminary data were also reported for a site in Newark, NJ. These results were based on comparisons of indoor concentrations, obtained using dichotomous samplers for collection and ion chromatography (IC) for analysis, with average annual surface accumulations that were obtained by collecting water extracts of zinc and aluminum structural surfaces and then analyzing by IC. In this paper we report the complete results for the Newark site and for a new site in Neenah, Wisconsin. The deposition velocities are based on average annual surface accumulations derived from approximately 500 extractions of zinc and aluminum surfaces as well as indoor concentrations measured for an annual cycle at weekly intervals at both the Newark and Neenah sites. The results demonstrate that deposition velocities for each of the major ions are similar, regardless of location, and can be used to predict surface accumulation rates and that variations in surface accumulations are not attributable to variations in airborne concentrations.

INTRODUCTION

While the composition and concentrations of particulate matter in outdoor environments have been studied extensively through air monitoring networks in the United States and elsewhere, until very recently little attention has been given to indoor/outdoor relationships and especially to the relationship between indoor airborne concentrations and indoor surface accumulation rates. In our first studies in this area, we reported on measurements of airborne concentrations and surface accumulations of ionic substances at electronic equipment installations in Wichita (Kansas), Lubbock

(Texas), and Newark (New Jersey)[1-3]. Airborne substances were collected at a convenient central location in the equipment room and outdoors on the roof of the equipment building using dichotomous samplers. Ionic substances that accumulated on surfaces were sampled using filter paper extraction[4]. Samples of fine and coarse particulates (0.1 - 2.5 and 2.5 - 15μm aerodynamic diameter, respectively) were collected on Teflon filters at weekly or twice weekly intervals for several months. The separation of particles into fine and coarse fractions bears directly on understanding the accumulation of ionic substances on horizontal and vertical surfaces, as discussed in earlier work[1-3]. These studies showed that: (1) reliable indoor/outdoor concentration ratios for the major ions could be estimated, in the absence of indoor sources, from readily available air handling system parameters; and (2) measured deposition velocities for each of the dominant ions in fine or coarse particles were similar, regardless of location, and could be useful in predicting surface accumulation rates at locations where indoor concentrations were known or could be reasonably estimated.

Further measurements at Newark and at a new location in Neenah, Wisconsin have now been completed. In these new studies, in addition to using dichotomous samplers to collect airborne particles, compact personal air sampling pumps have also been used. In both cases airborne samples were collected on Teflon filters. In contrast with the dichotomous samplers, the small size and low flow rates of these personal sampling pumps allow a large number of them to be used without appreciably altering the airborne concentration. They can also be placed in close proximity to the surface sampling locations. The intention of these additional studies was to: (1) obtain further information on the variation of the deposition velocities of the major ions among locations with varying climates and environments; (2) examine the possibility that the known variations in surface accumulations could be attributed to variations in local concentrations; and (3) determine if the data obtained using the dichotomous samplers are representative of concentrations across the entire equipment room for each sampling interval.

EXPERIMENTAL

Details of the measurements of surface accumulations rates using filter paper extractions and airborne concentrations using dichotomous samplers are discussed elsewhere[1-4].

Localized collection of airborne particles was accomplished using personal air sampling pumps operated at a nominal flow rate of 3 l/min. These pumps were modified to run continuously. Samples were collected on 37 mm Teflon filters. Total suspended particulate matter (TSP) was based on filter weighings to the nearest 0.01 mg. Ion concentrations were measured by ultrasonically extracting the filters in 25 ml of purified water and then analyzing the extract using an ion chromatograph equipped with standard columns and using standard eluents.

RESULTS AND DISCUSSION

The average accumulation rates of the major anions at Newark for each of ten equipment frames are compared in Figures 1 & 2. The data for sulfate and nitrate are the averages for zinc and aluminum surfaces. Sulfate and nitrate accumulations are predominantly a result of particle deposition. The data for chloride are for aluminum surfaces only, because it was shown previously[1] that chloride accumulation on zinc surfaces is due to the combined effects of particle deposition and reaction with

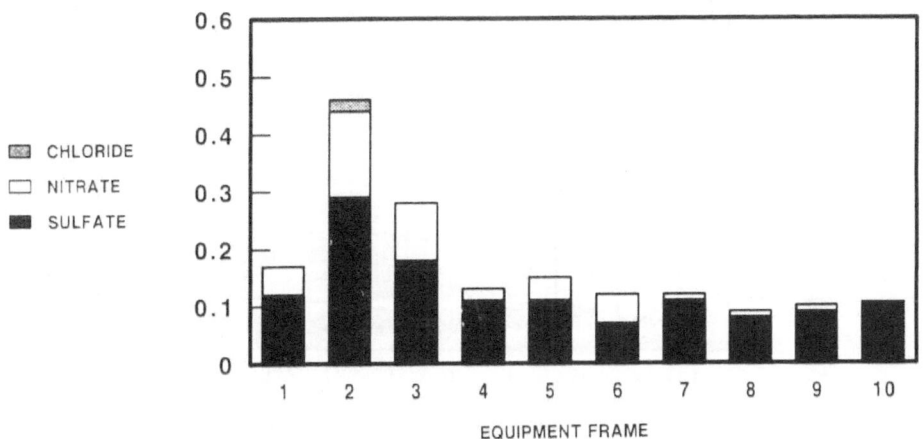

Figure 1.   Average accumulations on vertical surfaces at Newark ($\mu g/cm^2$ yr).

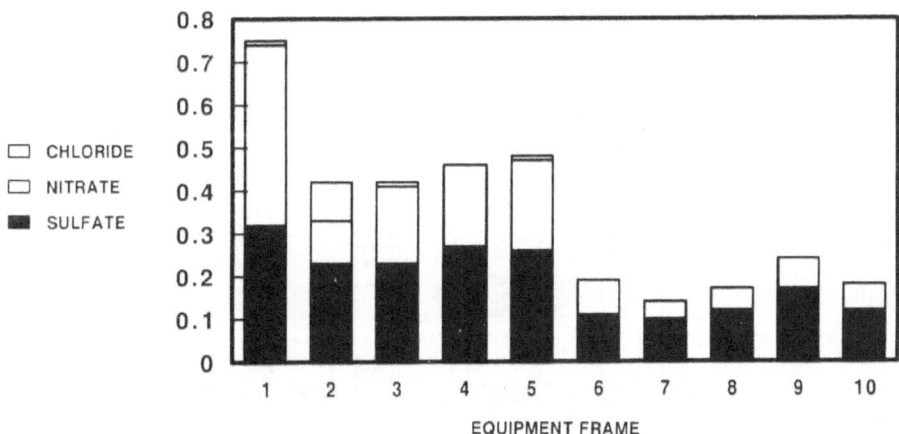

Figure 2.   Average accumulations on horizontal surfaces at Newark ($\mu g/cm^2$ yr).

corrosive chlorine-containing gases. Aluminum does not accumulate appreciable amounts of chloride through reactions with chlorine-containing gases. From Figures 1 and 2 it is evident that the sulfate and nitrate accumulation rates tend to track. The chloride accumulation was usually either very low or below the detection limit. The chloride data provide a useful measure of chloride contamination but evaluation of correlations involving chloride is not appropriate.

The average accumulation rates for the anions and cations measured at Neenah are given in Figures 3 and 4. Again, the chloride data are exclusively for aluminum surfaces. At Neenah the tendency for major ions to track was very strong for accumulation on horizontal surfaces. Except for chloride, the ions in coarse particles, which dominate deposition on these surfaces, appear to be uniformly proportioned across the equipment room. For vertical surfaces many of the ion accumulations were low or undetected for at least a few sampling locations. In view of the large errors associated with such measurements, high correlations are not expected.

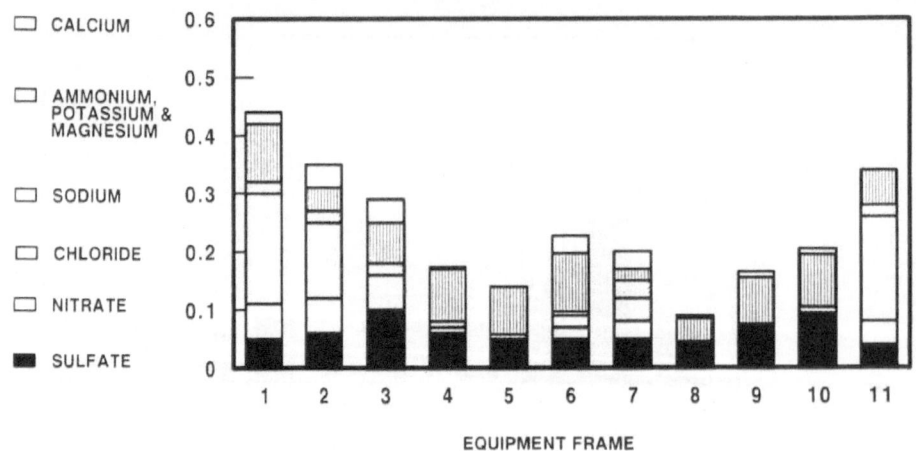

Figure 3. Average accumulations on vertical surfaces at Neenah (μg/cm^2 yr).

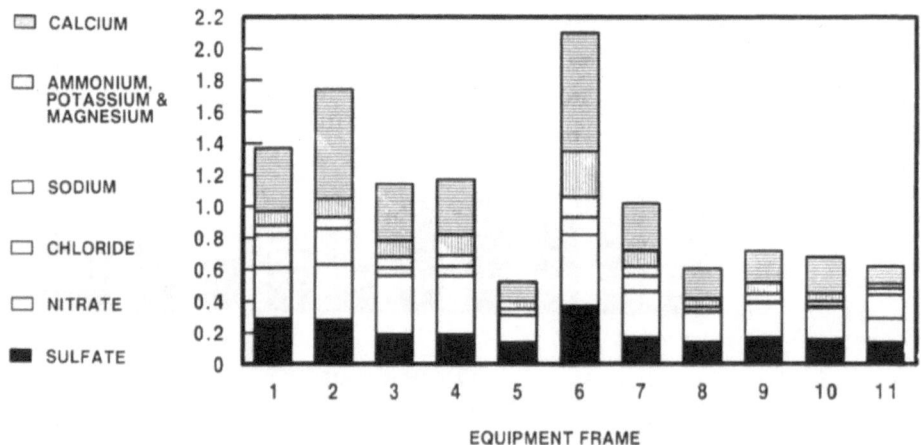

Figure 4. Average accumulations on horizontal surfaces at Neenah (μg/cm^2 yr).

Figure 5 gives the overall airborne concentrations of sulfate and chloride at Newark, as determined using the personal sampling pumps. Figure 6 gives overall airborne concentrations for all the ions analyzed at Neenah. With the exception of chloride, the airborne concentrations of the ions measured at Neenah and Newark for the various sampling locations span very narrow ranges. For chloride, the analytical error inherent in determining very low concentrations is undoubtedly responsible for the wider distribution of concentrations than was seen for other ions. Comparison of the average airborne concentrations obtained using the personal sampling pumps with concentrations obtained using the dichotomous sampler are given in Table I. The clear conclusions from this data are that the airborne concentrations of all components of the particles are remarkably uniform across the equipment room and that results from a single dichotomous sampler placed at a single sampling point are representative of airborne concentrations across the entire equipment room. It should be noted that, due to the high efficiency filtration at the Neenah location, the fraction of TSP contributed by coarse particles is very small. The large majority of the water soluble ionic material collected by the personal sampling pumps is contributed by fine particles.

Table I. Comparisons of Total Airborne Concentrations
($\mu g / m^3$)

	Dichotomous Sampler Particle Collection	Personal Sampling Pump Particle Collection
Newark		
TSP	3.7	3.7
Sulfate	0.81	0.96
Chloride	0.004	0.004
Neenah		
TSP	2.2	2.4
Sulfate	0.77	0.82
Chloride	0.002	0.002
Sodium	0.009	0.009
Ammonium	0.23	0.27
Potassium	0.007	0.011
Magnesium	0.003	0.006
Calcium	0.010	0.012

Table II. Variation in Deposition Velocities of Ions Associated
With Particles at Newark
(cm/sec)

Frame	------------ Fine ------------		------------ Coarse ------------	
	Sulfate	Chloride	Sulfate	Chloride
1	0.004	<0.06	0.3	0.1
2	0.010	0.63	0.2	2.8
3	0.005	<0.06	0.2	0.1
4	0.004	<0.05	0.3	<0.05
5	0.004	<0.05	0.3	0.1
6	0.002	<0.11	0.1	<0.1
7	0.004	<0.08	0.1	<0.08
8	0.003	<0.11	0.1	<0.1
9	0.003	<0.06	0.2	<0.06
10	0.004	<0.16	0.1	<0.1
Avg.	0.004	<0.14	0.2	<0.3

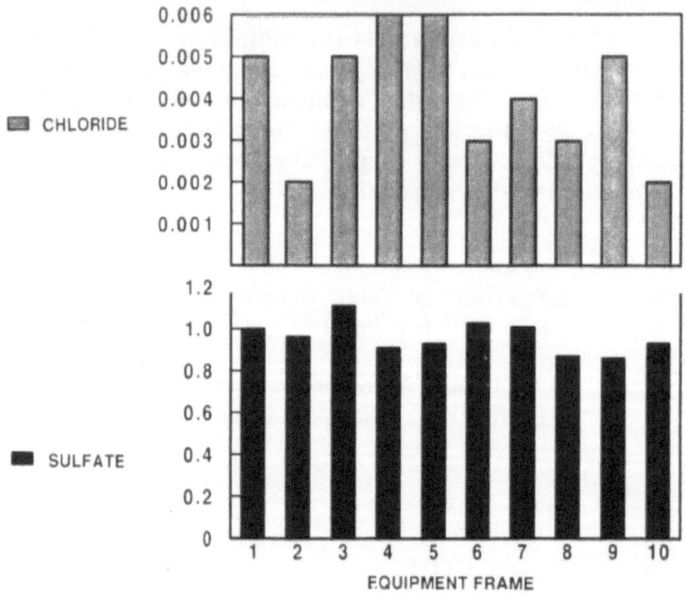

Figure 5.   Average airborne concentrations of anions at Newark ($\mu g/m^3$).

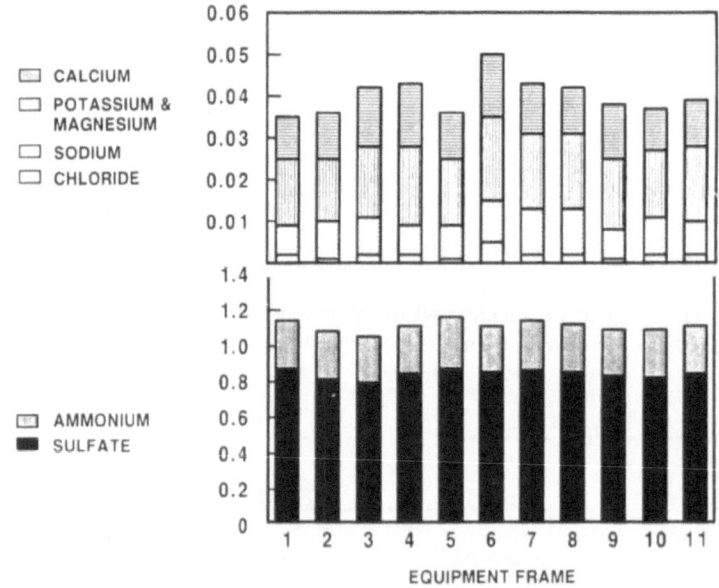

Figure 6.   Average Airborne concentrations of ions at Neenah ($\mu g/m^3$).

At both Newark and Neenah the broad ranges in surface accumulation rates contrast sharply with the narrow ranges in airborne concentrations for the major ions associated with fine and coarse particles.   Comparison of airborne ion concentrations with surface accumulation rates of the ions is best illustrated by the deposition velocities (calculated as the ratio of the surface accumulation rate to the airborne concentration) at each sampling location.   The deposition velocities for Newark are given in Table II.   Those for Neenah are given in Table III.

Table III. Variation in Deposition Velocities of Ions Associated With Particles at Neenah
(cm/sec)

			Fine Particles			
Frame	Sulfate	Chloride	Sodium	Potassium	Magnesium	Calcium
1	0.002	1.9	0.13	0.02	0.05	0.14
2	0.003	8.2	0.10	0.01	0.10	0.26
3	0.004	<0.03	0.10	0.02	0.09	0.20
4	0.002	<0.03	0.06	0.003	0.03	0.005
5	0.002	<0.06	0.03	<0.003	0.02	<0.006
6	0.002	0.5	0.03	0.003	0.03	0.09
7	0.002	0.9	0.12	<0.003	0.10	0.18
8	0.002	<0.03	0.01	<0.003	0.03	0.03
9	0.003	<0.06	0.03	0.10	0.02	0.05
10	0.004	<0.03	0.05	<0.003	0.09	0.07
11	0.002	5.7	0.11	0.01	0.02	<0.006
			Coarse Particles			
Frame	Sulfate	Chloride	Sodium	Potassium	Magnesium	Calcium
1	0.2	6.6	0.4	0.9	0.6	2.3
2	0.2	14.6	0.4	1.8	0.7	3.6
3	0.1	1.6	0.4	1.3	0.5	1.5
4	0.1	1.6	0.5	1.6	0.6	1.3
5	0.1	0.2	0.2	0.3	0.3	0.6
6	0.3	1.4	0.6	3.2	0.9	2.9
7	0.1	3.2	0.2	1.1	0.6	1.4
8	0.1	0.2	0.1	0.5	0.3	1.0
9	0.1	0.4	0.3	0.6	0.5	0.9
10	0.1	0.3	0.2	0.3	0.5	1.3
11	0.1	4.7	0.2	0.3	0.2	0.6

In calculating these deposition velocities, the airborne concentrations of ions were apportioned into fine and coarse particle categories from data collected using dichotomous samplers. The percentages of the total ion concentrations that were contributed by fine particles at Newark were: sulfate - 97; and chloride - 50. The same percentages for Neenah were: sulfate - 95; chloride - 50; sodium - 70; ammonium - 95; potassium - 90; magnesium - 50; and calcium - 45. The ranges in the deposition velocities were appreciable. At Newark and Neenah the ranges for sulfate in fine particles were 0.002-0.010 and 0.002-0.004 cm/sec, respectively, while the range for sulfate in coarse particles at both locations was 0.1-0.3 cm/sec. For many of the ions the surface accumulation rates were very small or below detection limits. For these the deposition velocities are given as upper limits. The deposition velocities determined for Newark using personal sampling pumps to collect airborne particles compare quite well with the deposition velocities obtained previously using measurements from only one dichotomous sampler to represent the airborne concentrations across the equipment room[2,3].

Since the variations in surface accumulations are not caused by variations in the airborne concentrations, they must be due to variations in the deposition process. If there were no turbulence or force fields acting on the particles, the effects of which have been characterized in several recent modeling studies,[5-7] the deposition rate to all surfaces would be controlled by a combination of impaction and convective diffusion. Since the large majority of the TSP at Newark and Neenah is attributable to fine particles, impaction will be a significant factor for only a small portion of the TSP. Variations in convective diffusion effects for fine particles are expected to be small In an equipment room where the natural flow of air is progressively disrupted by encountering surfaces with complex geometries, the turbulence will be variable, and variations in deposition velocities are not surprising. Particles passing near a surface can also be affected by electrostatic and thermal effects. A detailed quantitative discussion of these factors is not possible without precise characterization of the turbulence, surface electrical potentials, and surface temperature. However, the available qualitative information on these factors is useful for gauging the relative importance of those factors.

Electrostatic effects are unlikely to be an important factor because the zinc and aluminum surfaces are structural members of equipment frames that are commonly grounded. Power dissipation produces appreciable variations in surface temperatures at these locations. The aluminum surfaces usually enclose heat releasing components while the zinc surfaces are associated with structural surfaces that remain close to room temperature. Particle deposition to cooler surfaces is expected to be enhanced. The deposition velocities given in Tables I and II are computed as the average for zinc and aluminum surfaces, except as noted for chloride. The deposition velocities are usually slightly less for zinc than for aluminum surfaces. If thermal effects are significant the opposite is expected. However, other effects, including the observation that air flow rates tended to be lower near zinc surfaces, which are deeper within the equipment frames than aluminum surfaces, may be more important. The reduced air flow at zinc surfaces is consistent with lower deposition velocities at these surfaces. Clearly, thermal effects are not the dominant contributor to the variations in deposition velocities at these locations.

CONCLUSIONS

The results of this study demonstrate that: (1) the deposition velocities for each of the major ions contained predominantly in fine particles are similar regardless of location, and can be used to predict surface accumulation rates at unknown locations; (2) variations in surface accumulations are not attributable to variations in the airborne concentrations; and (3) airborne indoor concentrations of species contained in particles are sufficiently constant across equipment rooms to conclude that results obtained using a single dichotomous sampler are representative of the concentrations across the entire equipment room. From this study it is clear that minimizing and controlling the rate of degradation of electronics caused by surface accumulation of ionic substances requires that attention be given not only to the quality of the air filtration but also the air flow characteristics at surfaces. Substantial reductions in the accumulation rates can be realized by reducing turbulence at surfaces.

REFERENCES

1. J. D. Sinclair, L. A. Psota-Kelty, in "Proceedings of the International Congress on Metallic Corrosion", Published by National Research Council of Canada, NRCC 23163, Volume 2, p. 296, Toronto, Canada, June 3-7, 1984.

2. J. D. Sinclair, L. A. Psota-Kelty, C. J. Weschler, *Atmos. Environ.*, *19*, 315 (1985).

3. J. D. Sinclair, L. A. Psota-Kelty, C. J. Weschler, *Atmos. Environ.*, *3*, 461 (1988).

4. J. D. Sinclair, *Anal. Chem.*, *54*, 1529 (1982).

5. W. W. Zararoff, G. R. Cass, *J. Aerosol Sci.*, *18*, 445 (1987).

6. H. Fissan, J. R. Turner, *J. Aerosol Sci.*, *18*, 623 (1987).

7. B. Y. H. Liu, B. Fardi, K. H. Ahn, "Proceedings of 33rd Annual Technical Meeting of the Institute of Environmental Sciences," San Jose, California, May 4-8, 1987.

ADHESION OF ASH PARTICLES ON HEAT TRANSFER SURFACES IN COAL COMBUSTION

APPLICATIONS: MECHANISMS AND IMPLICATIONS

R. Nagarajan*

Department of Mechanical and Aerospace Engineering
West Virginia University
Morgantown, WV 26506-6101

The useful lifetime of rotor blades and stator vanes in gas
turbines burning pulverized coal may be limited by the
build-up of thick, strongly-adherent, insulating
multiphase deposits composed of captured ash particles
and condensed liquid 'glue'. The rate of growth of the
deposit layer is governed by the competing dynamics of
ash particle deposition and deposit erosion due to oncoming
particles. A model for ash deposition developed previously
has been extended here to account for deposit erosion due
to particle impact. Sticking coefficient (ratio of captured
to incident particle mass) predictions made using the present
theory are compared against corresponding measurements in the
DOE-METC combustion/deposition entrained reactor (CDER), a
test facility designed to simulate the deposition characteristics of a cooled turbine blade surface. Inferences are
drawn regarding the deposition mechanisms predominating at
different gas and surface temperatures, and the role
deposit erosion plays in determining the rate of deposit
build-up.

INTRODUCTION

The direct utilization of coal in combustion gas turbines offers the
substantial advantages of low-cost fuel availability and economical
equipment operation for power generation and locomotion. However, before
coal and coal-derived gases and liquids can be considered viable fuels
for gas turbines, certain technological challenges need to be met:
Incombustible inorganic material in the coal matrix ('ash') will
survive the combustion process, and impact on turbine blades in the
form of supermicron ash particles. A fraction ('sticking fraction') of
incident particle mass will adhere to the surface. The strongly-bonded
ash deposit can build to a thickness of a few millimeters within a matter
of a few hours of turbine operation. The associated loss in aerodynamic
and heat-transfer efficiency can severely curtail the economical
lifetime of the turbine engine. Deposition-related problems, such as
blade erosion and corrosion, may reduce blade life as well.
These environmental concerns constitute so-called 'barrier issues' which

*Present Address: IBM Corporation, 5600 Cottle Rd., San Jose, CA 95122.

need to be addressed on a priority-basis. This paper describes an on-going program of investigation into ash deposition mechanisms in coal combustion applications. By means of a combined experimental and theoretical modeling approach, a framework is being developed with which to evaluate several coals with respect to their fouling tendencies, to evaluate several selective coal-cleaning/additive techniques with respect to their effectiveness, and to determine a set of optimum operating conditions that would minimize ash deposition.

The central concept of 'self-regulated' sticking of ash material in the presence of condensed liquid glue was formulated by Rosner and Nagarajan[1] and used in the context of deposit growth on convection-cooled heat-exchanger tubes in industrial boilers. This preliminary low-temperature sticking model considered only sodium sulfate as glue material, and took silica to represent a coal ash particle in terms of physical and chemical properties. The glue phase was allowed to form only via vapor diffusion/ deposition of the sulfate precursors, or by thermophoresis of nucleated sulfate aerosols in the combustion product stream. This 'first-cut' analysis was later extended to incorporate more complex ash and glue chemistry, and to account for other high-temperature sources of liquid glue characteristic of gas turbine operation -- e.g., glassy interfacial films of alkali silicates and aluminosilicates on the surfaces of ash particles, fused oxides in the sintering regime, etc. The different deposition regimes and prevailing deposition mechanisms have been delineated in Figure 1. The computational procedure involved, and comparisons of model predictions with relevant experimental deposition data obtained at DOE-METC have been presented in a series of publications[2-7]. In this paper, some recent refinements to the ash-adhesion theory are presented, and their implications for ash capture illustrated by means of sample calculations.

REFINEMENTS TO ASH DEPOSITION THEORY: EFFECT OF DEPOSIT EROSION

The initial ash deposit that forms on a cooled turbine blade surface is typically a loosely-bonded aggregate of particles which depends on molecular-level attraction forces -- such as van der Waals forces, electrostatic attachment, etc.[8] -- and any 'trapping' by surface irregularities of the oxidized metal to provide the necessary adhesion. The combined action of these forces will not be sufficient to hold supermicron sized ash particles onto blade surfaces, especially in the presence of combustion gas-dynamic shearing, unless enhanced by liquid-phase adhesion and chemical and mechanical bond-formation at the ash deposit/metal oxide interface. This may account for the frequently-observed initial induction period of high particle reflectivity when the cooled surface is free of any deposit at the start. The exploitation and extension of such low-deposition induction periods constitutes an important aspect of the present program of research.

The viscous dissipation and surface energy associated with the liquid provide effective sinks for the kinetic energy of the impinging ash particle; in addition, the liquid phase in a porous deposit preferentially segragates to the inter-particle contact areas. Because of this 'liquid bridging' effect, even a small quantity of liquid in the deposit -- say, less than 1 percent by volume -- is sufficient to significantly increase deposit strength, thermal (and electrical) conductivity, and stickiness. Here it is useful to introduce two dimensionless parameters that are fundamental to ash deposition: the sticking coefficient ($\underline{s}$), which is the ratio of captured particle mass to incident particle mass, and the erosion co-efficient, $\underline{e}$, which is the ratio of the number of particles ejected from the deposit to the number of particles impacting on the deposit. With increasing liquid inventory in the deposit, $\underline{s}$, would increase, and $\underline{e}$ would decrease.

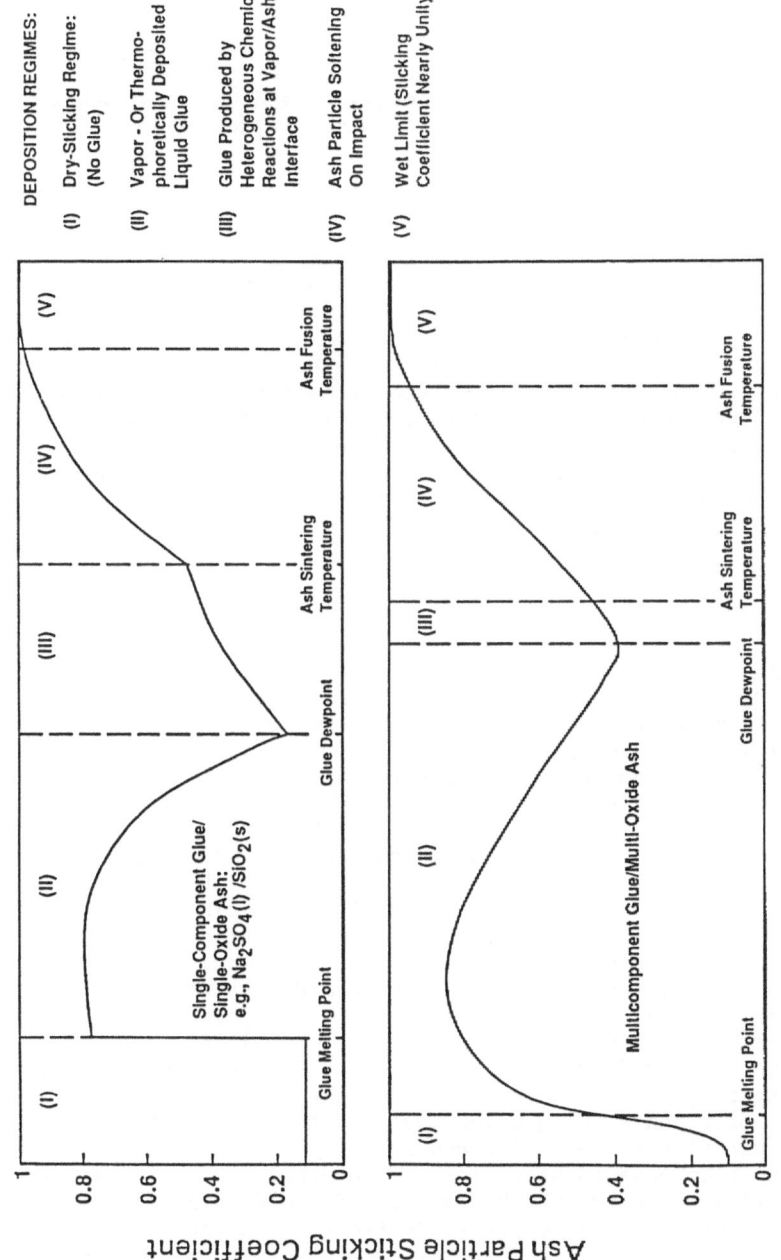

Figure 1. Liquid-Enhanced Capture of Impacting Ash Particles: Delineation of Deposition Regimes.

The net rate of growth of the ash deposit layer is governed by the competing dynamics of particle sticking and particle removal (by large particle impact-induced deposit erosion). The evolution of the deposit thickness, Δ, may be mathematically represented by the single non-dimensional ordinary differential equation:

$$\frac{d(\Delta/a)}{d\tau} = \underline{s}\,(\theta_W(\Delta)) - \underline{e}\,(\theta_W(\Delta)) \tag{1}$$

where a is a characteristic target dimension, and θ_W the dimensionless deposit outer (hot gas-side) surface temperature defined as:

$$\theta_W(\Delta) \equiv T_W/T_g \tag{2}$$

with the reference temperature, T_g, being the gas temperature. Equation 2 is an algebraic supplementary relation obtained by means of a "nested annulus" energy balance in which the deposit is subdivided into annular elements in order to allow for variable thermal conductivity radially within the growing deposit. The relationships between the sticking and erosion coefficients, $\underline{s}$ and $\underline{e}$, and deposit surface temperature are derived from their corresponding dependences on included glue volume fraction, Φ_G (which is, in essence, a function of deposit temperature, provided other variables, such as pressure, fuel-to-air ratio, gas temperature, etc. are held constant). θ_W, s and e are allowed to change with dimensionless time:

$$\tau \equiv (-\overset{\circ}{m}_p'')_{s=1} \cdot t/(a\rho_d) \tag{3}$$

where $(-\overset{\circ}{m}_p'')_{s=1}$ is the particle arrival rate (deposition rate corresponding to a sticking coefficient of unity), t is real time, and ρ_d is the deposit density in terms of particle mass per unit volume. The particle arrival rate itself is not allowed to vary with time, i.e., aerodynamic "action-at-a-distance" effects associated with very thick deposits have been neglected in the present analysis.

The dependences of $\underline{s}$ and $\underline{e}$ on the particle-to-glue deposition rate are sketched in Figure 2, which graphically represents the solution procedure used to estimate steady-state values for the particle deposition rate, sticking coefficient and erosion coefficient. The coupling between these parameters may be explained as follows: The inertial impaction deposition rate, $-\overset{\circ}{m}_p''$ depends linearly on $(\underline{s}-\underline{e})$, but $\underline{s}$ and $\underline{e}$ themselves depend on the inventory of glue available to each particle comprising the deposit surface layer. Therefore, for a given glue deposition rate, $-\overset{\circ}{m}_G''$, $\underline{s}$ and $\underline{e}$ are not known in advance, but must be obtained from the solution of a nonlinear equation, as sketched in Figure 2. Rational interim dependences of $\underline{s}$ and $\underline{e}$ on Φ_G have been incorporated in the model based on limited experience, intuition and necessary limiting behavior. While these necessarily have to be improved, the present mechanistic description of the underlying factors which are expected to determine $\underline{s}$ and $\underline{e}$ is in many ways preferable to attempts to "extract" sticking coefficients from limited sets of experimental observations (see, e.g. reference 9).

It is assumed for the sake of computational convenience (and in the absence of contradictory evidence) that the erosion coefficient $\underline{e}$ is of the same functional form as the sticking coefficient $\underline{s}$, albeit in an inverted sense. Thus, as $\underline{s}$ tends to unity, $\underline{e}$ would tend to zero; and as $\underline{s}$ approaches its "dry" value, s_{dry} (which may be as low as 0.0001), $\underline{e}$

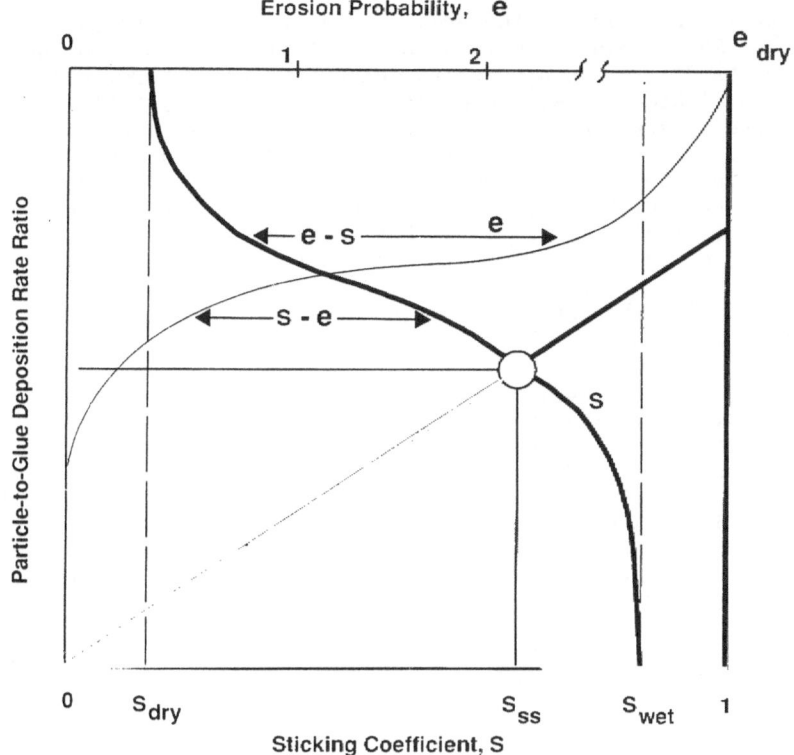

Figure 2. Inertial Particle Capture in the Presence of Simultaneous 'Glue' Deposition (After Rosner and Nagarajan, 1987): Estimation of Steady-State Sticking Coefficient and Erosion Probability.

reaches its maximum value, e_{max}, defined as the number of particles ejected per impacting particle from a dry, unsintered deposit. Pending the development of a plausible theory to estimate e_{max}, it is treated as a variable parameter, and studies are conducted to gauge the effect of variations in e_{max}. $\underline{e}$ may then be related to e_{max} as:

$$\underline{e} = \left(\frac{s_{wet} - \underline{s}}{s_{wet} - s_{dry}} \right) \cdot e_{max} \tag{4}$$

where s_{wet} is the sticking coefficient corresponding to complete inundation of the deposit by the glue liquid; it is usually taken to be unity.

MODEL PREDICTIONS

Introduction of the concept of glue-regulated deposit erosion, due to particle impaction at sufficiently high relative kinetic energies, has opened the way for understanding the conditions necessary for the phenomenon of "asymptotic fouling," a cessation of net deposit growth sometimes observed in boiler applications. The cause for this may be traced to a dynamic balance between glue-induced particle capture and removal events. Figure 3 displays the calculated effects of erosion on the net growth rate of an ash layer on a superheater tube in a pulverized-coal fired power station. These calculations correspond to a gas mainstream temperature of 1,558 K; gas approach velocity of 14.3 m/s, particle density of 2,500 kg/m, particle diameter of 20 micrometers, particle mass loading in the gases of the order of 10^{-3}, particle-to-glue burden ratio of about 150, and a steam temperature of 719 K within the tube. The glue liquid, $Na_2SO_4(1)$ was assumed to deposit by trace vapor concentration-diffusion (thick lines; dew point = 1,506 K), or by nucleated-droplet thermophoresis (thin lines). $SiO_2(s)$ was taken to represent coal-ash material. Under these conditions, as the deposit grows, its outer surface gets hotter, and the vapor deposition rate of glue drops accordingly. Thus, with time, the sticking coefficient decreases, and the erosion coefficient increases. For an assumed value of 2 for e_{max}, the "asymptotic fouling" behavior is seen here in the case of thermophoretically deposited glue, whereas vapor-deposited glue displays an approach to the same condition wherein the deposit layer thickness tends to a time-independent value corresponding to the intersection s=e in Figure 2.

Asymptotic fouling, highly desirable in terms of boiler/turbine operation, is attainable only under conditions where vapor diffusion or thermophoresis are the only modes of glue formation. However, coal-ash deposition involves high-temperature regimes (Regions III-V in Figure 1) where the rate of formation of the glue is kinetically controlled, and actually increases with temperature. In addition, the presence of coal-ash mineral impurity constituents decreases the initial deformation temperature of ash to values well below the melting temperature of pure silica (about 2,000 K). For instance, the initial deformation temperature of Arkwright Pittsburgh bituminous coal-ash (48 percent silica by mass) is 1,465 K, and the ASTM fluid temperature is 1,656 K; for some lignite coals, the fluid temperature is as low as 1,500 K. For these coals, the sticking coefficient would typically increase with temperature when the initial surface temperature exceeds about 1,200 K. This would eventually result in "catastrophic fouling" wherein ash begins to accumulate at ever-increasing rates, forcing a shutdown of the turbine (or boiler) within a matter of hours. Deposit erosion, in this case, becomes less and less

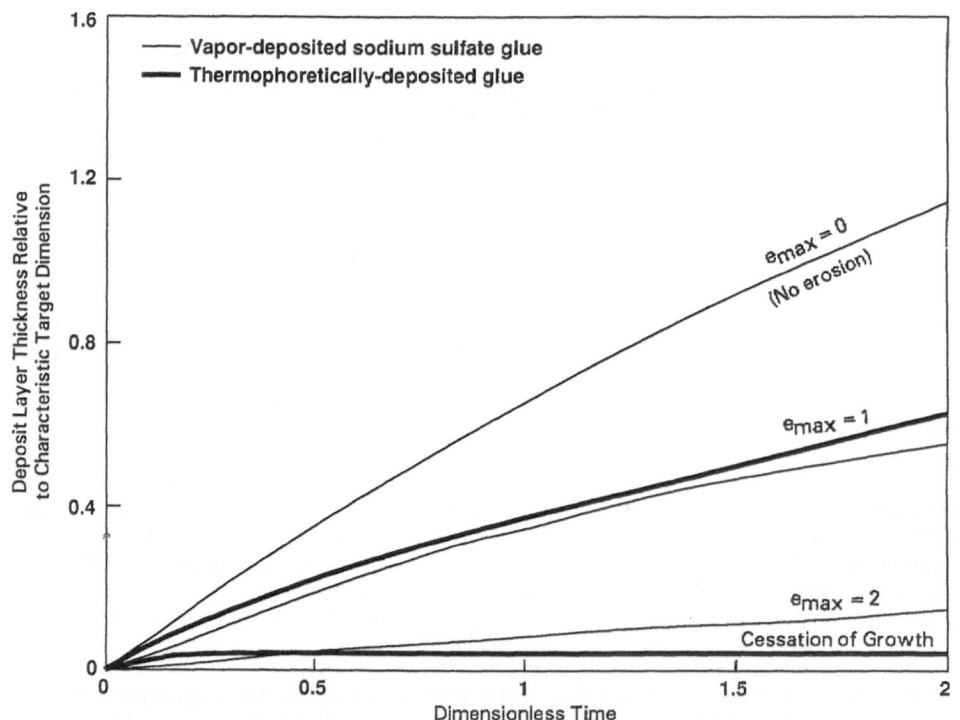

Figure 3.   Predicted Effect of Erosion on Deposit Layer Growth Under Boiler Simulation Conditions.

significant as the deposit builds up, unless the deposit gets to be so thick that it starts to affect particle trajectories and gas streamlines to a noticeable extent. Illustrative calculations incorporating erosion have been performed for the atmospheric-pressure combustion of Arkwright Pitts-burgh bituminous coal in the DOE-METC combustion/deposition entrained reactor, and are presented here in Figure 4. The gas temperature for these runs was taken to be 1,573 K, and the corresponding initial target surface temperature was 1,300 K. While consideration of erosion does lead to lower estimates of deposit growth, the approach to "catastrophic fouling" is clearly evident even for e_{max} values of up to 3. Given the high velocity of impact (approximately 300 m/s) encountered in turbine sections, 3 may not be an unrealistically high estimate for e_{max}; the actual value may, indeed, be even higher. There is no way to quantify this rather crucial parameter at present; this shortcoming will be redressed in follow-up work.

In Figure 5, the "effective" sticking coefficient, $\underline{s}$-$\underline{e}$, is plotted against time for various e_{max} values. The approach to the "wet" limit is obviously delayed by the erosion process which acts to retard the growth rate of the deposit. Negative values for $\underline{s}$-$\underline{e}$ imply impact-induced erosion of the target surface itself in the absence of an insulating deposit layer. The interference may be drawn here that many deposition-control strategies are liable to result in increased erosion and corrosion of the blade surface; additive techniques that rely primarily upon the physical dislodging effect of the added material in shaking the deposit loose are especially suspect in this regard. Additives that counteract deposition principally via chemical "gettering" reactions with condensible glue precursors are more benign with respect to surface degradation.

In the next section, predictions of the theoretical model are compared with relevant experimental data, and inferences are drawn regarding the role erosion actually plays in determining deposit evolution characteristics.

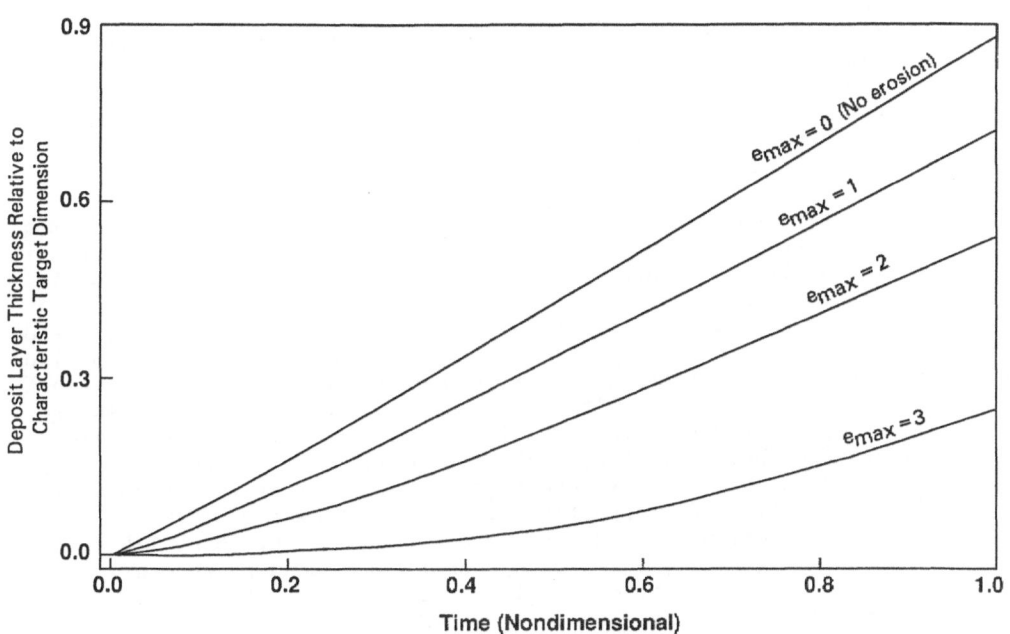

Figure 4. Predicted Effect of Erosion on Deposit Layer Growth Under Turbine Simulation Conditions.

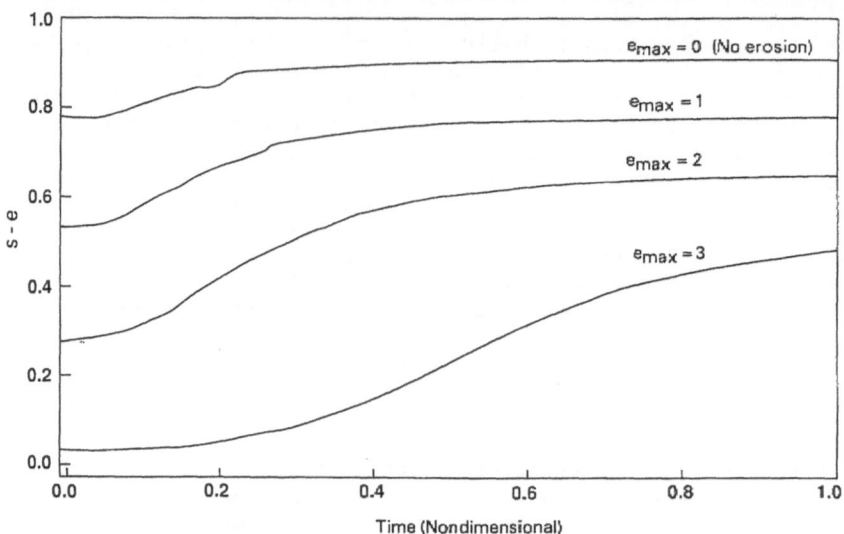

Figure 5.   Predicted Effective Sticking Coefficient (s-e)
Under Turbine Simulation Conditions.

DEPOSITION DATA MEASUREMENT AND COMPARISONS WITH MODEL PREDICTIONS

Deposition tests at DOE-METC are currently performed in a pressurizable, optically accessed high-temperature, high-velocity laboratory-scale combustion/deposition entrained reactor (CDER). The CDER is designed to simulate the combustion conditions of an actual gas turbine, with the deposition target assembly being representative of the leading edge of a turbine blade. Coal particles are entrained in dry air and introduced into the reaction chamber as a second stream. For a more detailed description of the experimental setup and test procedure, the reader is referred to reference 10. Coal particles burn and devolatilize as they flow down the drop-tube furnace. Unburned ash particles, along with the combustion product gases, are accelerated through a nozzle located at the bottom of the tube and impinge on a cylindrical platinum target held normal to the gas/particle stream. The target is cooled by means of an opposing (annular) jet of cooling air directed at the underside of the target. This allows the temperature of the target to be controlled independently of the gas temperature, and, thus, enables the simulation of deposition behavior of actively cooled engine components.

Target temperature is measured in situ during deposition by radiation pyrometry. Particle delivery rates are estimated by replacing the platinum target with a filter designed to capture all particles incident upon it. By monitoring weight gain as a function of time for both the filter and target under identical conditions, ash arrival rates and deposition rates may be determined; their ratio yields the sticking coefficient. In Figures 6 through 8, measured integral sticking coefficients are plotted against coolant flow rate; these results correspond to the atmospheric-pressure combustion of Arkwright Pittsburgh bituminous coal with the combustion gas temperatures being 1,573 K, 1,473 K, and 1,373 K, respectively. These gas temperatures correspond to uncooled target temperatures of about 1,300 K, 1,200 K, and 1,100 K, respectively. Coolant flowrate of 20 1/min results in a reduction of target surface temperature of about 125 K. The solid lines in Figures 6, 7 and 8 represent corresponding theoretical predictions of the sticking coefficient. In Figure 6, the influence of deposit erosion is displayed as well.

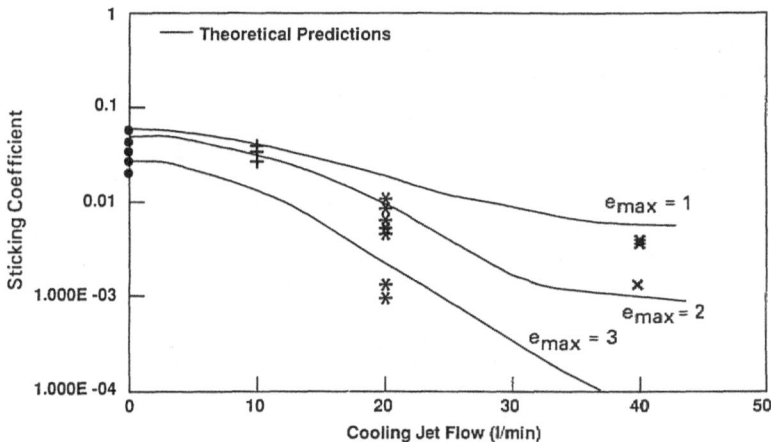

Figure 6. Sticking Coefficient Variation with Target
Surface Cooling: Gas Temperature = 1573K.

For an initial (uncooled) target surface temperature of 1,300 K, both
data and theory indicate a loss of particle adherence with increasing surface
cooling. Close quantitative agreement between theory and data is observed
when e_{max} is set to 2; e_{max} values exceeding 2 lead to sticking coefficient
underpredictions for coolant flow rates exceeding 20 1/min. From Figures
7 and 8, it is clear that there is a wide discrepancy between observations
and predictions at the lower gas and surface temperatures. At 1,473 K gas
temperature, the sticking coefficient is seen to increase slightly with
increased cooling; according to the present model, the sticking coefficient
should have decreased up to about 20 1/min coolant flow, and increased with
further cooling. At 1,373 K gas temperature, the sticking coefficient is
seen to be virtually independent of cooling, whereas theory predicts a
more complex dependency. In addition, theory predicts that for the uncooled
target, the sticking coefficient will increase monotonically with increasing gas temperature; however, the sticking coefficient corresponding
to 1,373 K gas temperature and an uncooled target is observed to be higher
than that at 1,473 K or 1,573 K.

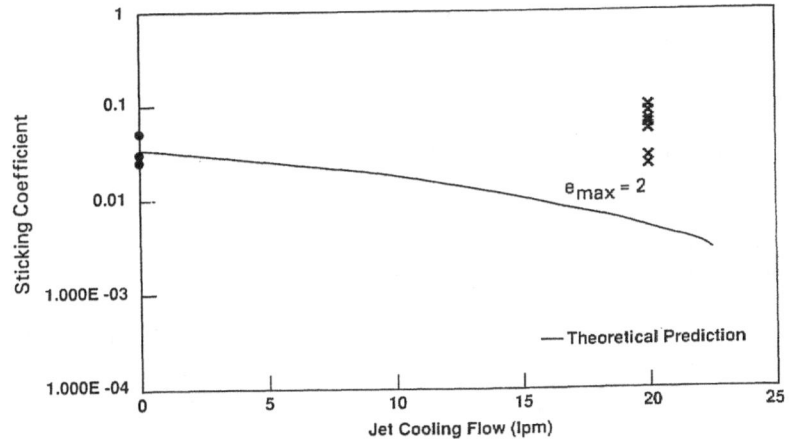

Figure 7. Sticking Coeffcient Variation with Target
Surface Cooling: Gas Temperature = 1473 K.

These apparent contradictions are principally the consequence of an assumption inherent in the present model -- viz. the neglect of hetero-geneous/homogeneous nucleation phenomena within the gas boundary layer. It is likely that such phase change processes occur to a greater extent at lower gas and target surface temperatures. This would affect the nature of the "glue" liquid delivery mechanism, as well as the dynamics of resulting supermicron particle encounters with the deposit. Ash particles would already possess a surface film of glue liquid under these conditions; this would imply, in general, higher sticking coefficients and lower erosion coefficients than predicted by the present theory. This would also imply a relative insensitivity to surface temperature, since deposition is essentially controlled by gas phase phenomena. The 1,573 K temperature situation is more in line with the present hypothesis that deposit surface temperature is the crucial parameter that determines particle capture probability. This "shift" in the deposition mechanism between low-temperature and high-temperature deposition regimes is an interesting phenomenon that will be taken up in more detail in future work.

SUMMARY

In this paper, the consequences of deposit erosion and boundary-layer nucleation on ash deposit growth rates in coal combustion environments have been examined. Erosion is shown to be of importance during the initial stages of deposit evolution, when the particles are not strongly bonded together. The present analysis has one adjustable parameter -- number of particles ejected from a dry deposit per oncoming particle -- which needs to be determined through well-defined and well-controlled laboratory tests under dynamic conditions. Glue uptake by particles in the gas layer is shown to contribute substantially to their adherence characteristics, resulting, in general, in higher, surface temperature-independent sticking coefficients. While the latter subject has been dealt with qualitatively here, a quantitative prediction of ash deposition rates in the presence of simultaneous homogeneous/heterogeneous nucleation is planned for the future. It is hoped that the current mutually supportive experimental and theoretical program of deposition research will help develop guidelines for effective deposition control in coal-fired heat engines.

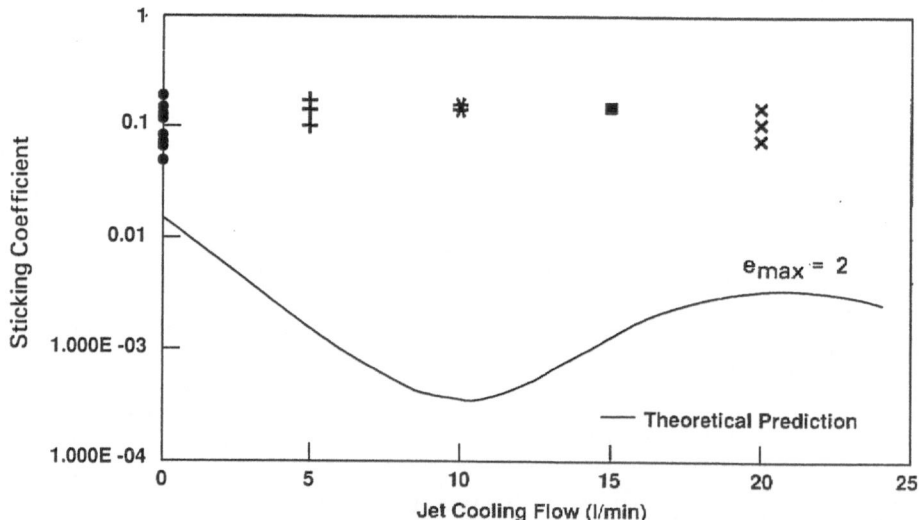

Figure 8.   Sticking Coefficient Variation with Target
Surface cooling: Gas Temperature = 1373 K.

REFERENCES

1. D. E. Rosner and R. Nagarajan, in "Heat Transfer," AIChE Symp. Series No. 83, R. W. Lyczkowski, editor, pp. 285-296, AIChE, 1987.

2. J. S. Ross, R. J. Anderson, and R. Nagarajan, Energy and Fuels, 2, 282-289 (1988).

3. R. Nagarajan and R. J. Anderson, Effect of coal constituents on the liquid-assisted capture of impacting ash particles in direct coal-fired gas turbines, ASME Paper No. 88-GT-192 (1988).

4. J. S. Ross, M. Ramezan, R. J. Anderson, and R. Nagarajan, Prediction and in situ measurement of the thermal conductivity of multiphase fouling deposits formed in direct coal-fired combustors, ASME Paper No. 87-WA/HT-6 (1987).

5. M. Ramezan, J. S. Ross, R. J. Anderson, and R. Nagarajan, in "Proc. Fourth Annual Pittsburgh Coal Conference," Pittsburgh, Pennsylvania, September 28 through October 2, 1987, pp. 905-915.

6. R. Nagarajan, R. J. Anderson, S. D. Woodruff, and C. Meyer, Coal ash mineral composition and its relation to ash deposition: A theoretical and experimental study, paper presented at the Engineering Foundation Conference on Mineral Matter and Ash Deposition from Coal, Santa Barbara, California, February 21 through 26, 1988.

7. R. Nagarajan and R. J. Anderson, High temperature ash deposition mechanisms: A theoretical treatment (in preparation, 1988).

8. E. Raask, "Mineral Impurities in Coal Combustion: Behavior, Problems, and Remedial Measures," Hemisphere Publishing Corp., 1985.

9. R. K. Ahluwalia, K. H. Im, and R. A. Wenglarz, Flyash adhesion in simulated coal-fired gas turbines, ASME Paper No. 88-GT-135 (1988).

10. R. J. Anderson, C. T. Meyer, and R. A. Dennis, A combustion/deposition entrained reactor for high temperature/pressure studies of coal and coal minerals, paper to be presented at the 1988 Annual AIChE Meeting, Washington, D.C., November 27 through December 2, 1988.

FACTORS AFFECTING ADHESION OF DRUG PARTICLES TO SURFACES

IN PHARMACEUTICAL SYSTEMS

P. J. Stewart

Department of Pharmacy
University of Queensland
St. Lucia, Queensland
Australia 4067

The mechanism of adhesion of drug powders on the
surface of polymer coated glass beads was studied. Glass
beads (500 μm) were air suspension coated using a 5%
hydroxypropylmethylcellulose phthalate solution in
dichloromethane: methanol solvent (50:50). Specific
particle size fractions of several sulphonamide powders were
prepared by fluid energy milling and sonic sifting;
distributions were characterized by laser diffraction.
Average charge to mass ratios on particle detachment were
measured using an air stream Faraday cage. Total adhesion
was determined by a centrifugal method and characterized by
an S50 value, i.e. the speed required to detach 50% of the
drug particles. Electrical interactions probably caused by
triboelectrification during the preparation of the
interactive system contributed significantly to the initial
drug adhesion. Adhesion decreased with time and was well
correlated with the charge to mass ratio decrease. The rate
of decrease in adhesion increased with increasing relative
humidity. Adhesion did not occur uniformly over the polymer
surface with local multilayer adsorption occurring when the
drug concentration was increased. When the interactive
systems were prepared at higher relative humidity conditions
(60% RH), adhesion between the drug particles and surface
was significantly decreased; non-electrical interactions
contributed to the adhesion process under these conditions.

INTRODUCTION

Adhesion in pharmaceutical systems has been known for many years
but quantification of adhesion phenomena has occurred only in recent
times. The degree of adhesion of drugs to surfaces has important
implications in the formulation and processing of solid dosage forms.
For example,
  (a) the flow properties of drug powders will be influenced
       by their cohesiveness;
  (b) the degree of interaction between drug and excipients can
       greatly influence the homogeneity of a powder mix;

(c) the uniformity of powder filling procedures will depend
     on the cohesiveness of the powder mixture;

(d) sticking of powder to the walls of the processing
     equipment causes material loss and drug homogeneity
     problems;

(e) the compressibility of some tablet granulations will be
     determined by the adhesion characteristics of lubricants;

(f) the dissolution times of tablets and capsules will be
     affected by the degree of interaction of hydrophobic
     lubricants to granule surfaces.

     The purpose of this paper is to review the research undertaken at
the Department of Pharmacy, University of Queensland and to report on
the mechanism and degree of interaction of drug particles on a model
polymer coated carrier and the factors which influence the interaction.

RESULTS AND DISCUSSION

1. Quantification of Adhesion

     The centrifugal method (Figure 1) allowed the determination of the
adhesion profile which was a logarithmic normal function when the
percent of drug remaining on the carrier analysed by u-v spectroscopy
was regressed against the square of the speed of rotation[1]. The profile
could be characterized by the S_{50} , i.e. the speed required to dislodge
50 percent of adherent particles, and σ , i.e. the geometric standard
deviation of the distribution. In these experiments the total degree of
adhesion of the drug particles in the interactive systems was measured
by the S_{50} parameter.

     The air stream Faraday cage (Figure 1) determined the charge
generated on detachment of the drug particle from the surface of the

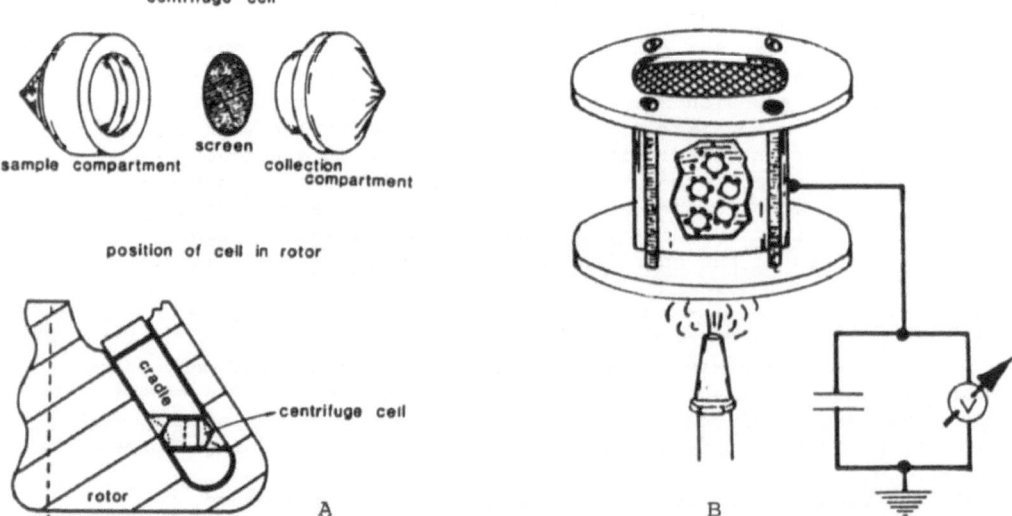

Figure 1.   Design of (A) the centrifuge cell for total adhesion
            measurement and (B) the air stream Faraday cage.

carrier in the interactive systems which was calculated as the average charge to mass ratio or tribo (T) in microcoulombs per gram of adherent powder, i.e. $T = Q/M$, where Q is charge magnitude measured and M is the mass of powder detached. The concentration of powder detached (C%) was expressed as 100M/W, where W is the weight of mixture after powder detachment[2].

2. Effect of Ambient Humidity during Blending

Interactive mixes of sulphapyridine powder ($d_v=27.2$ μm, $\sigma=1.1$ μm) and a hydroxypropylmethylcellulose phthalate coated glass carrier were used to observe the adhesion tendency of drug particles when prepared at humidity conditions of 26, 40, 50, 65, and 80% at constant temperature of 25 °C (Figure 2). The carrier and drug were equilibrated over silica gel and blended within the environmental chamber at the required relative humidity and temperature. The sample of mixture was immediately taken for adhesion measurement after preparation. S_{50} values of five replicate mixtures were determined at each humidity level.

The relationship between S_{50} and relative humidity at which the mixes were prepared is shown in Figure 3[3]. S_{50} values were highest at a relative humidity of 26%, showed a rapid and significant decrease (analysis of variance; P<0.001) in adhesion tendency at 50% R.H. and reached a minimum at 65-80%. The reduction of S_{50} with increasing relative humidity during blending can be explained by considering the forces involved in the adhesion process. Four types of forces are primarily important in the adhesion of small particles to surfaces; molecular forces, electrical forces including electrical double layer

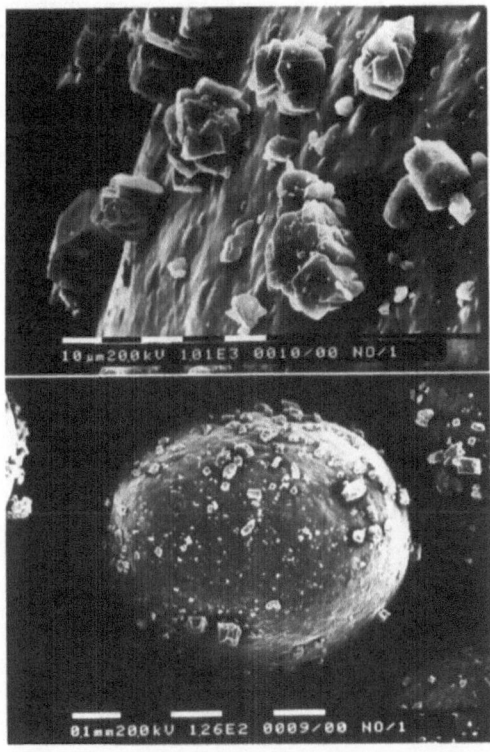

Figure 2. Scanning electron micrograph of the interactive mix showing the sulphapyridine particles adhering to the HPMCP coated glass beads

formation at contact points and the Coulombic interaction of the
electrical charge distributed over the particle surface due to previous
electrifications, capillary forces due to the formation of liquid
bridges, and interfacial forces including salt bridge formation,
mechanical interlocking of particles and sintering effects[4,5].

While the molecular force is effective over the whole relative
humidity range, the electrical double layer interaction contribution
occurs up to intermediate humidity conditions until charge leakage
begins at the interface. Coulombic interaction which will occur at low
to intermediate humidity level will increase the adhesion of interactive
system, depending on the degree of static electrification of the
adherents and the treatments of the materials. At high humidity levels
the capillary force should become the main component of adhesive forces.
The critical humidity between 40-50% in Figure 3 at which the rapid
reduction of S_{50} occurred corresponded with the humidity range at which
sudden increases in electrical conductivity of the materials and
surrounding atmosphere have been observed. There was little evidence
that the capillary interaction dominated at relative humidities between
65 and 80%. If substantial capillary interaction occurred at relative
humidities between 65 and 80%, an increase in the adhesion tendency
would be observed following the reduction of S_{50} at intermediate
humidities. Such a result was not unexpected since the adhesion
measurements were performed immediately after interactive system
preparation when the liquid bridges between surfaces were probably not
well established[11].

3.   Effect of Storage Relative Humidity

Interactive systems of sulphapyridine were prepared by blending
for 10 minutes at relative humidity and temperature of 26% and 25 °C,
respectively. The interactive mixes then were placed in crucibles and
stored in desiccators which were conditioned to various relative
humidities (6, 33, 53, 75, and 94%) by saturated salt solutions at
constant temperature (25 °C). Samples of mixtures were then taken at
time intervals for adhesion measurements. Determination of the degree
of adhesion of the interactive mixes was performed over an 11 day
period.

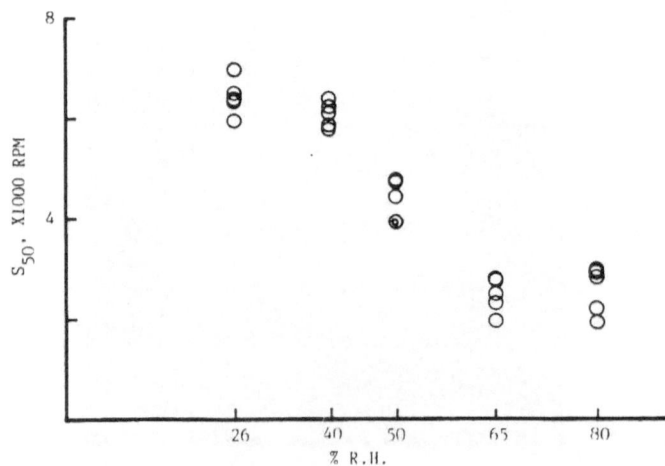

Figure 3.   Effect of relative humidity during blending on the adhesion
(S_{50}) of a sulphapyridine interactive mix, replicate runs.

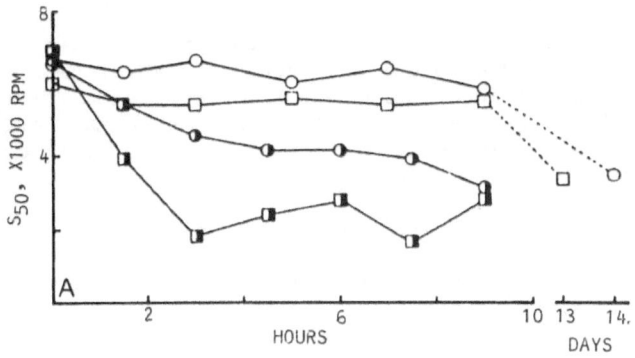

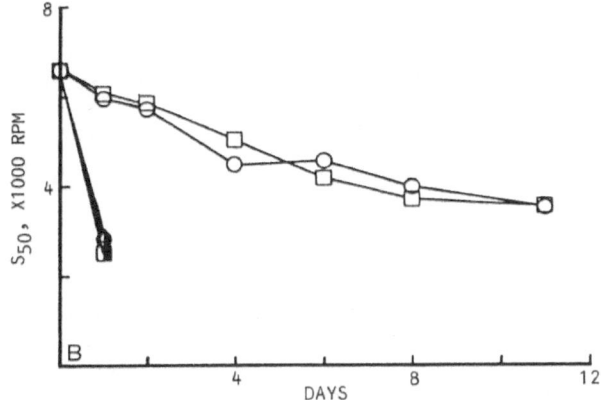

Figure 4. Effect of relative humidity on the adhesion (S_{50}) of sulphapyridine interactive mixes during storage (A) over 9 hours and (B) over 11 days.

○ =6%RH; □ =33%RH; ◐ =53%RH; ◧ =75%RH and △ =94%RH

Figure 4 shows the relationship between S_{50} of the sulphapyridine interactive mixes and time of storage at different humidity conditions[3] S_{50} values at time zero were obtained for the freshly prepared mixes before storage. The S_{50} of the mixes kept at 53, 75, and 94% R.H. were markedly reduced after one day's storage whereas S_{50} of the mixes kept at 6 and 33% showed a gradual decrease over the 11 days. The S_{50} profiles of the mixes kept at 6 and 33% decreased at about the same rate. The S_{50} reduction curve always retains a certain asymptotic level as the other force components, i.e. molecular forces, capillary forces will always exist when the electrical interactions are diminished by charge decay.

4. Correlation Between Total Adhesion and Charge Decay

Since the capillary force will be relatively ineffective under dry condition or at low relative humidity (i.e. 33%), the force components assisting particle adhesion at low to medium relative humidity will be as follows[4-6]

$$F_{AD} = F_{NEL} + F_{EL}$$

where F_{NEL} is the nonelectrical force component and includes the molecular force, salt bridge formation, mechanical interlocking, etc, and F_{EL} is the electrical force component which consists of interactive forces caused by the electrical double layer at the contact points (F_C) and the Coulombic interaction of the electrical charge distributed over the particle surface due to previous electrifications (F_{IM}).

During storage of the interactive systems studied both the total adhesion and the average charge-to-mass ratio decreased[7]. Correlation of the total adhesion force and the electrical forces quantitatively is not possible. For example, accurate estimations of the total adhesion force using the centrifugal technique is hindered by lack of knowledge of the particle size distribution removed from the carrier at each centrifuging time and by particle orientation effects during detachment[1]. In addition, calculation of the electrical forces requires a knowledge of the drug-carrier separation distances and areas of contact. The following theoretical treatment of the force components in these interactive systems can be achieved using the mathematical expressions derived for each of the force components[8] -

$$F_{AD} = F_{NEL} + F_{IM} + F_C$$

$$d^3\pi\rho\omega^2 1/6 = F_{NEL} + Q^2/x^2 + 2\pi Q^2/A$$

$$\omega^2 = 6F_{NEL}/d^3\pi\rho 1 + [6/x^2\pi d^3\rho 1 + 12/Ad^3\rho 1] Q^2$$

where d is the adhered particle diameter, ρ is the particle density, ω is the angular rotating velocity, 1 is the distance between the particle-carrier interface and the axis of rotation during centrifugation, Q is the charge on the particle, x is the distance between the centres of the charges during coulombic interaction between the charged particle and its image, and A is the contact area between the particle and substrate during contact electrification.

The square of the rotational speed is, therefore, related to the square of the charge of the drug powder. The physical meaning of the charge Q in the derived equation is different for the image and contact force interaction. Q is the charge on the particle during Coulombic interaction between the precharged particle and a neutral surface and is the charge developed when a particle is detached from a surface after interaction by contact electrification. In the air stream Faraday cage both these charges will be measured and the parameter Q can be represented by the average charge-to-mass ratio. Since the interactive systems contain a distribution of adherent particle sizes and since the intrinsic adhesion force is also distributed in intensity, the detachment of particles cannot be reflected by a single angular velocity. In the developed centrifugal method the " average " detachment velocity is represented by the S_{50} value. Such a value can be used to compare the degree of interaction in mixtures provided the standard deviation of the forces in the system remains constant.

Figure 5 shows the correlation between the square of the S_{50} and the square of the average charge to mass ratio[7]. Several observations were made -

(a) Good correlation existed between the S_{50} as a measure of total adhesion and the square of the average charge to mass ratio for the interactive mixtures shown, i.e. the correlation coefficients for the sulphapyridine and succinylsulphathiazole interactive mixtures were 0.98 and 0.93 with significance at the 99% and 95% level, respectively

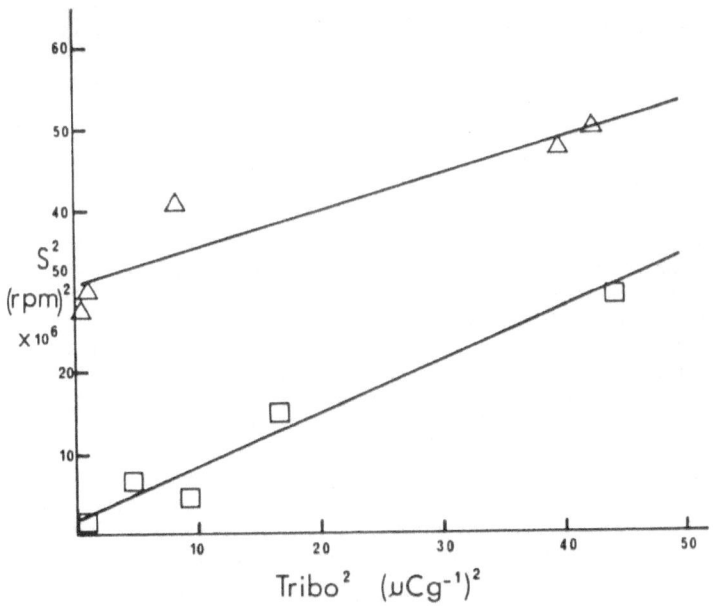

Figure 5. Correlation between the square of the S_{50} and the square of the average charge-to-mass ratio for the sulphapyridine ($\square$) and succinylsulphathiazole ($\triangle$) interactive mixes.
————————— , linear least-squares regression.

(b) The intercept value is a complex function involving the non electrical force component, particle size and density and is different for the two interactive mixes, i.e. 3.08×10^7 and 1.74×10^6 for the succinylsulphathiazole and sulphapyridine interactive mixes respectively. Since the sulphapyridine and succinylsulphathiazole possess similar densities (1.35 and 1.43 $g\ cm^{-3}$ respectively) and similar particle size distributions (volume mean diameter of 27.2 μm, sd = 1.1 μm and 23.4 μm, sd = 1.6 μm, respectively), the different intercept values for the two mixes reflect different nonelectrical degrees of interaction. The succinylsulphathiazole, therefore, probably has a greater molecular interaction with the polymer coated carrier than the sulphapyridine since the degree of capillary interaction and other surface forces would be expected to be low in the humidity conditions of the study.

(c) Changes in the slope of the regression will be a function of the particle density and size distribution, the area of contact between the drug particles and the carrier surface and the distance between particle centre and its image at the interface. The slightly differing slopes obtained in Figure 5 (i.e 4.52×10^5 and 6.54×10^5 for the succinylsulphathiazole and sulphapyridine interactive mixtures respectively) reflect the contributions of these parameters in the two mixes. While the particle size and density are similar for the drug powders and will not influence the slope significantly, small particle size differences will affect x^2, and the particle shape and orientation at the surface, the deformation characteristics of the powder and the degree of impact during mixing will affect the area of contact.

5. Effect of Moisture Content of the Powder

The cumulative effects of the previous history of a powder reflect its physical properties as well as its interactive characteristic[9]. The most likely contaminant of powder during storage is the moisture vapour of the ambient atmosphere. Since the contact electrification process is a surface phenomenon, the moisture contamination of the surface of contacting bodies will alter its nature of charge generation and the rate at which it is dissipated. The investigation of the effect of relative humidity conditions during material storage on the S_{50} of the interactive mix was carried out.

The carrier and drug powders were stored for 24 hours in a desiccator at controlled temperature (25 °C) and relative humidity (26, 43, 53, 75, and 84%) using saturated salt solutions. Four replicate interactive systems were prepared at each relative humidity by blending for 10 minutes.

Figure 6 and 7 shows the relationships between the degree of adhesion of succinyl sulphathiazole and sulphmerazine powders and the relative humidities at which the powders were stored before blending[3]. The adhesion tendency of succinyl sulphathiazole powder mixes was significantly reduced when the relative humidity conditions were increased (analysis of variance; $P < 0.001$). This behaviour conformed to the general trend of the effect of humidity on electrostatic interactions. The S_{50} values of the mixes prepared from the powder stored at 75% R.H. were in the same order as the minimum S_{50} reduction of the mixes during storage.

Unexpected results for the adhesion tendency of sulphamerazine interactive mixes were observed (Figure 7). The adhesion tendency of sulphamerazine powder was significantly increased when powders were stored at 53% R.H. and above (analysis of variance; $P < 0.001$).

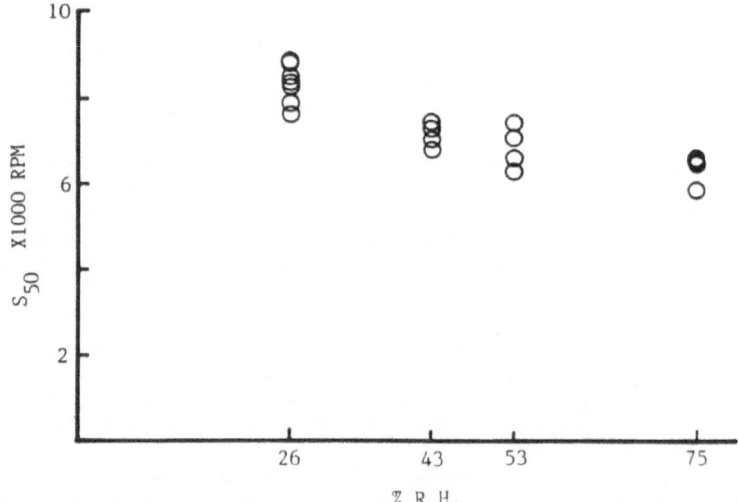

Figure 6. Effect of relative humidity preconditioning on the adhesion (S_{50}) of succinylsulphathiazole interactive mixes, replicate runs.

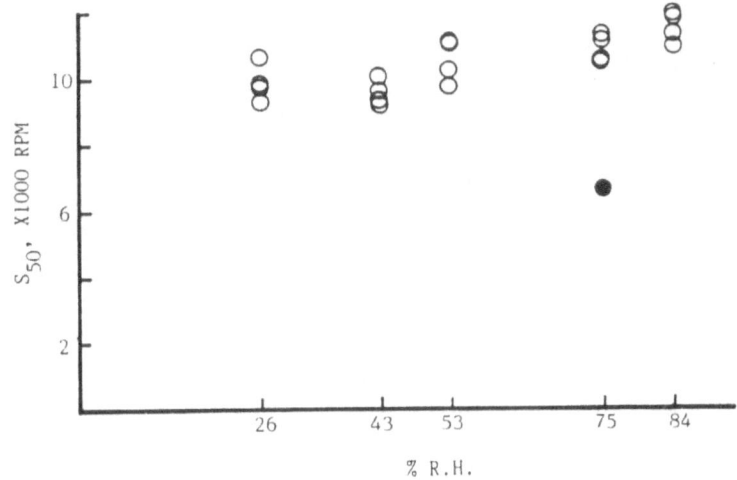

Figure 7. Effect of relative humidity preconditioning on the adhesion (S_{50}) of sulphamerazine interactive mixtures, replicate runs. ● =S_{50} of mixture stored at 75% RH for 24 hours.

The capillary force would be expected to be low in this adhesion system as the hydrophobic surface property of materials, in particular the sulphamerazine powder would not form adsorbed moisture film layers thick enough to induce such capillary interaction. Adsorbed water vapour could have two effects on the surface charge[10]. Although it is well known that adsorbed water will increase surface conductance hence cause rapid charge dissipation, the adsorbed moisture vapour may also provide more charging sites on a particle. This effect possibly occurred on sulphamerazine particle with the water vapour associated with the surface localizing at various spots and behaving as additional charging sites resulting in an increase of electrostatic interactions.

This effect was tested by observing the adhesion tendency of an interactive mixture which was prepared from the materials stored at 75% R.H. for 24 hours and then stored at the same relative humidity. The result of adhesion measurement showed a decrease in S_{50} to 6672 rpm (initially the S_{50} was 11224 rpm, shown by closed circle in Figure 5). Such a decrease was in agreement with the data in Figure 4 reflecting that charge decay had occurred.

6. Influence of Particle Concentration

All drug materials showed a reduction in their adhesion capabilities to the carriers when the amount of drug powder in the mixtures increased[1] (Figure 8-10). The collision effect and formation of particulate layers were considered to be the cause of the decrease in S_{50}.

The collision effect could occur by the impingement of detached particles with neighbouring particles which were in the path of the particle leaving the carriers. Such an effect would cause premature dislodgement of other more strongly bound particles. Particles would be removed from the substrate by different methods depending of the directions of the forces on the particles. Particles will be directly detached if the forces were applied normal to the surface of substrate, or they will be slid off under application of tangential forces, or will

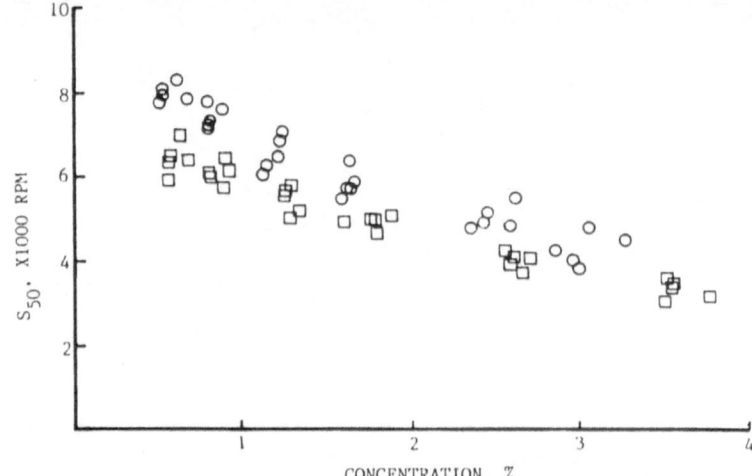

Figure 8.   The effect of particle size and concentration on the adhesion (S_{50}) of sulphapyridine interactive mixes. ☐ d_v=15.5 μm; ○ d_v=27.2 μm.

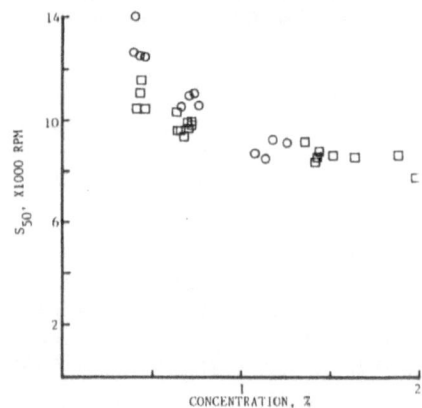

Figure 9.   The effect of particle size and concentration on the adhesion (S_{50}) of sulphamerazine interactive mixes. ☐ d_v=14.5 μm; ○ d_v=17.7 μm.

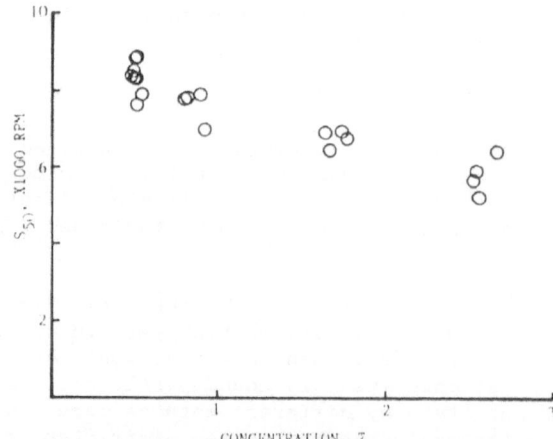

Figure 10.   The effect of particle concentration on the adhesion (S_{50}) of succinylsulphathiazole interactive mixtures.

be rolled off by simultaneous shear and tension forces. Because of the spherical shape of the interactive system employed in this experiment, particles at different locations on the carrier surface experienced the centrifugal forces at different angles relative to the substrate surface. This condition promotes the rolling and sliding actions of detached particles leaving the carriers. Consequently, any drug particles under removal by sliding or rolling off the carrier will tend to collide with other particles resulting in multiple dislodgements of particles. When the amount of drug powder in mixture increases, decrease in interparticle distance occurs with a greater chance of particle collisions and a further reduction in the S_{50}.

Figure 11 shows the scanning electron micrographs of interactive mixes containing sulphapyridine powder at concentrations of 0.9% (a), 1.9% (b), 2.7% (c), and 3.1% (d). The tendency of the adherent drug particles to form particle layers increased with drug concentration. The particle layer formation observed for the higher drug concentrations occurred despite free space still available on carrier surface (monolayer saturation was about 12 percent).

The layer formation caused a decreasing adhesion tendency of drug particles by hindering direct contact with carrier surface. The binding capability of the outer particle layer relied on the interaction with its own species and might be assisted by long range attractive forces with the carrier. The outer particle layer could, therefore, form strong autoadhesion bonding but, when the inner particles were

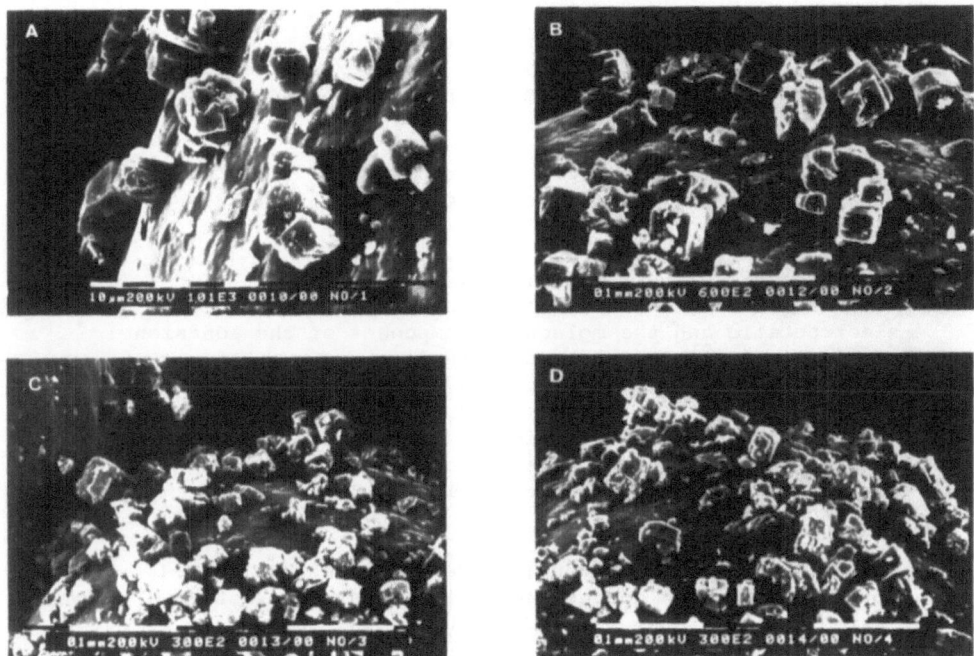

Figure 11. Scanning electron micrographs of the sulphapyridine interactive mixes (a,0.9%; b,1.9%; c,2.7%; d,3.1%)

dislodged, the whole aggregate would be removed from the carrier. Both collision and layer formation effects could cause reduction in the S_{50} by initiating the multiple dislodgement of drug particles. However, the collision effect was unlikely to be effective until the number of adherents on the carrier became so dense that interparticle distances were equal to the path travelled by removed particles during rolling or sliding before dislodgement. In addition, layer formation effects occurred more frequently with increase in drug concentration.

The other possible drug segregation mechanism that could have occurred in these studies was the multiple detachment of drug particles due to the collision effect and particle layer formation. Surface roughness of pharmaceutical excipient carriers could facilitate the formation of particle layers. Fine powder was likely to fill up the carrier surface irregularities in agglomerates rather than covering the surface in a particulate manner. Although uneven surfaces could adsorb more drug powder, more strongly than the less irregular carrier surfaces, multiparticle detachment in the form of particle layers or agglomerates might also occur if the binding force between inner particles contacting with the carrier surface was less than the external accelerative force.

REFERENCES

1   P.Kulvanich and P.J. Stewart, Fundamental
     considerations in the measurement of adhesional forces
     between particles using the centrifuge method. Int. J.
     Pharm., 35, 111-120 (1987).

2   P.Kulvanich and P.J. Stewart, An evaluation of the
     air stream Faraday cage in the electrostatic charge
     measurement of interactive drug systems. Int. J. Pharm.,
     36, 243-252 (1987).

3   P.Kulvanich and P.J. Stewart, Influence of
     relative humidity on the adhesive properties of a model
     interactive system. J. Pharm. Pharmacol., 40, 453-458
     (1988)

4   H.Krupp, Particle adhesion: Theory and
     experiment. Adv. Colloid Interface Sci., 1, 111-239
     (1967)

5   B.V.Derjaguin, Yu.P.Toporov, V.M. Muller, and
     I.E. Aleinikova, On the relationships between the
     electrostatic and the molecular component of the adhesion
     of elastic particles to a solid surface. J. Colloid
     Interface Sci., 58, 528-533 (1977).

6   I.E.Aleinikova, B.V.Derjaguin and Yu.P.Toporov, The
     electrostatic component of adhesion of dielectric
     particles to a metal surface. Kolloidn. Zh., 30,
     177-182 (1968).

7   P.Kulvanich and P.J.Stewart, Correlation between
     total adhesion and charge decay of a model interactive
     system during storage. Int. J. Pharm., 39, 51-57
     (1987).

8   A.D.Zimon,"Adhesion of Dust and Powder,"2nd ed.,

pp.108-119, Consultants Bureau, New York,1982.

9   E.N.Hiestand, Powders: Particle - particle
        interactions.   J. Pharm. Sci., 55, 1325-1344 (1966)

10   G.A.Turner and M.Balasubramanian, The frequency
        distributions of electrical charges on glass beads.
        J.Electrostat., 2, 85-89 (1976).

11   W.J.Whitfield, A study of the effects of relative
        humidity on small particle adhesion to surfaces,
        in "Surface Contamination: Genesis, Detection
        and Control", K.L.Mittal, editor, Vol.1, pp. 73-81,
        Plenum Press, New York, 1979.

PART II.   PARTICLE DETECTION, ANALYSIS AND CHARACTERIZATION

THE ROLE OF INFRARED AND RAMAN MICROSPECTROSCOPIES IN THE

CHARACTERIZATION OF PARTICLES ON SURFACES

Kenneth J. Ward and David R. Tallant

Materials Characterization Department 1823
Sandia National Laboratories
Albuquerque, NM   87185

The nondestructive identification of particles on surfaces can be accomplished quickly and definitively using vibrational microspectroscopy. Both Fourier-transform infrared (FT-IR) and Raman microspectroscopies are vibrational spectroscopy methods which are now commercially available. These two techniques are strongly complementary for several reasons. Both methods detect vibrational energies between atoms in molecules. However, the quantum mechanical selection rules differ such that infrared spectroscopy is more sensitive to asymmetric molecular distortions and Raman spectroscopy to symmetric modes. This means that weak bands in infrared spectra are often strong in Raman spectra and vice versa. Infrared spectroscopy directly measures absorption of incident infrared radiation. However, Raman spectroscopy observes the same vibrational frequencies by measuring the differences in energy between incident and scattered visible light. This means that infrared microspectroscopy has a diffraction limitation of about 10 micrometers while the diffraction limit for Raman microspectroscopy is near 1 micrometer, so Raman microspectroscopy can be used for smaller particles. Furthermore, because visible light detectors are used, Raman spectroscopy can detect vibrational frequencies as low as 50 wavenumbers. However, detection of infrared frequencies in this signal-limited application requires the use of more sensitive infrared detectors which have low frequency cutoffs at about 750 wavenumbers. These advantages of Raman microspectroscopy, the smaller particle capability and the increased frequency coverage, are offset by the low quantum efficiency of the Raman effect and sometimes by interference due to fluorescence. These differences between the two methods lead to the general observation that infrared spectroscopy more readily identifies organic contaminants, while Raman spectroscopy is usually better for inorganic species. Numerous examples of the application of these methods to materials analysis problems will be presented to illustrate these principles and to demonstrate the broad utility of vibrational microspectroscopy to particle identification.

INTRODUCTION

The papers included in this proceedings volume reveal that particle characterization is an arduous and complex process. Simply decreasing the numbers and sizes of airborne particulates to levels which are acceptable for a particular class of clean room or type of processing application is not sufficient. Analyses of airborne particulates do not provide a complete picture of the myriad of contaminants which inevitably migrate to undesirable locations on production components. These other types of contaminants include materials originating from processing equipment or from the process itself, such as lubricants, plasticizers, adhesives and encapsulants. Also, human debris such as skin, hair, cosmetics and clothing fibers are difficult particulates to identify. In general, the full range of contaminants in a processing environment are both particulate and non-particulate in nature. The location and characterization of this diverse variety of contaminants requires a broad spectrum of analytical instrumentation and expertise. This paper will review one type of solution to these problems, the techniques of infrared and Raman microspectroscopies. These methods are extremely versatile and supply definitive identifications for almost all types of contaminants. For situations involving unknown non-metallic contaminants, these analytical methods are an excellent first choice. The uses and limitations of these methods will be discussed in detail in this paper.

Let us examine how these techniques can be simultaneously versatile and definitive. Infrared and Raman microspectroscopies are versatile because of the diverse sampling applications that are possible. They are definitive because they provide chemically specific (i.e. legally defensible) identifications of materials. Gases, liquids and solids are all easily sampled. For processing applications, solids are the most common substrates. These methods can be used for contaminants on solids which are in the form of particles, droplets, smears, fibers, crystals, inclusions, bubbles and additives. Important characteristics of the techniques are that they are nondestructive and are readily accomplished at normal temperatures and pressures. If desired, these methods are also easily applicable to situations involving extremes of temperature and/or pressure if desired. Samples of almost any size can be accommodated, from a micrometer to a meter or more. The techniques can be used to identify many types of organic and inorganic contaminants. These methods are not useful for metals and metallic alloys, for coatings with thicknesses of less than ten monolayers, for dilute solutions of less than millimolar concentrations, and for complex mixtures of contaminants. Thus the primary limitation of these methods is sensitivity, not versatility. Hardware improvements to the instrumentation are continuously reducing these limitations, for example a grazing angle microscope has been developed by Spectra-Tech Inc. (Stamford, CT) for use in the analyses of near monolayer thin film coatings.

Infrared and Raman microspectroscopies are extremely specific in that each chemical species in a particular environment exhibits a unique and distinctive vibrational spectrum. These spectra are physical properties of a substance like the boiling point or density. Unlike many physical properties, different species do not exhibit the same vibrational spectra. Even many conformational isomers of the same compound can be identified, which cannot be accomplished by a technique as specific as mass spectroscopy. In fact, mass spectroscopy may have difficulties with some structural isomers, which are always distinguishable using vibrational spectroscopy. However, there are complications which limit the specificity of these methods. If the contaminated area of interest is a mixture, the resulting spectrum will be a mixture of the spectra of the individual molecular components of the mixture. This renders identification of the individual contaminants of interest more difficult. Methods to solve this complication will be discussed in the sampling section of this paper. All of the reasons outlined above illustrate that infrared and Raman microspectroscopies are versatile techniques for all types of sampling applications in addition to being extremely specific chemical identifiers.

The field of vibrational microspectroscopy is now experiencing an explosion in growth. Although the method was first developed in the 1940's for dispersive infrared spectrometers[1,2] and in the 1970's for Raman microspectroscopy,[3,4] the poor sensitivity of early instrumentation limited the utility of the methods. Modern computer-driven instruments with high sensitivity detector systems have now changed the situation completely. Two recent books have been entirely devoted to the subject of infrared microspectroscopy.[5,6] Many papers have focused upon the general utility of these vibrational microspectroscopic methods to industrial problem solving.[7-12] Early applications included work which identified industrial processing contaminants by FT-IR microspectroscopy[7] and by Raman microspectroscopy.[8] A variety of applications using both methods were summarized in two papers.[9,10] In another article, FT-IR microspectroscopy was applied to the identification of numerous industrial processing contaminants.[11] A general review of the technique of infrared microspectroscopy was the feature article in an issue of Analytical Chemistry.[12]

In addition, these methods have been applied to numerous specialized applications in the fields of microelectronics,[13-22] polymer science,[23-41] ceramics,[8,42-44] corrosion,[45,46] forensics,[47-50] art conservation,[51,52] paper manufacturing,[53] mineralogy,[54-58] coal utilization,[59,60] explosives,[61,62] electrochemistry[10,63,64] and biology.[65,66] In the microelectronics industry, early work focused on the localization of carbon, oxygen and dopant contents and thicknesses on silicon wafers using infrared methods.[13] These subjects have been recently revisited.[14] Organic and fluorescent processing contaminants have been identified,[15] and packaging contaminants have been studied,[16] using infrared microspectroscopy. Failure analyses have been described for circuit boards,[17] and for disk media.[18] The mapping of contaminants on microelectronic devices using infrared methods have been reported by numerous authors. Buried layers in optoelectronic materials were mapped,[19] and the mapping of GaAs has been accomplished.[20,21] Raman microspectroscopy was used to map the structure of silicon after laser damage.[22]

In the field of polymer science, thin film coatings have been studied.[23,24] Numerous types of fibers have been identified using infrared microspectroscopy,[25,26] and orientations of crystalline structures in fibers were determined via polarization studies.[27-29] Inclusions in polymer substrates have been identified in numerous papers.[30-33] Multilayer polymer coatings have been cross-sectioned before analysis,[24,34-36] and even imaged.[37,38] The group at the Molecular Microspectroscopy Laboratory at Miami University have published three papers[39-41] detailing the precautions that must be considered when identifying suspected polymer contaminants. One such difficulty is distinguishing between common polyamide materials like proteins and nylon, as well as cellulosics such as cotton and paper.[39] Another problem is in the area of dust analysis,[40,41] which can result from improper handling of materials prior to microspectroscopic examination.

The structure of ceramics have been investigated using Raman microspectroscopy. Studies of alumina, zirconia and silicon nitride have been reported.[8,42] The silica system has also been examined.[43,44] For corrosion chemistry, ferrous metals[45] as well as the corrosion of materials used in the oil industry,[46] have been monitored using Raman microspectroscopy. Infrared microspectroscopy is experiencing an explosion of use recently in the field of forensics, for drug analyses,[47,48] and for fiber,[48-50] paint chip[48,49] and gunshot residue[49,50] identifications. In the area of art conservation, infrared microspectroscopy has been applied to determine the authenticity of paintings[51] and of a painted manuscript.[52] Infrared microspectroscopy has been applied to the characterization of paper products.[53] Mineralogy has been the subject of numerous publications, including work related to identification of inclusions,[54-56] of individual grains in mixtures[57] and of a new borate mineral.[58] Infrared[59] and Raman[60] microspectroscopy have been applied to the analysis of coals. Explosives have also been examined using these methods: penetration into propellent grains by infrared,[61] and nitro compounds by Raman[62] microspectroscopy. An electrochemical cell has been developed which allows the acquisition of in-situ Raman microspectroscopic data.[10,63] Lithium/thionyl chloride battery chemistry has been investigated using Raman microspectroscopy.[64] Finally, in the field of biochemistry, agricultural systems were examined by infrared microspectroscopy,[65] and the structure of gallstones were determined using both infrared and Raman microspectroscopy.[66] From this summary, the broad range of disciplines which can benefit from vibrational microspectroscopy technology is clear.

In this paper, numerous aspects and applications of infrared and Raman microspectroscopies will be presented in the following manner. First, the theory of both techniques will be briefly discussed in order to understand the relative strengths and weaknesses of each method. As will be shown, the techniques are extremely complementary. Quite often the definitive solution to a problem is to apply both infrared and Raman microspectroscopies to the same region of the same sample. Following the theory section, the experimental methods used in our laboratory in terms of both equipment and typical conditions will be presented. A discussion of sampling techniques will then be given, with particular emphasis on infrared sampling procedures since sampling methods for Raman microspectroscopy are usually extremely simple. A method to detect the presence of mixtures in microscopic samples will also be presented. A broad range of applications will then be discussed to exhibit the variety of problems that have been solved with these methods. A first step in the utilization of these methods is the development of a spectral library. Some comments regarding the construction and use of spectral libraries will be made. Categories of applications to be discussed include the contamination of microelectronic devices, contaminants on circuit boards, contaminants from microelectronic processing equipment, and problems due to various epoxies. In addition, some miscellaneous problems involving pump motors and meteors will serve to illustrate the broad utility of the methods. Finally, some of the future directions foreseen for the field of vibrational microspectroscopy and a brief summary of chemical analyses using these techniques will be given.

THEORY

Infrared Microspectroscopy

Infrared and Raman are types of vibrational spectroscopy, used to identify the energies of vibrations of atoms, or groups of atoms, in molecules. However, the mechanism and method by which these vibrational energies are detected is quite different for the two techniques as shown in Figure 1. This leads to various strengths and weaknesses for each method. Infrared spectroscopy is accomplished by direct measurement of absorbed incident infrared light which has the same energy as a molecular

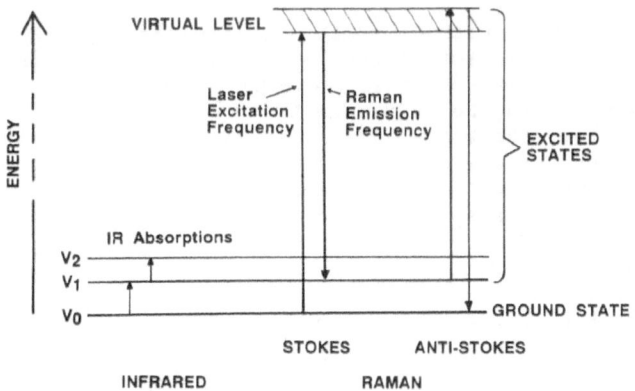

Figure 1.   Molecular vibrational energy level (ν_0 - ν_2) diagram
comparing the mechanisms for infrared absorption and Raman
scattering.

vibration (Figure 1). The sample is irradiated with infrared light of
various frequencies. Then the absorption of infrared radiation is
detected by measuring the difference in intensity between the incident
light and the light which is transmitted through or reflected from the
sample. This absorption is measured over a range of frequencies, and
this set of absorbances as a function of frequency is known as a
spectrum. As discussed previously, since this spectrum is characteristic
of a particular molecular species, unknown compounds can be identified.

In addition to the coincidence of energy of a molecular vibration
with the incident infrared frequency, there are quantum mechanical
selection rules which must be met for a particular vibrational mode to
result in the absorption of infrared radiation. One such selection rule
is that the molecular motion involved in the vibration must result in a
change in the permanent molecular dipole. An asymmetric vibration
results in a greater dipole moment change than a symmetric vibration of
the same type for the same molecule. Thus asymmetric molecular
vibrations, in general, lead to more intense infrared absorption than
symmetric vibrations. This requirement of a change in the permanent
molecular dipole means that homonuclear species, such as amorphous carbon
(soot) and diatomic gases such as nitrogen and oxygen, cannot be detected
by infrared spectroscopy.

The fact that infrared microspectroscopy involves direct measurement
of infrared absorption leads to several other limitations for the method,
which have been extensively discussed by Messerschmidt.[67-70] Very
sensitive infrared detectors are required for measurement of absorption
when using a very small incident light aperture. These narrow band
mercury-cadmium-telluride detectors typically have low frequency response
cutoffs in the range of 800-650 cm^{-1}. Therefore infrared
microspectroscopy cannot be used for numerous species which only have
absorptions below these values. In particular, since the frequency of
vibrational absorption decreases with increasing mass of the vibrating
group, this means that many inorganic species cannot be identified by
infrared microspectroscopy. Furthermore, since the diffraction limit at
750 cm^{-1} is 13 micrometers, infrared microspectroscopy cannot be used

below this sample size without spectral distortion. Thus, if more sensitive infrared detectors become available with cutoffs below 750 cm^{-1}, they cannot be used without increasing the diffraction limitation. In summary, along with the requirements for sample preparation which will be discussed below, there are three main limitations of infrared microspectroscopy. These are the detection of homonuclear species, the detector cutoff which often makes inorganic identifications difficult, and the relatively poor spatial resolution of about 10 micrometers.

Raman Microspectroscopy

In contrast to infrared spectroscopy, Raman spectroscopy measures vibrational energies by a completely different process. As displayed in Figure 1, a laser light source is used to perturb molecules to some combination of electronic, vibrational and rotational excitation equal to the laser frequency. The excited state generally does not correspond to a distinct quantum mechanical state of the molecule. This state is known as a virtual level. The absorbed energy is then reemitted (scattered) as another photon. The vast majority of the time the scattered photon is of the same energy as the laser excitation frequency, which is called Rayleigh scattering. However, about 10^{-6} of the absorbed photons are scattered at a different frequency from the excitation, which is the Raman effect. The difference in energy between the excitation and the emission corresponds to a vibrational energy as measured directly in the infrared. The energy axis for Raman spectra is plotted as the difference in energy between the laser excitation and the emission frequencies, and this difference in energy is known as the Raman shift. Thus the Rayleigh band is centered at a Raman shift of zero. The Raman shift is presented in wavenumbers (cm^{-1}), which corresponds to the vibrational absorption energy measured directly by infrared spectroscopy. The intensity axis for Raman spectroscopy is emission intensity (photon flux) per unit time, usually counts/second.

Figure 1 shows that there are two types of Raman scattering. When the molecule is in the vibrational ground state at the time of laser absorption, Raman emission occurs at frequencies less than the excitation frequency. This is termed a Stokes shift. Since some molecules are initially in an excited vibrational state, Raman emission also occurs at frequencies greater than that of the laser frequency. These are known as anti-Stokes Raman shifts. The vibrational energy differences measured by Raman spectroscopy are the same for Stokes and anti-Stokes scattering, so the Raman spectrum is a mirror image on either side of the Rayleigh band. The only difference between the two types of Raman scattering is the initial state of the molecule. Since molecules in the vibrational ground state are more prevalent at lower temperatures, Stokes lines are more intense than anti-Stokes lines at room temperature. The relative intensities of the Stokes and anti-Stokes lines can then be used to obtain an in-situ, noninvasive measurement of temperature.[71]

As with infrared spectroscopy, various quantum mechanical selection rules govern which Raman bands are allowed. For a Raman frequency to be observed, the molecular vibration must result in a change in the induced dipole moment, which depends upon the polarizability of the molecule. This is in contrast to the change in the permanent molecular dipole which is required for infrared absorption. The polarizability of a molecule is an ellipsoid which indicates the distortion of the electron cloud when a point charge approaches the molecule. In practice, this means that Raman spectroscopy is much more sensitive to symmetric vibrations than to asymmetric ones, exactly the opposite of infrared spectroscopy. This is one reason the two methods are so complementary.

The different process by which Raman spectroscopy is accomplished leads to several strengths and weaknesses for the method compared to infrared spectroscopy. One primary difference is that both the laser excitation and the Raman emission can be visible frequencies. Thus, the diffraction limit for Raman microspectroscopy is on the order of 0.5 micrometers, the same as for visible light microscopy. This is about an order of magnitude better than for conventional infrared microspectroscopy. It also means that photomultiplier tube detectors can be utilized for photon counting, so that vibrational energies down to 50 cm^{-1} and sometimes less can readily be detected. The only limitation to the low energy end of the vibrational spectrum in Raman spectroscopy is interference due to the intense Rayleigh line. The fact that Raman spectroscopy can detect these low frequency vibrational modes means that Raman spectroscopy is better suited for the detection of molecules containing atoms of higher mass, in other words, inorganic compounds. This fact again points to the complementarity of infrared and Raman spectroscopies.

A disadvantage to working in the visible is the use of an intense laser as the excitation source which sometimes leads to sample volatility and/or decomposition. Furthermore, laser plasma lines may lead to spectral interpretation errors. These lines are readily filtered with a small monochromator or a line filter placed between the laser and the sample. The low quantum efficiency of the Raman effect, where only on the order of 10^{-6} of the incident photons are Raman scattered in the absence of resonance or surface enhancement, means that acquiring a Raman spectrum requires more time to attain the same signal-to-noise level than for an FT-IR spectrum. Another major disadvantage for Raman microspectroscopy is fluorescence. This process has a much higher quantum efficiency than the Raman effect. Emission frequencies are produced in the same spectral region as Raman scattering. This has the effect of obscuring the narrow, distinctive Raman bands for a particular molecule with a broad, often uninformative background. In fact, this fluorescence is so efficient that this broad background is often only due to a minor impurity in the sample of interest. Fluorescence is probably the single most important limitation to the general utility of Raman microspectroscopy.

EXPERIMENTAL

Infrared Microspectroscopy

A schematic of the infrared microspectrometer is shown in Figure 2. The heart of the system is a Nicolet 7199 Fourier transform infrared spectrometer interfaced to a Spectra-Tech (Stamford, CT) IR-PLAN microscope. A collimated infrared beam is passed out of the spectrometer and a flipper mirror either directs the beam into the microscope or allows the beam to pass through the microscope interface to the gas chromatography interface. The microscope interface is equipped with a narrow band mercury-cadmium-telluride (MCT) detector having a 0.25mm element with a low frequency cutoff near 750 cm^{-1}. The IR-PLAN microscope has been designed[68-70,72] to include some important new features for infrared microscopes. The microscope is an adaptation of an Olympus (Lake Success, NY) optical microscope, which has a superior optical quality to microscopes which have been used for infrared microspectroscopy in the past. This microscope has a four position objective turret above the stage, with 4X and 10X glass viewing objectives. For infrared spectroscopy, totally reflecting, silvered Cassegrainian objectives of 15X and 32X power are used. The objectives provide an image of the sample stage at an aperture above the objective.

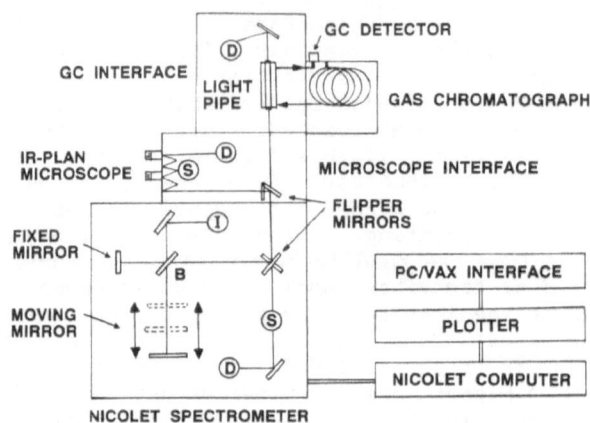

Figure 2. Schematic diagram of a microscope/GC/FT-IR spectrometer.
For the abbreviations, B is a beamsplitter, I is an
infrared source, D are detectors and S are sampling areas.

At this image plane, four blade masking is used to isolate the sample
region of interest. Either transmission or reflection sampling is
available. For transmission work, a 10X condenser below the stage
provides a second image of the sample plane at a lower remote aperture, a
process known as redundant aperturing. This feature reduces stray light
from passing around the sample and therefore increases spatial and
photometric accuracy.[69,70] The microscope is equipped with 4" x 5"
Polaroid automatic exposure photographic capability as well as a color
television viewing system for discussion of contaminant locations.

Data are acquired on the Nicolet FT-IR spectrometer at 4 cm^{-1}
resolution (1.97 cm^{-1}/data point) using a mirror velocity of 2 cm/sec.
Interferograms are then transformed using Happ-Genzel apodization. The
number of scans is varied between 256 and 4096 depending upon the
aperture size(s). Using a two minute data collection time, 256 scans,
and apertures which image to a 100 micrometer diameter, this system
results in open beam (no sample) transmission data having a
signal-to-noise ratio of about 3500:1 measured peak-to-peak across the
2500-2600 cm^{-1} region. In reflectance, using a gold slide surface under
the same experimental conditions, the signal-to-noise is roughly 2000:1.
This factor of two difference is due to the fact that in reflectance,
half of the beam cross section is used to send the IR beam into the
objective using a pick-off mirror. The other half of the cross section
is used to get the 180° reflected beam back out of the microscope to the
detector. In transmittance, the entire beam is utilized for both sample
illumination and collection of transmitted light.

Raman Microspectroscopy

The Raman microspectrometer system consists of a SPEX (Edison, NJ)
Raman spectrometer using a 0.85m double monochromator equipped with
holographic gratings. A schematic diagram of the instrument is displayed
in Figure 3. The spectrometer is interfaced to a Zeiss (New York, NY)
20T microscope. Although the original SPEX design had the microscope
directly bound to the monochromator entrance slit, the system was

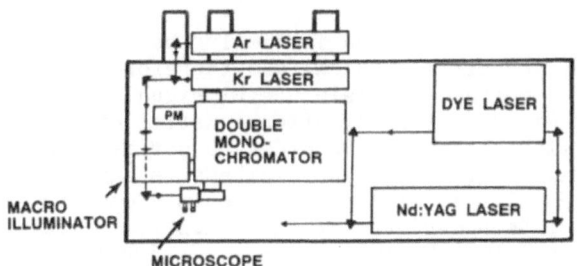

Figure 3.   Schematic diagram of a Raman/microprobe/fluorescence
             spectrometer. PM denotes the photomultiplier detector.

modified to mount the microscope onto the floating optical table and
couple the optics of the two components. Both Argon and Krypton lasers
(manufactured by Coherent, Inc., Palo Alto, CA) are used to provide
monochromatic excitation. The lasers are focused through the microscope
to a 1-2 micrometer spot size. Either 10X, 40X or 80X glass objectives
are used. After collecting the scattered light in a 180° geometry, the
light is passed through the monochromator. The slits of the
monochromator are set to give about 5 cm^{-1} resolution. Detection is
accomplished via a water cooled RCA 31034 photomultiplier tube operating
in the photon counting mode. Data are collected, using a DEC (Merrimack,
NH) MINC 11/23 minicomputer, at 2 cm^{-1}/data point with a sample dependent
dwell time per data point of 1-4 seconds.

SAMPLING METHODS

Infrared Microspectroscopy

     Infrared microsampling is usually more difficult than for Raman
microspectroscopy. Microsampling for infrared microspectroscopy is
complicated by the fact that data can be acquired either by reflectance
off the sample or transmission through the sample. Reflectance sampling
is almost always straightforward, a matter of simply focusing the
infrared beam on the sample. For thick organic samples, front surface
specular reflection which accounts for about 4% of the incident radiation
can sometimes be utilized. This decrease in signal limits the utility of
the method,[73] along with the spectral distortion due to dispersion in
refractive index effects which accompany specular reflection. This
effect is due to the change in refractive index of the sample which
occurs near an absorption band. The effect can be partially overcome by
a Kramers-Kronig transformation of the data as promoted by Hill and
Krishnan.[74] However, this treatment does not strictly hold for FT-IR
microspectroscopy due the range of specular angles sampled in a
Cassegrain objective and the impossibility of infinite mirror travel.
For thinner films on infrared reflective substrates, infrared
reflection-absorption spectroscopy (IRAS) can be utilized without the
magnitude of the dispersion in refractive index effect which accompanies
specular reflectance sampling. This technique is particularly effective
for samples which have thermally decomposed on a metallic substrate
resulting in the formation of a stain which is extremely difficult to
remove and sample by transmission methods.

     The primary difficulty for transmission FT-IR microspectroscopy sampling is that infrared absorption is simply too quantum efficient. Most samples as received are too thick, and therefore absorb too much of the incident infrared energy to obtain useful signal at the detector. Therefore, infrared microsampling is often a matter of crushing, pressing or otherwise distorting the sample in order to get it thin enough to transmit infrared light. The desired thickness is sample dependant, although usually a pathlength of less than ten micrometers is ideal. Numerous methods have been devised for this relatively tedious process of sample preparation.[24,26,33,35,75] Most of them require the use of a low magnification (10-100X), long working distance visible microscope, which is a required device in any microspectroscopy laboratory.

For liquids, attenuated total reflectance (ATR) sampling has been a major factor in the development of infrared spectroscopy, particularly as applied to the analyses of aqueous solutions.[76] In this technique, the infrared beam is multiply reflected through an infrared transmitting crystal such as zinc selenide. The solution of interest is placed in contact with this crystal. When the infrared beam reflects off the faces of the crystal the evanescent wave will slightly penetrate into the solution. The penetration depth depends upon the reflectance angle and upon the refractive index difference between the crystal and the solution. This penetration depth is much less than can be obtained in thin film capillary cells. Thus, strong absorptions in the liquid are no longer totally absorbing, and the entire spectrum may be analyzed. Of course, the price of this decrease in effective pathlength is a decrease in sensitivity. One such ATR accessory is the CIRCLE cell, made by Spectra-Tech Inc. which uses cylindrical internal reflectance for better sensitivity. This sampling accessory simplifies routine analyses of liquids. Obvious potential applications for this method include the sampling of degreasing solutions, plating baths and polymer coating processes among others.

Solids can be thinned for transmission infrared microspectroscopy by a variety of mechanical methods.[24,35,75] One such method is the slicing of a very thin wedge shaped section from a larger piece of the sample using a scalpel. This technique is known as skiving. Skived samples or fibers may be further flattened by using a wheel type roller device or by simply rolling the sample under the blunt part of a needle probe next to the tip. The probe rolling method can also be used to dilute crystalline samples in a few crystals of KBr. This microscopic KBr pellet will reduce the amount of specular reflected light from the crystal of interest and the accompanying dispersion in refractive index effect.

For polymer samples, a reproducible method of skiving is to use a microtome. Although expensive and time consuming to use, microtomes provide excellent samples, particularly for multilayered samples or of samples where depth profiles are of interest.[37,38] One should always microtome in the direction such that a given point on the blade is always in contact with a particular depth of the sample, in order not to drag layers through each other by slicing down into the depth of the sample. Microtomed samples are most useful in the 2-10 micrometer thickness range depending on the material and whether minor additives are of interest. Trial and error is required to determine the optimum thickness for a particular sample. Since the resulting samples have well defined thicknesses, this opens up possibilities for quantitative infrared microspectroscopy. A less expensive method for polymer sample preparation is a simple screw press using sodium chloride or calcium fluoride windows. One simply places a small piece of malleable material between two windows. Then the screw press is used to exert and maintain pressure on the sample while observing under the preparatory microscope. If the sample is not malleable it will not appear to get larger and will

124

simply imbed itself in the soft salt windows. In this case, more drastic procedures are required. A miniature diamond anvil cell is now available (Precision Diamond Optics, Tucson, AZ) with narrow enough clearance to work with the IR-PLAN microscope. The 1 mm diamond faces of this device are used in conjunction with bolts to provide higher pressures and coerce many tough samples to flatten.[39,77] This device is also rather expensive.

The above discussion is intended to provide an introduction to the methods involved in infrared microsampling. The key to successful microsampling is experience. Keep trying. Patience, practice and making mistakes appear to be required. In deciding whether to use reflectance or transmission sampling methods, remember that a sample is affected the least by reflectance sampling, which only requires placing a sample on the stage. Minimal manipulation of microscopic samples is desirable because the sample could be lost, destroyed or otherwise changed. Furthermore, there is always the possibility of introducing new contamination from sample handling. This may result in the misidentification of the real problem.[39-41] However, the infrared microspectroscopist should always attempt to do transmission work if the sample is expendable. If the sample is scarce, as most microscopic samples are by definition, first use reflectance sampling. Then use sample preparation methods to obtain a transmission spectrum. A transmission spectrum will almost always give a more readily identifiable match to a suspected contaminant or a compound contained in a spectral library. This is because specular reflectance sampling leads to distortion of spectral bandshapes due to the dispersion in refractive effect discussed previously. Thicker samples are much more effectively sampled in transmission after thinning than by having to depend upon front surface reflection. In summary, sample preparation for infrared microspectroscopy requires a veritable toolbox full of various dental hardware and sample compression devices, along with a good quality preparatory microscope.

Raman Microspectroscopy

For Raman microspectroscopy, sampling methods are generally quite simple. Since all spectra are collected in a 180° reflectance arrangement, sampling often consists only of focusing the 1 micrometer diameter laser beam on the region of interest. Increases in sample thicknesses up to around 20 micrometers in thickness usually lead to more intense Raman bands. This is in contrast to infrared microspectroscopy where many samples thicker than 10 micrometers are totally absorbing, which causes degradation of signal-to-noise ratios. For Raman microspectroscopy, further sample thickness increases beyond 20 micrometers do not lead to intensity gains, however signal is not degraded by thicker samples as occurs for infrared spectroscopy. Although Raman sampling is often straightforward, there are important exceptions.

Fluorescence, and laser heating leading to sample evaporation or decomposition, are complications which decrease the utility of Raman microspectroscopy. Fluorescence is best minimized by using a laser frequency at which the sample does not absorb. When this is not feasible, fluorescence can often be reduced by laser irradiation (bleaching) of the sample for periods of time up to several hours. This may reduce the fluorescence to sufficiently low levels so that Raman bands can be observed on top of the fluorescence background. For some samples, degradation by the laser beam cannot be avoided even at extremely low power levels. The normal approach to sample volatilization is to reduce laser output gradually in order to maintain Raman intensity as much as possible. In addition, one must sample new areas of the sample as volatilization is induced.

One method to increase Raman intensity in reflectance microsampling is to attempt to obtain surface enhancement of the Raman effect, known as SERS spectroscopy. This technique may be useful when the sample intimately contacts a roughened gold or silver surface, and the laser is focused onto the sample/metal interface. In practice, the effect is difficult to obtain. For Raman microspectroscopy in general, despite the complications discussed above and the quantum inefficiency of the Raman effect, sampling techniques are almost always simpler than for infrared microspectroscopy.

It should be remembered that there are numerous pitfalls which await the spectroscopist who enters the world of the microscopist. This is equally true for Raman and infrared microspectroscopies. These types of problems fall into two categories, artifact sampling and errors in identifications. In the area of artifact sampling, the introduction of further contamination via sample handling has already been mentioned. In addition, the occasional loss of a scarce contaminant should not be catastrophic, since a true processing contaminant will manifest itself on numerous substrates over time. One should be cautious of spending vast amounts of time on identifications of contaminants which are not reproducible in the processing environment. Single occurrence contaminants are often extremely difficult to pinpoint in the processing facility even if identifications are made. These problems are often found to be due to sample handling between the facility and the acquisition of the microspectroscopy data. The question of hermeticity of supposedly sealed packages must be answered before microsampling can begin. Be sure to leak test suspected failed devices before disassembly.

Another difficulty arises when the contaminant of interest is a mixture. A single spectrum with features representing multiple components may easily be misidentified. One method to reduce this risk is to sample multiple areas of the contaminant to check the homogeneity of a deposit. This should be done routinely since mixtures are almost always a potential complication in contaminant identifications. If the ratios of bands in the spectrum change as a function of spatial location, the contaminant is probably a mixture. The group of spectral features whose intensities change together can often be associated with a single component for identification purposes.

In layered samples, for which vibrational microspectroscopy is ideally suited, sample mixing by incorrect microtoming or skiving procedures must be avoided. Spatial mixing via diffraction effects in the infrared must also be considered. Infrared absorption will occur in regions of the sample that visually appear to be masked, due to the diffraction of the infrared beam at the aperture(s). Thus significant amounts of infrared energy impinge on the sample beyond the apertured area, leading to spectral features in the resulting infrared spectrum from areas of the sample beyond the region of interest.[70]

In the area of identification and interpretation of anomalous spectral features, there is no substitute for experience. However, a large bookshelf of reference spectra is certainly quite helpful. Thermal degradation products of polymer materials are especially difficult to identify. For infrared microspectroscopy, inorganic species are often difficult as well, due to the low frequency detector cutoff near 750 cm^{-1}. A minor organic residue in an inorganic which is infrared inactive in the detectable region can lead to a faulty identification. For Raman microspectroscopy, laser cooking of samples and thermal decomposition from the production environment both result in amorphous carbon. The contaminant problem itself usually had an organic origin. Two types of materials often lead to erroneous interpretation. Polyamides from animal products such as wool, silk, hair and skin closely resemble nylon and each other. However, a carefully constructed spectral library can be used to distinguish these species on the basis of the N-H amide stretching near 3300 cm^{-1} region and the C-H stretch region from 2800-3000 cm^{-1} as shown in Figure 4. However, use of the C-H region to

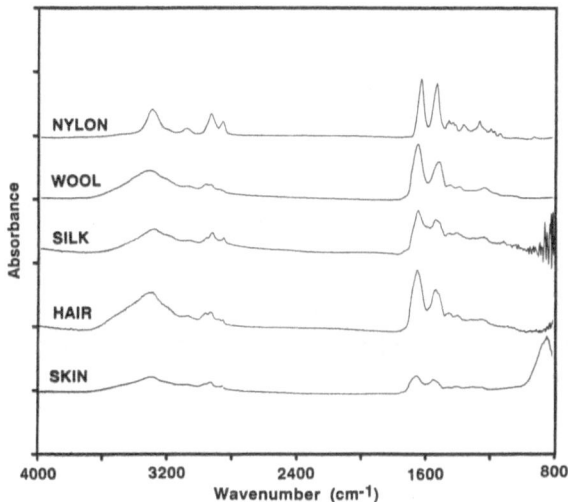

Figure 4.   FT-IR transmission spectra of nylon, wool, silk, hair and skin particles.

distinguish these materials must be approached with caution, since hair from different people, for example, have different ratios of the C-H stretch frequencies in this region.[78] Natural products have a broader N-H stretch band than a synthetic polyamides such as nylons.[79] Optical microscopy can be used to differentiate wool from human hair and natural from synthetic fibers.[80]

Similarities also exist for cellulosic materials like cotton, rayon, paper and wood. Again, these species can be distinguished upon careful examination of their spectra as shown in Figure 5. In addition, differentiation may be possible because wood contains lignin,[41] and paper is often filled with kaolin to provide luster.[41] Overall, the problems discussed above can be surmounted with caution and an excellent set of library reference spectra. Once these problems are recognized, the satisfaction of problem solving for processing applications can begin.

APPLICATIONS

The potential applications for infrared and Raman microspectroscopies for particles of greater sizes than the diffraction limitations are essentially infinite. Since almost all types of organic and inorganic contaminants may be identified in nearly all types of sampling environments, the possibilities are nearly limitless. As discussed previously, mixtures are probably the most difficult complication. The applications that will be touched upon here include

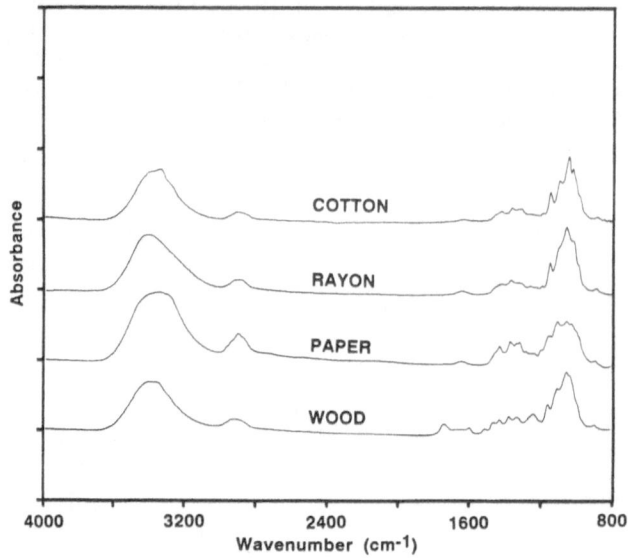

Figure 5.   FT-IR transmission spectra from cotton, rayon, paper and wood particles.

the identification of contaminants on microelectronic devices and packages, contamination on or from processing equipment, contamination of circuit boards, contamination by epoxies, and two miscellaneous interesting problems involving motors and meteors.

The infrared transmission spectrum of a brown fiber observed on a bond pad of a microelectronic package and the library reference spectrum of a cotton fiber are shown in Figure 6. Although there are differences in the relative intensities of the spectral features in the contaminant and reference spectra, the brown fiber is clearly identifiable as a cellulosic material. This fiber was thought to have originated from a clean room worker wearing a cotton garment beneath their protective clothing. Various types of fibers have been observed on microelectronic devices and packages. After establishing a spectral library of the various garments from our packaging facility, fibers can be identified as originating from hats, gowns or booties. Often these species are decomposed due to subsequent thermal excursions of the microelectronic devices. Figure 7 shows the presence of a decomposition product of nylon as well as of amorphous carbon in the Raman spectrum of a particle on a package seal rim. Nylon is the material used for clean room gowns at our microelectronic packaging facility. Figure 8 illustrates the identification of a deposit on an IC die which was determined to be due to a photoresist material from its Raman spectrum. Changes in the processing operation were made to ensure complete masking of the die during photoresist application.

Another problem manifested itself as droplets appearing on a large lot of IC dies after the wafer saw step. The infrared transmission spectra shown in Figure 9 reveal that the culprit was a contaminated house compressed air line used to blow silicon debris from the wafers, and was not due to lubricants used for the wafer saw. Filtered bottled

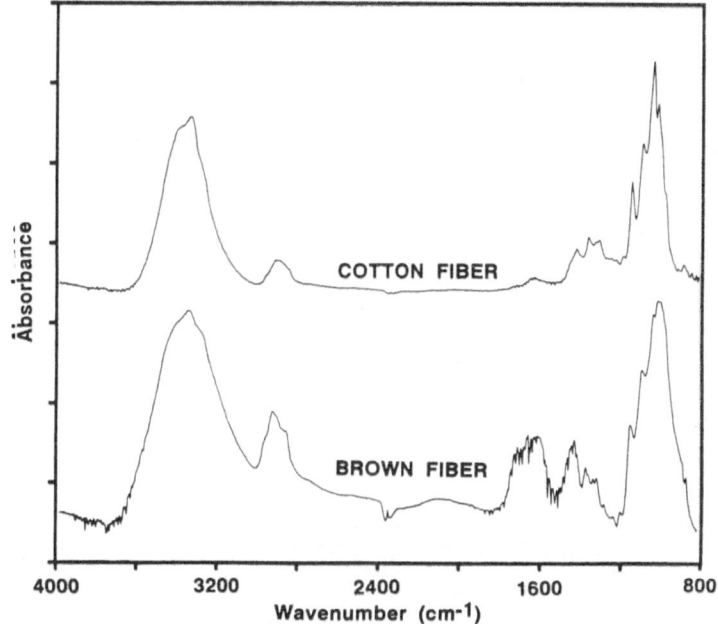

Figure 6.   FT-IR transmission spectrum of a reference cotton fiber and reflectance spectrum of a brown fiber on an IC bond pad.

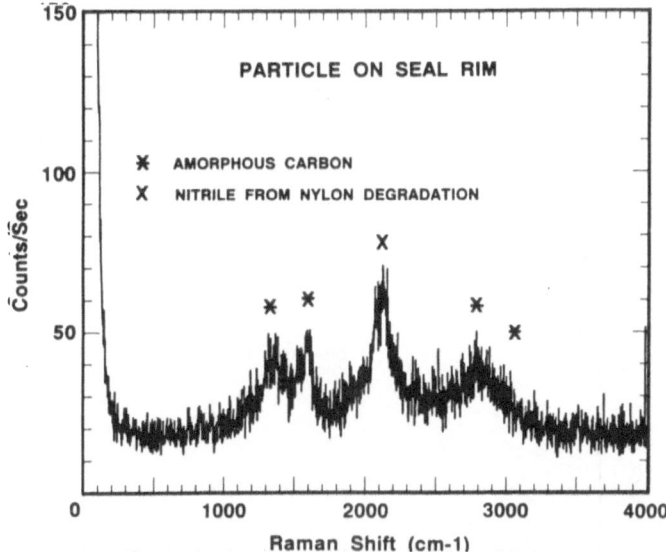

Figure 7.   Raman spectrum of a mixture of thermally decomposed nylon and amorphous carbon from a particle on the seal rim of an IC package.

nitrogen is now connected to the wafer saw for this operation. A cleaning procedure for this lot which had oil contamination was recommended and implemented. After cleaning, some dies exhibited

129

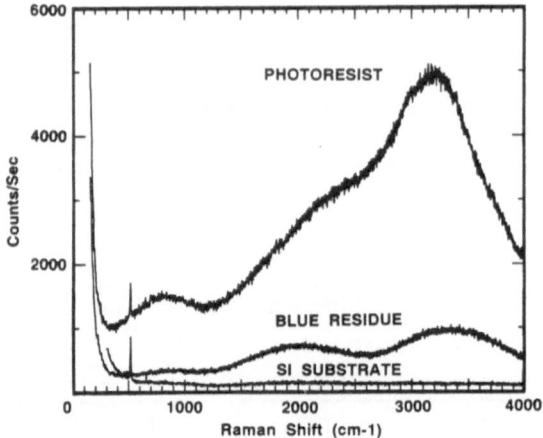

Figure 8.   Fluorescence/Raman spectra of photoresist, of an IC die residue and of a clean area of the IC die.

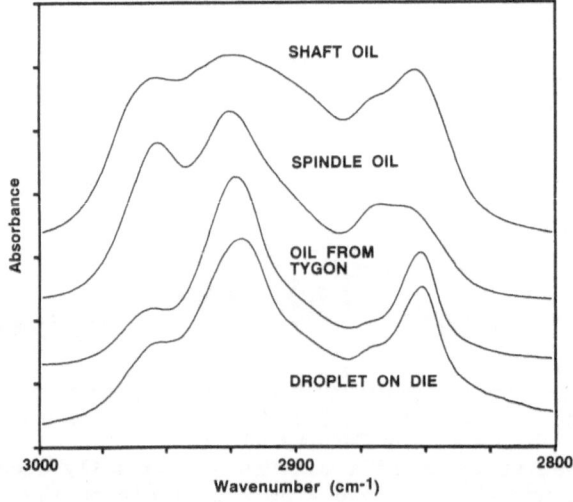

Figure 9.   FT-IR transmission spectra of lubricants used for IC wafer saw shaft and spindle, from a Tygon tube used for nitrogen blow off, and reflectance spectrum of an oil droplet on an IC die.

needle-like crystals. The Raman spectrum of one of these needles shown in Figure 10 is identifiable as calcium carbonate. It turns out that this problem was traced back to a faulty deionization apparatus for water used in the recommended cleaning procedure.

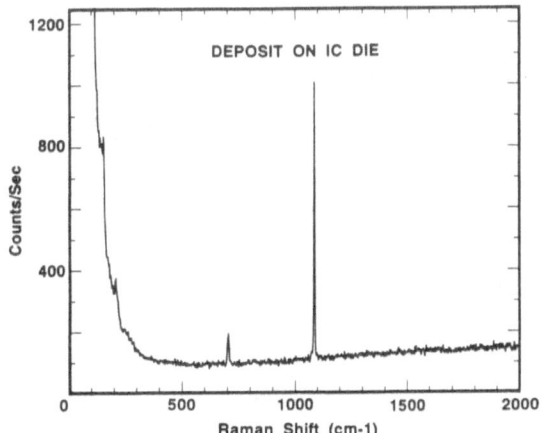

Figure 10.   Raman spectrum of a hydrated calcium carbonate particle left on an IC die after cleaning.

Contamination from process equipment can assume many forms. Figure 11 shows the infrared reflectance spectrum of a residue found in a dehumidifying bake-out oven. Investigation revealed that the thermocouple wires in this oven had recently been replaced. A controlled thermal decomposition of the polymer insulation from a new thermocouple wire provided the reference spectrum shown in the figure.

In another study, a ferrite insert in a metallic component exhibited a white precipitate formation. The infrared microspectrum shown in Figure 12 was in this case able to make a positive identification of an inorganic species, calcium sulfate. This component had been improperly exposed to a sulfuric acid plating solution instead of the less corrosive citric acid bath intended.

Contamination occurs on circuit boards from a variety of sources as well. The most common species observed are residual flux or photoresist. The fluorescence/Raman spectrum of a photoresist residual was shown in Figure 8. An infrared spectrum of rosin based flux (abietic acid) from a printed wiring assembly is shown in Figure 13. This board was returned to the manufacturer for further cleaning. Another deposit observed on a circuit board consisted of alternating black and white blotches, identified from the Raman spectra shown in Figure 14 as both the rutile and anatase phases of TiO_2, and amorphous carbon. This resulted from the exposure of the board in an oven which had been used to bake out epoxy which was filled with TiO_2. The amorphous carbon resulted from epoxy decomposition. Neither of these species could have been identified with infrared microspectroscopy. The reasons for this are infrared detector cutoff in the case of TiO_2, and the infrared inactivity due to selection rule in the case of amorphous carbon.

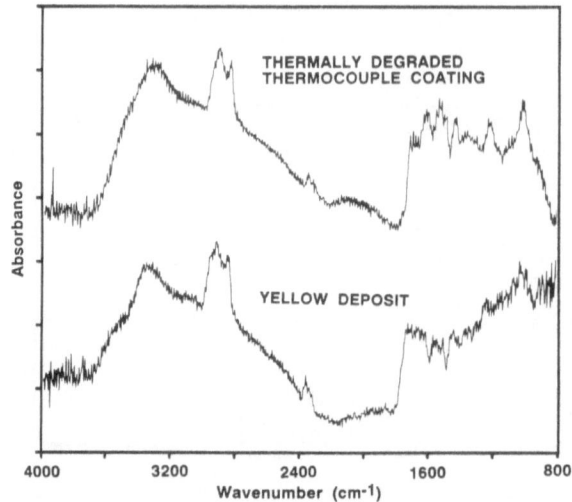

Figure 11.   FT-IR reflectance spectra of thermally decomposed
thermocouple insulation and of a deposit from a bake-out
oven.

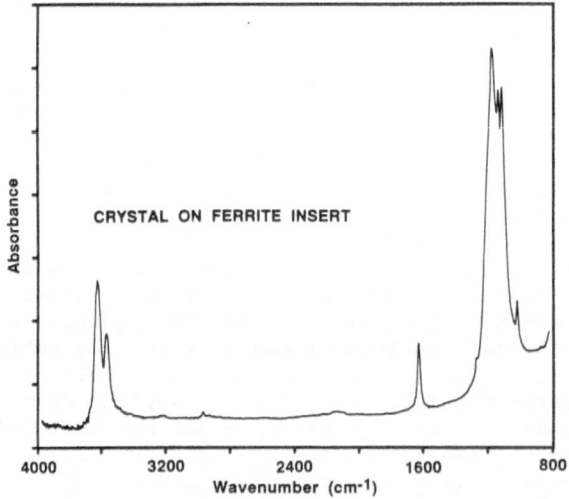

Figure 12.   FT-IR reflectance spectrum of a hydrated calcium sulfate
crystal on a ferrite surface.

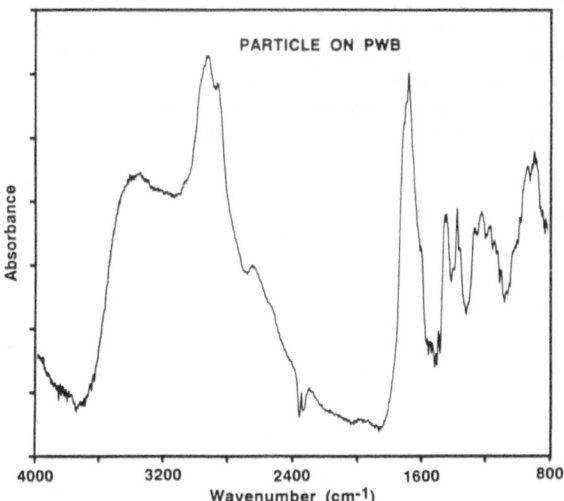

Figure 13. Infrared reflectance spectrum of rosin based flux residue on a printed wiring board.

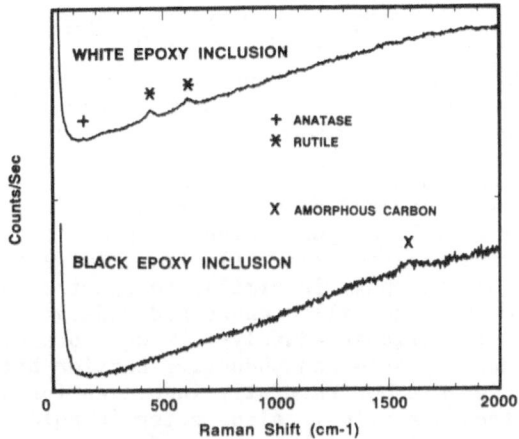

Figure 14. Raman spectra of titanium dioxide phases and of amorphous carbon from inclusions in the epoxy coating of a circuit board.

An evasive problem we have encountered on circuit boards involved a white deposit observed unevenly coated over many of the solder dipped leads and solder connections. The reflectance infrared spectrum of Figure 15 was used to identify this contaminant as a long chain aliphatic carboxylate salt. This material was then positively identified using mass spectroscopy as a mixture of lead and tin di-n-octanoate. The problem was found to have originated from residual octanoic acid present in the antistatic packaging bags used to store these circuit boards before assembly. The active antistatic agent in these bags is octylamide. Octanoic acid can result either from hydrolysis of octylamide or from residual octanoic acid used in octylamide synthesis. The octanoic acid readily reacted with the lead/tin solder, resulting in weakening of the solder joints. Users of antistatic bags should be extremely cautious of this problem.

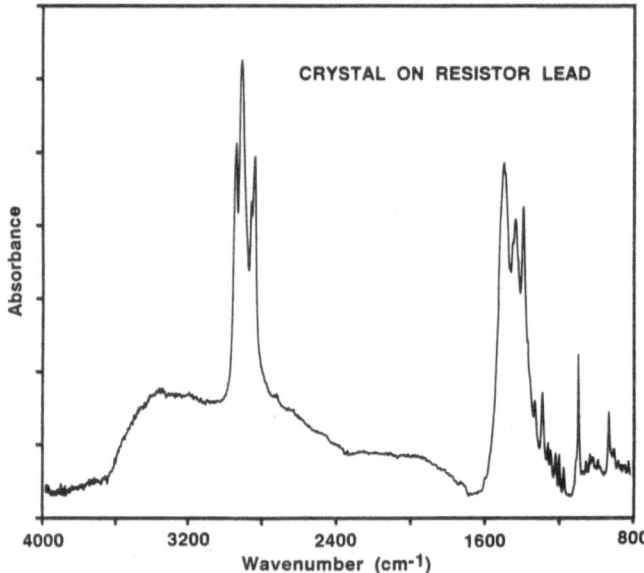

Figure 15.   FT-IR reflectance spectrum of octanoate salts from corrosion product on resistor lead due to contamination of circuit boards from antistatic bags.

Other contamination problems involving epoxies have been observed. A reddish residue was observed on an IC package adjacent to the IC die. Although a 70% silver filled epoxy which was grey in color was used to attach the die to the package, the reflectance infrared spectrum of the organic component of the epoxy is similar to spectrum of the contaminant, as shown in Figure 16. The filled epoxy had undergone separation of the silver filler from the organic adhesive during storage. This problem could provide an unacceptable nonconductive barrier between the package and the die. Epoxies are now carefully inspected for separation before use in the die attach operation. Also, eutectic gold die attachment is now used whenever possible, which is part of a general trend to eliminate organic materials from microelectronic processing operations.

Two other interesting problems illustrate the flexibility of these techniques for other than microelectronic applications. A particle was isolated from pump oil used in a defective vacuum system. The Raman microspectrum of this particle only displayed the 1350 and 1600 cm^{-1} peaks characteristic of amorphous carbon. However, the reflectance infrared microspectrum of this particle clearly revealed the presence of

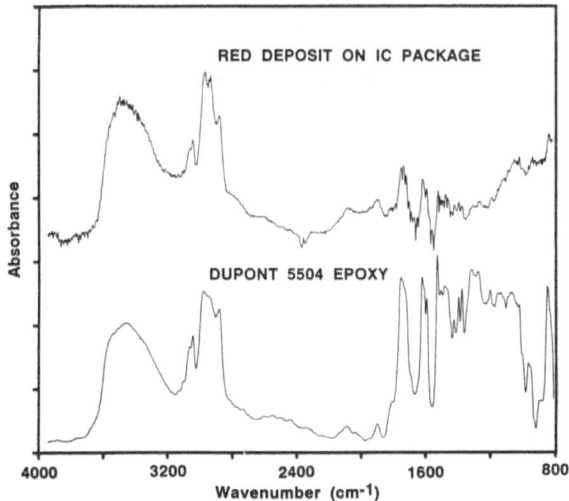

Figure 16.   FT-IR reflectance spectra of a red deposit observed on an IC
             package adjacent to the die and of Dupont 5504 epoxy used to
             attach the die to the package.

Teflon, Figure 17. The carbon from pump oil breakdown had preferentially
deposited on the Teflon fragment. Fortunately, the infrared beam was
able to penetrate the carbon and determine the real contamination problem
due to the insensitivity of infrared spectroscopy to amorphous carbon.
Mechanical degradation of a Teflon gasket was found to be responsible for
this problem.

      A meteorite provided another interesting diversion. A rare sample
of a Lodran meteorite was examined by Raman microspectroscopy. Grains
were observed in this mineral using polarized light microscopy of about
ten different species. Raman microspectroscopy was able to identify all
of the nonmetallic phases except one. The spectrum of a new crystalline
phase of (sodium, potassium, aluminum)silicate, contained within a phase
of olivine, was obtained and is shown in Figure 18. Line shape analysis
using Raman microspectroscopy data was used to show that this phase was
crystalline rather than a glass or a loaded phase of another mineral.

      For all of these applications, the techniques of Raman and infrared
microspectroscopies brought new information to bear on the problem at
hand. At Sandia National Laboratories, vibrational microspectroscopy is
usually the first method of choice for the identification of completely
unknown microscopic contaminants because of the identification
specificity and broad utility of the methods.

FUTURE DIRECTIONS

      As the techniques of infrared and Raman microspectroscopies mature
for routine analytical problem solving such as contaminant
identification, new areas of research using these methods are being·

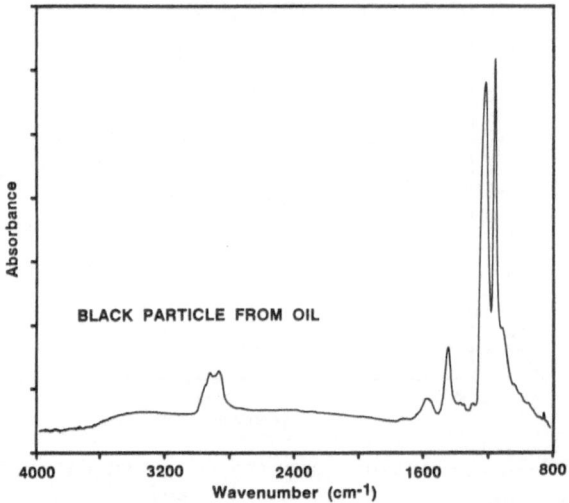

Figure 17.   FT-IR transmission spectrum of a particle isolated from pump oil, showing the presence of Teflon.

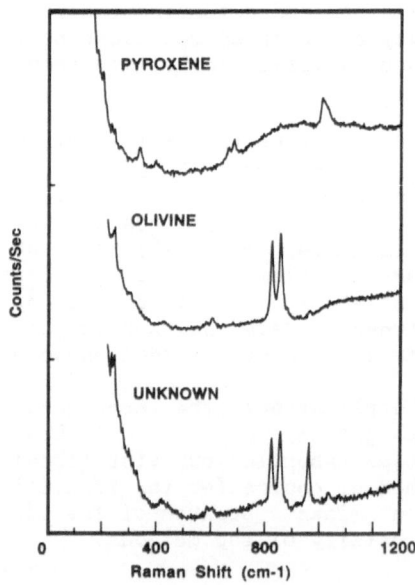

Figure 18.   Raman spectra of some of the microscopic phases from the Lodran meteorite.

developed. The most fundamental of these is the continued expansion of infrared libraries and improvements in spectral search algorithms. An exciting new development in this field is the neural network method developed for infrared spectroscopy at Sprouse Scientific Systems (Paoli, PA). This technique greatly reduces search times relative to the spectrum by spectrum search methods so often used in the past.

The expansion of spectral libraries, particularly for industrial processing applications, must include thermal degradation spectra for polymers. The broad range of temperatures experienced by devices in numerous processing applications makes spectral identifications of contaminants extremely difficult when library spectra in commercially available libraries are all of thermally unstressed materials. These thermal degradation effects have been the largest barrier to successful identifications in our laboratory. We are attempting to expand infrared spectral libraries in this area.[81,82] An example of this work and of the dramatic spectral changes which can accompany thermal decomposition is shown in Figure 19 for a piece of phthalate plasticized polyvinylchloride clean room glove material.

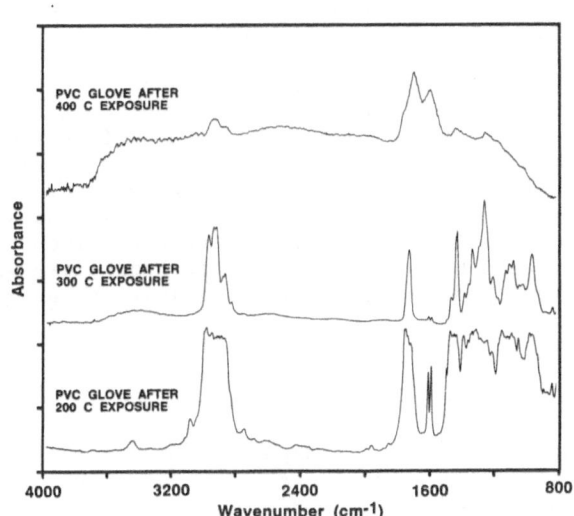

Figure 19.   FT-IR reflectance spectra from three stages of thermal decomposition of phthalate plasticized PVC clean room gloves.

Quantitative microspectroscopy is another field that is just beginning to be developed. The difficulties in obtaining useful data have been well documented.[67,70] However, single frequency methods have been used successfully with Raman microspectroscopy to quantify carbonate[83] and sulfate[56] in aqueous solutions, and in infrared microspectroscopy for polymer blends.[84] There is obviously an opportunity to extend this work using full-spectrum multivariate statistical methods such as principal components analysis and classical and partial least-squares methods.[85] Polymer doping profiles and local defects in microelectronic coatings such as borophosphosilicate glass (BPSG) are potential applications.[86]

In the area of chromatography detection, large strides have already been made by Peter Griffiths' group at U.C. Riverside,[87,88] and by Sid Bourne of the Digilab Division, of Bio-Rad Inc., Cambridge, MA. The combination of the use of focusers for column effluents, low temperature trapping and microscopic IRAS detection has dramatically lowered detection limits for GC/FT-IR beyond the capabilities of light pipe systems.[87] These ideas have also been applied to the detection of supercritical fluid chromatography (SFC),[88] along with a study of high performance liquid chromatography (HPLC) and thin layer chromatography (TLC) chromatography fractions.[89]

Image analysis is another exciting area for vibrational microspectroscopy applications. Although quite time consuming, the benefits from mapping contaminants on devices are evident. Mapping has been used with Raman microspectroscopy to analyze GaAs,[20] Si[22] and ferrous metals.[45] Infrared stimulated emission has been used to map GaAs,[21] single crystal silica,[90] silicon nitride[91] and silicon.[92] An emissivity correction algorithm[91] and a method to increase contrast by double illumination[92] have been reported. Using standard infrared microspectroscopy, Harthcock has shown that functional group analysis can be used for contaminant mapping.[37,38] However, in all of the above work, single frequency analyses have been used for data reduction. The use of Gram-Schmidt reconstruction[93] of the infrared data should improve signal-to-noise for this data significantly.

Finally, the area of superresolution is now being explored.[69,94] This technique was originally developed for the far-IR region,[94] and involves placing an aperture smaller than the diffraction limit closer to the sample than the diffraction limit, which is not a simple task.[95] This will provide improved spatial resolution beyond the infrared diffraction limitation. For microelectronics processing applications, the ability to obtain a well defined infrared spectrum from a 1 micrometer size contaminant would be very useful. This capability would even further develop the complementarity between infrared and Raman microspectroscopies.

SUMMARY

The field of vibrational microspectroscopy is fast moving, exciting and extremely useful to a broad range of disciplines. It has been shown that a wide variety of sampling conditions and different types of contaminants can readily be accommodated. In particular, the identification of contaminants involved in microelectronics processing has been emphasized. For contaminant characterization in clean room applications, one should be concerned about the prevention of both particulate and nonparticulate contaminants. Nonparticulate contaminants such as organic coatings are extremely difficult to monitor in ambient air by conventional particle detection methods. However, these vibrational microspectroscopic methods can identify both types of contaminants easily and yet with extreme specificity. It has been our experience that all materials used in clean rooms seem to eventually find their way onto production devices. The high temperature processing of components can create thermally decomposed materials. This makes identification and backtracking to the source of the problem much more difficult. Better spectral libraries are required to overcome this problem.

The positive side is that the complementarity of infrared and Raman spectroscopies allows the identification of almost all types of potential contaminants. Sampling is usually quite simple, yet caution must be observed to prevent artifact sampling or the introduction of other contaminants. Furthermore, the complexity of considering the absorbances at thousands of frequencies simultaneously in a vibrational spectrum may lead to interpretation errors even after valid data are collected. The presence of mixtures of contaminants in a single spectrum may also lead to identification errors. Caution in jumping to interpretive conclusions is always prudent. However, because infrared and Raman microspectroscopies are easy to use, rapid and are relatively inexpensive, the general utility of these methods to a range of laboratories and processing facilities is clear.

ACKNOWLEDGEMENTS

The authors would like to acknowledge the contributions of Karen Higgins, Celeste Case and Alejandro Pimentel of Sandia National Laboratories for acquiring some of the data presented in this paper.

REFERENCES

1. R. Barer, A. R. H. Cole and H. W. Thompson, Nature, 163, 198 (1949).
2. V. J. Coates, A. Offner and E. H. Siegler, J. Opt. Soc. Am., 43, 984 (1953).
3. G. J. Rosasco, E. S. Etz and W. A. Cassatt, Appl. Spectrosc., 29, 369 (1975).
4. M. Delhaye and P. Dhamelincourt, J. Raman Spectrosc., 3, 33 (1975).
5. P. B. Roush, editor, "The Design, Sample Handling and Applications of Infrared Microscopes", A.S.T.M., Philadelphia, PA, 1987.
6. R. G. Messerschmidt and M. A. Harthcock, editors, "Infrared Microspectroscopy", Marcel Dekker, New York, 1988.
7. R. Cournoyer, J. C. Shearer and D. H. Anderson, Anal. Chem., 49, 2275 (1977).
8. F. Adar, Microbeam Anal., 16, 67 (1981).
9. M. Mehicic, M. A. Hazle, R. L. Barbour and J. G. Grasselli, Microbeam Anal., 20, 68 (1985).
10. N. R. Smyrl, R. L. Howell, D. M. Hembree and J. C. Oswald, in "Infrared Microspectroscopy", R. G. Messerschmidt and M. A. Harthcock, editors, p. 211, Marcel Dekker, New York, 1988.
11. E. V. Miseo and L. W. Guilmette, in "The Design, Sample Handling and Applications of Infrared Microscopes", P. B. Roush, editor, p. 97, A.S.T.M., Philadelphia, PA, 1987.
12. J. E. Katon, G. E. Pacey and J. F. O'Keefe, Anal. Chem., 58, 465A (1986).
13. D. J. Zearing and V. J. Coates, Proc. S.P.I.E., 276, 249 (1981).
14. K. Krishnan, in "Infrared Microspectroscopy", R. G. Messerschmidt and M. A. Harthcock, editors, p. 139, Marcel Dekker, New York, 1988.
15. J. N. Ramsey and H. H. Hausdorff, Microbeam Anal., 16, 91 (1981).
16. K. Madden, B. Bergin, N. Klymko and J. N. Ramsey, in "Infrared Microspectroscopy", R. G. Messerschmidt and M. A. Harthcock, editors, p. 129, Marcel Dekker, New York, 1988.
17. P. L. Lang and J. E. Katon, Microbeam Anal., 21, 47 (1986).
18. J. C. Schearer and D. C. Peters, in "The Design, Sample Handling and Applications of Infrared Microscopes", P. B. Roush, editor, p. 27, A.S.T.M., Philadelphia, PA, 1987.
19. B. Sartorius and M. Rosensweig, J. Appl. Phys., 60, 3401 (1986).
20. P. Kidd, G. R. Booker and D. J. Stirland, Appl. Phys. Lett., 51, 1331 (1987).
21. P. Dobrilla and J. S. Blakemore, J. Appl. Phys., 61, 1442 (1987).

22. P. M. Fauchet, Scanning Electron Microsc., 1986(I), 425 (1986).
23. W. Herres and G. Zachmann, Fresenius Z. Anal. Chem., 319, 701 (1984).
24. M. A. Harthcock, in "The Design, Sample Handling and Applications of Infrared Microscopes", P. B. Roush, editor, p. 84, A.S.T.M., Philadelphia, PA, 1987.
25. P. L. Lang, J. E. Katon, J. F. O'Keefe and D. W. Schiering, Microchem. Journal, 34, 319 (1986).
26. E. G. Bartick, in "The Design, Sample Handling and Applications of Infrared Microscopes", P. B. Roush, editor, p. 64, A.S.T.M., Philadelphia, PA, 1987.
27. D. B. Chase, in "Infrared Microspectroscopy", R.G. Messerschmidt and M. A. Harthcock, editors, p. 93, Marcel Dekker, New York, 1988.
28. J. W. Brasch and A. Lustiger, in "Infrared Microspectroscopy", R. G. Messerschmidt and M. A. Harthcock, editors, p. 103, Marcel Dekker, New York, 1988.
29. S. L. Hill and K. Krishnan, in "Infrared Microspectroscopy", R. G. Messerschmidt and M. A. Harthcock, editors, p. 115, Marcel Dekker, New York, 1988.
30. X. Jouan and J. L. Gardette, Polymer Commun., 28, 329 (1987).
31. D. W. Schiering, in "Infrared Microspectroscopy", R. G. Messerschmidt and M. A. Harthcock, editors, p. 229, Marcel Dekker, New York, 1988.
32. H. J. Humecki, Solid State Technol., 28, 309 (1985).
33. H. J. Humecki, in "Infrared Microspectroscopy", R. G. Messerschmidt and M. A. Harthcock, editors, p. 51, Marcel Dekker, New York, 1988.
34. M. A. Harthcock, L. A. Lentz, B. L. Davis and K. Krishnan, Appl. Spectrosc., 40, 210 (1986).
35. F. M. Mirabella, in "The Design, Sample Handling and Applications of Infrared Microscopes", P. B. Roush, editor, p. 74, A.S.T.M., Philadelphia, PA, 1987.
36. R. Kellner, G. Fischboeck and C. Minich, Mikrochim. Acta, 1986(I), 271 (1986).
37. M. A. Harthcock and S. C. Atkin, Microbeam Anal., 22, 173 (1987).
38. M. A. Harthcock and S. C. Atkin, Appl. Spectrosc., 42, 449 (1988).
39. P. L. Lang, J. E. Katon, D. W. Schiering and J. F. O'Keefe, Polymer Mater. Sci. Eng., 54, 381 (1986).
40. P. L. Lang, J. E. Katon, A. S. Bonanno and G. E. Pacey, in "Infrared Microspectroscopy", R. G. Messerschmidt and M. A. Harthcock, editors, p. 41, Marcel Dekker, New York, 1988.
41. P. L. Lang, J. E. Katon and A. S. Bonanno, Appl. Spectrosc., 42, 313 (1988).
42. F. Adar and D. R. Clarke, Microbeam Anal., 17, 307 (1982).
43. L. C. Klein, C. Nelson and K. L. Higgins, Mat. Res. Soc. Symp. Proc., 32, 293 (1984).
44. R. K. Janssen and D. M. Krol, Appl. Opt., 24, 275 (1985).
45. D. J. Gardiner, D. J. Littlejohn and M. Bowden, Appl. Spectrosc., 42, 15 (1988).
46. F. Purcell and R. Heidersbach, Microbeam Anal., 20, 57 (1985).
47. J. P. Beauchaine, K. D. Kempfert, M. P. Fuller and R. J. Rosenthal, Microbeam Anal., 22, 187 (1987).
48. J. P. Beauchaine, J. W. Peterman and R. J. Rosenthal, Mikrochim. Acta, 1988(I), 133 (1988).
49. J. C. Shearer, D.C. Peters and T. A. Kubic, Trends Anal. Chem., 4, 246 (1985).
50. T. A. Kubic, Microbeam Anal., 21, 463 (1986).
51. J. C. Shearer, D. C. Peters, G. Hoepfner and T. Newton, Anal. Chem., 55, 874A (1983).
52. B. Guineau, J. Forensic Sci., 29, 471 (1984).

53. A. J. Sommer, P. L. Lang, B. S. Miller and J. E. Katon, in "Infrared Microspectroscopy", R. G. Messerschmidt and M. A. Harthcock, editors, p. 245, Marcel Dekker, New York, 1988.

54. J. Debessy, C. Beny, N. Guilhaumou, P. Dhamelincourt and B. Poty, J. de Phys., Colloq., C2, 811 (1984).

55. B. Wopenka and J. D. Pasteris, Appl. Spectrosc., 40, 144 (1986).

56. K. L. Higgins and C. L. Stein, Microbeam Anal., 21, 31 (1986).

57. J. P. Dybwad, L. M. Logan and K. P. Zinnow, Amer. Mineral., 59, 604 (1974).

58. E. S. Etz, D. E. Newbury, P. J. Dunn and J. D. Grice, Microbeam Anal., 20, 60 (1985).

59. P. D. Green, C. A. Johnson and K. M. Thomas, Fuel, 62, 1013 (1983).

60. D. Brenner, A.C.S. Symp. Ser., 252, 47 (1985).

61. E. Varriano-Marston, J. Appl. Polymer Sci., 33, 107 (1987).

62. F. W. S. Carver, D. P. Wyndham and J. T. Sinclair, J. Raman Spect., 16, 332 (1985).

63. D. M. Hembree, J. C. Oswald and N. R. Smyrl, Appl. Spectrosc., 41, 267 (1987).

64. M. C. Dhamelincourt, F. Wallert and J. P. Lelieur, J. Power Sources, 20, 69 (1987).

65. M. P. Fuller and R. J. Rosenthal, in "Infrared Microspectroscopy", R. G. Messerschmidt and M. A. Harthcock, editors, p. 153, Marcel Dekker, New York, 1988.

66. H. Ishida, R. Kamoto, S. Uchida, A. Ishitani, Y. Ozaki, K. Iriyama, E. Tsukie, K. Shibata, F. Ishihara and H. Kameda, Appl. Spectrosc., 41, 407 (1987).

67. R. G. Messerschmidt, Spectroscopy, 1(2), 16 (1986).

68. R. G. Messerschmidt, Microbeam Anal., 22, 169 (1987).

69. R. G. Messerschmidt, in "The Design, Sample Handling and Applications of Infrared Microscopes", P.B. Roush, editor, p. 12, A.S.T.M., Philadelphia, PA, 1987.

70. R. G. Messerschmidt, in "Infrared Microspectroscopy", R. G. Messerschmidt, and M. A. Harthcock, editors, p. 1, Marcel Dekker, New York, 1988.

71. D. A. Long, "Raman Spectroscopy", p. 84, McGraw-Hill, London, 1977.

72. T. Hirschfeld, Appl. Spectrosc., 30, 353 (1976).

73. J. A. Reffner, Pittsburgh Conference Paper 372 (1988).

74. S. L. Hill and K. Krishnan, Pittsburgh Conference Paper 373 (1988).

75. H. J. Humecki, in "The Design, Sample Handling and Applications of Infrared Microscopes", P.B. Roush, editor, p. 39, A.S.T.M., Philadelphia, PA, 1987.

76. P. T. McKittrick, B. E. Miller, N. D. Danielson and J. E. Katon, Pittsburgh Conference Paper 378 (1988).

77. J. E. Katon, P. L. Lang, D. W. Schiering and J. F. O'Keefe, in "The Design, Sample Handling and Applications of Infrared Microscopes", P. B. Roush, editor, p. 49, A.S.T.M., Philadelphia, PA, 1987.

78. P. L. Lang, private communication, 1988.

79. R. M. Scott and J. N. Ramsey, Microbeam Anal., 17, 239 (1982).

80. W. C. McCrone, R. G. Draftz and J. G. Delly, "The Particle Atlas", Ann Arbor Science Publishers, Ann Arbor, Michigan, 1967.

81. K. J. Ward, Pittsburgh Conference Paper 376, 1988.

82. K. J. Ward, Pacific Conference Paper 237, 1988.

83. E. S. Etz, Microbeam Anal., 16, 73 (1981).

84. R. Kellner and C. Weigel, Mikrochim. Acta, 1988(I), 163 (1988).

85. D.M. Haaland, Anal. Chem., 60, 1193 (1988).

86. D. M. Haaland, Anal. Chem., 60, 1208 (1988).

87. K. H. Shafer, P. R. Griffiths and R. Fuoco, J. High Resol. Chrom., Chrom. Comm., 9, 124 (1986).

88. S. L. Pentoney, K. H. Shafer and P. R. Griffiths, J. Chromatogr. Sci., 24, 230 (1986).

89. U. Bode and H. M. Heise, Mikrochim. Acta, 1988(I), 143 (1988).

90. F. Dacol and S. Utterback, Scanning Microsc., 1, 1045 (1987).
91. J. W. Shepherd, Proc. SPIE Int. Soc. Opt. Eng., 780, 130 (1987).
92. A. Castaldini and A. Cavallini, Proc. SPIE Int. Soc. Opt. Eng., 897, 55 (1988).
93. J. A. de Haseth and T. L. Isenhour, Anal. Chem., 49, 1977 (1977).
94. G. A. Massey, J. A. Davis, S. M. Katnik and E. Omon, Applied Optics, 24, 1498 (1985).
95. H. J. Sloane and R. J. Obremski, Appl. Spectrosc., 31, 506 (1977).

INFRARED AND RAMAN MICROSPECTROSCOPY: AN OVERVIEW OF THEIR USE IN THE

IDENTIFICATION OF MICROSCOPIC PARTICULATES

Patricia L. Lang[*], Andre' J. Sommer and J. E. Katon

[*]Department of Chemistry, Ball State University, Muncie
IN 47306

Molecular Microspectroscopy Laboratory, Department of
Chemistry, Miami University, Oxford, OH 45056

The recent coupling of infrared and Raman
spectrometers with microscopes has provided the
analyst with two very powerful new tools for
identification of microscopic particulates. Unlike
other analytical tools, these microspectroscopic techniques
yield molecular, as opposed to elemental,
information. Both infrared and Raman
microspectroscopy are reviewed with respect to
their historical development and their respective
advantages, limitations and capabilities. The two
methods are then discussed in terms of their
complementarity in solving chemical identification
problems, and it is shown that both methods are
often needed for complete sample characterization.
Finally, their application to the identification of
microscopic particulates is illustrated by two case
studies: identification of solder flux residue
contamination and characterization of various dust
particles.

INTRODUCTION

Microspectroscopy is a term which has been used to describe the
study of how light interacts with matter of microscopic dimensions.
The instrument which measures this interaction is often called a
microprobe or a microspectrometer. Although many different kinds of
microspectroscopic techniques exist, including electron/ion beam, fluorescence,
and UV-Visible, this paper will specifically overview the techniques of
infrared and Raman microspectroscopy. These techniques are of primary
importance to the modern analyst since they furnish the capability to
obtain selective, molecular and structural information on a compound.

The overview will include a brief history of infrared and Raman
microprobes, their capabilities and limitations, and the
complementarity of the two techniques. In addition, some of their
applications in the electronics industry for contaminant analyses are

given. Finally, the findings of a recent microspectroscopic study of dust found in an office/laboratory environment are reviewed.

HISTORY OF INFRARED AND RAMAN MICROSPECTROSCOPY

The concept of an infrared or a Raman microprobe is a simple one; it involves the coupling of a microscope to the respective infrared or Raman spectrometer. This coupling was successfully accomplished after a series of technological advances which occurred over three centuries.[1]

Both microspectroscopic techniques had their beginnings in the development of the light microscope by van Leeuwenhoek in the 17th century. However, an infrared microscope uses an objective made of reflecting surfaces while the conventional objective used for visible light microscopy is made of a refracting material such as glass. The reflecting objective, or Cassegrain, provides magnification without the absorption of the infrared radiation. Consequently, for an infrared microscope, the beginning may more suitably be attributed to Sir Isaac Newton who first made a reflecting microscope objective in 1672.[2,3] In Newton's day there was no good method for developing high quality (spherically corrected and coma-free) mirrored surfaces, and the reflecting microscope objective was virtually forgotten. Schwarzschild and Chretien independently solved these problems in 1905 and 1925, respectively,[2] thus paving the way for the development of the infrared microscope.

In the early 1920's infrared spectroscopy was gaining recognition primarily because of the efforts of W. W. Coblentz. Working at the National Bureau of Standards, Coblentz showed that infrared spectra were characteristic physical properties which could be used to identify a compound.

The coupling of the infrared microscope and spectrometer first occurred in the late 1940's, and the first commercially available infrared microscope, manufactured by the Perkin-Elmer Corporation, appeared in 1953.[4] Early investigators recorded spectra of single fibers, tissue sections, single crystals and blood smears.[5]

There was one major problem with these first generation microspectrometers. This problem was expressed well by an early investigator: "The lower limit of the cross-sectional area which can be studied is in many cases set ... by the signal-to-noise ratio (S/N) on the recorded spectrum."[5] An acceptable S/N was difficult to obtain on microscopic samples with dispersive spectrometers; consequently, the technique remained little used until the development of Fourier transform infrared (FTIR) spectrometers. FTIR spectrometers with their inherent multiplex advantage, fast data acquisition rate, and capability for signal-averaging of scans permitted good S/N to be readily obtained using microscope accessories.[6]

Accordingly, after FTIR spectrometers became available, microspectrometry regained popularity, and an infrared microscope designed for a FTIR spectrometer was introduced at the 1983 Pittsburgh Conference by Digilab.[6,7] This microscope had fixed focus optics with a variable aperture in the back focal plane of the Cassegrainian objective. This design allowed one to easily define the area under study by masking the image of the sample. Subsequently, infrared microscopes from other manufacturers became available. One manufacturer's microscope features two variable apertures, a feature which improves the photometric accuracy of the microspectrometer.[8]

Presently, all manufactured infrared microscopes have specular "reflectance" capabilities as well as transmission capabilities. Specular reflection allows one to obtain reflectance spectra of the smooth surface of a dielectric material. If the dielectric material is on a metal substrate and the dielectric material is thin enough, however, an absorption/reflection phenomenon dominates. The radiation is passed through the material at near normal incidence, is reflected by the metal substrate, and is passed back through the material.

A grazing angle reflectance microscope is also commercially available.[9] This microscope reflects the infrared radiation from the sample substrate at near grazing angles (10-30°, as measured from the plane of the microscope stage) allowing it to pass through a longer sample pathlength. Consequently, thinner films/samples can be examined than with standard FTIR microscopes. The grazing angle microscope has the advantage over the macroscopic accessory in that samples whose size is approximately 50 x 50 micrometers in dimension can be studied.[9]

The development of Raman spectroscopy began with the discovery of the inelastic scattering of photons by C. V. Raman in 1928. This physical phenomenon became known as the Raman effect. However, the Raman effect is so weak (only about one in every 1,000,000 scattered photons do so inelastically) that the development of powerful sources was required before the coupling of a microscope and Raman spectrometer was able to provide adequate S/N. This was accomplished with the advent of the laser. The intense, monochromatic radiation supplied by a laser source provided the needed enhancement in signal, and a commercially available Raman microprobe appeared in 1977.[10] For most samples, however, the Raman signal was still difficult to detect over instrumental noise and fluorescence. Concerning these early microprobes Tomas Hirschfeld wrote, "The existing Raman microprobe devices are thus compelled to work at the very edge of the state of the art."[11] In later generations of Raman microprobes, improvements were made to existing double monochromators and triple monochromators were developed to improve stray light rejection.[12] Recently, the development of multichannel detectors has provided the Raman microprobe with even greater S/N capabilities. For example, using a 100 micrometer entrance slit, an intensified diode array detector can provide a S/N comparable to a photomultiplier tube, but 500 times faster.[13] This gain is accomplished only with a sacrifice in either resolution or range, however.

ADVANTAGES, LIMITATIONS, AND CAPABILITIES

The peaks observed in an infrared or Raman spectrum result from transitions between vibrational energy levels. (In the case of a gas, rotational transitions are involved as well.) Both techniques are, therefore, sensitive to a molecule's structure and chemical composition. With the exception of optical isomers, each compound yields a unique spectrum which allows it to be "fingerprinted". Identification of an unknown can be obtained by matching its spectrum to a reference file, or functional group information can be obtained if the compound is not contained within the reference data base. Furthermore, infrared and Raman spectroscopy can be performed on a wide range of compounds, organic and inorganic, in the gas, liquid, or solid phase.

Minimum sample size and spatial resolution in the x-y plane are fundamentally limited by diffraction of the light employed in both techniques. The effect of diffraction is more severe in infrared microspectroscopy since the wavelengths of light employed for sample analysis are much longer than the wavelengths associated with Raman spectroscopy. The diffraction limited illumination diameter "D" of light focused through a lens or objective can be estimated from D = 1.22 L/N.A. where L is the wavelength of light employed and N.A. is the numerical aperture of the focusing or collecting lens.[14,15] Thus the limiting sample illumination diameter will be determined by the lowest wavenumber (longest wavelength) observed in a given spectrum. For a typical mid-infrared spectrum obtained with an objective of N.A.= 0.28 this diameter corresponds to 62 micrometers at 700 cm^{-1}. A similar calculation conducted for Raman analysis at 0.5145 micrometers yields a sample illumination diameter of approximately 2.2 micrometers.

In practice infrared spectra can be obtained on samples as small as 6 micrometers in diameter, however the spectra will exhibit artifacts due to diffraction by the sample. Diffraction by the sample usually manifests itself in the infrared spectrum as a sloping baseline rising toward lower wavenumbers (longer wavelengths). Figure 1. demonstrates this phenomenon in the spectrum of a 15 micrometer wide cellulose acetate slab. The slab was cut from a film of cellulose acetate and analyzed by barely overlapping the edges of the aperture with the sample. The quantitative effect of diffraction is a loss in photometric accuracy at the longer wavelengths (lower wavenumbers). The loss of photometric accuracy can be observed by comparing the spectrum of the slab to a that of the continuous film obtained under the same conditions. The relative intensities of bands located at 1368, 1229, and 1049 cm^{-1} in the spectrum of the slab are not representative of the true cellulose acetate spectrum as represented by the continuous film. At these longer wavelengths, light bends around the sample (diffracts) giving rise to false transmittance values. As the sample size decreases this effect will dominate the spectrum to the point where infrared absorptions characteristic of the sample are no longer observed.

Diffraction also limits the ability to obtain pure spectra from small particles near neighboring particles or from a small particle embedded in a support matrix. The diffraction limited illumination diameter discussed previously can, as a rule of thumb, be employed as the minimum distance between two particles needed to obtain the spectrum of one without spectral interference from the other. Consequently, two particles which are less than 62 micrometers apart, or a particle whose diameter is less than 62 micrometers and is embedded in a support matrix, may exhibit spectral interferences due to the neighboring particle or support matrix, respectively, as a result of diffraction. It should be noted that spectral interferences will not be linear but will be observed in accordance with diffraction i.e., greater impurity absorptions will be observed at longer wavelengths than at shorter wavelengths for a given sample. Messerschmidt[16] has recently demonstrated the problems associated with diffraction on spatial resolution by observing a 20 micrometer fiber embedded in polystyrene. Spectra obtained by not aperturing the sample, singly aperturing and redundant aperturing reduced the spectral impurity component due to the polystyrene support matrix.

Quantitation is difficult in either infrared or Raman microspectroscopy. In Raman spectroscopy, quantitation is difficult since the Beer-Lambert Law is not applicable to scattering phenomena. It is even more difficult for Raman microspectroscopy, since detailed geometrical and optical properties of microscopic samples are not readily determined. In infrared microspectroscopy, quantitation is

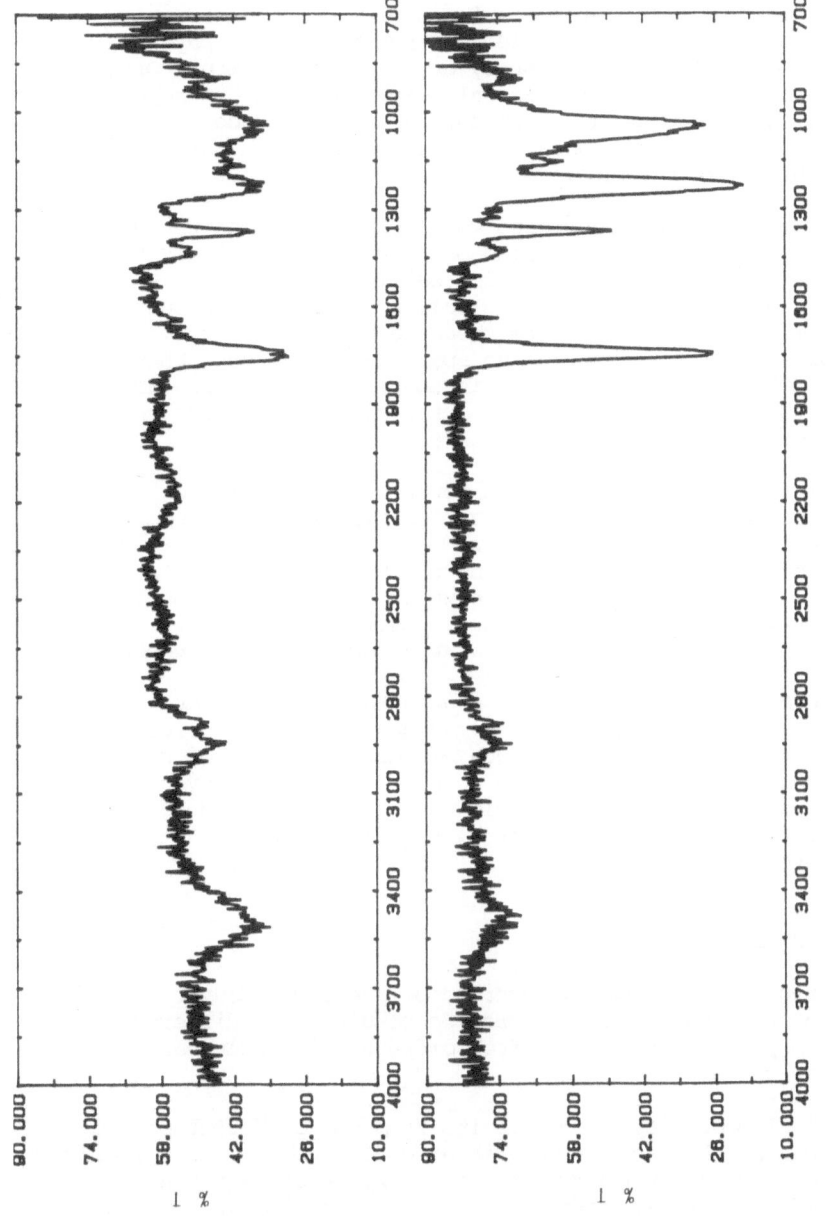

Figure 1. Infrared spectrum of a 15 micrometer wide cellulose acetate slab (top) and cellulose acetate film (bottom).

primarily difficult because the pathlength of microscopic particles is not easily controlled or measured.

There is generally little or no sample preparation for Raman microspectroscopy since the technique is based on a light scattering process.[17] For example, one is not troubled with flattening or thinning the sample in order to provide an adequate pathlength as is needed with most infrared microscopic samples.[15] Consequently, in Raman microspectroscopy the sample does not have to be physically altered, a process which can destroy crystal orientation. Often samples can be studied in situ as well with the Raman microprobe. Some examples of in situ sampling using the Raman microprobe have been the identification of pigments on ancient artifacts[18] and the study of fluid inclusions in crystals.[19]

One problem often encountered when using the Raman technique is that the extraordinarily weak Raman scattering is difficult to detect above fluorescence. This problem can be minimized using a Raman microprobe, however. Even though a sample fluoresces in bulk, one may focus on a microscopic spot where there is no fluorescence.[20] In addition, fluorescence can sometimes be spatially filtered with the microscope optics.[20] Finally, the intense, focused laser radiation delivered by the microscope objective often serves to photobleach the fluorescence.[20]

The disadvantage to using high power density radiation in Raman microspectroscopy is, of course, the possibility that thermal degradation of the sample will occur. Degradation from a focused laser source can be minimized by defocusing the beam, the use of a cold stage, or by immersing the sample in water or some other medium of high heat capacity.[21] Also, multi-channel detection allows one to either minimize the time exposed to the radiation or to use lower powers.[16]

Sample degradation from a hot source is typically not a problem in infrared microspectroscopy. The germanium-coated beamsplitters used in most interferometers filter out much of the near-infrared radiation which would heat the sample.[22] In both techniques, however, some heat from the source may cause the sample to thermally expand such that focus is lost.[23] This problem may be remedied by physically fixing the sample securely to a substrate.

COMPLEMENTARITY

Infrared and Raman microspectroscopy are complementary in that the spectral information which they provide results from different quantum mechanical phenomena as well as from different instrumental limitations.[15]

The complementarity which results from the different physical phenomena can be observed in the vibrational spectra of a polyethylene terephthalate (PET) fiber shown in Figure 2. Peaks which are strong in the infrared are usually weak in the Raman spectrum, and vice versa. The peak intensity differences result from different selection rules. Raman selection rules require that a change in polarizability occur, while infrared selection rules require that a change in dipole moment occur during a molecular vibration. These differences mean that symmetric molecular vibrations are more intense in Raman spectroscopy. Notice that the strongest band in the Raman spectrum at 1615 cm^{-1} is not observed in the infrared spectrum. This band is due to the aromatic carbon-carbon stretching vibration. Thus, if the techniques

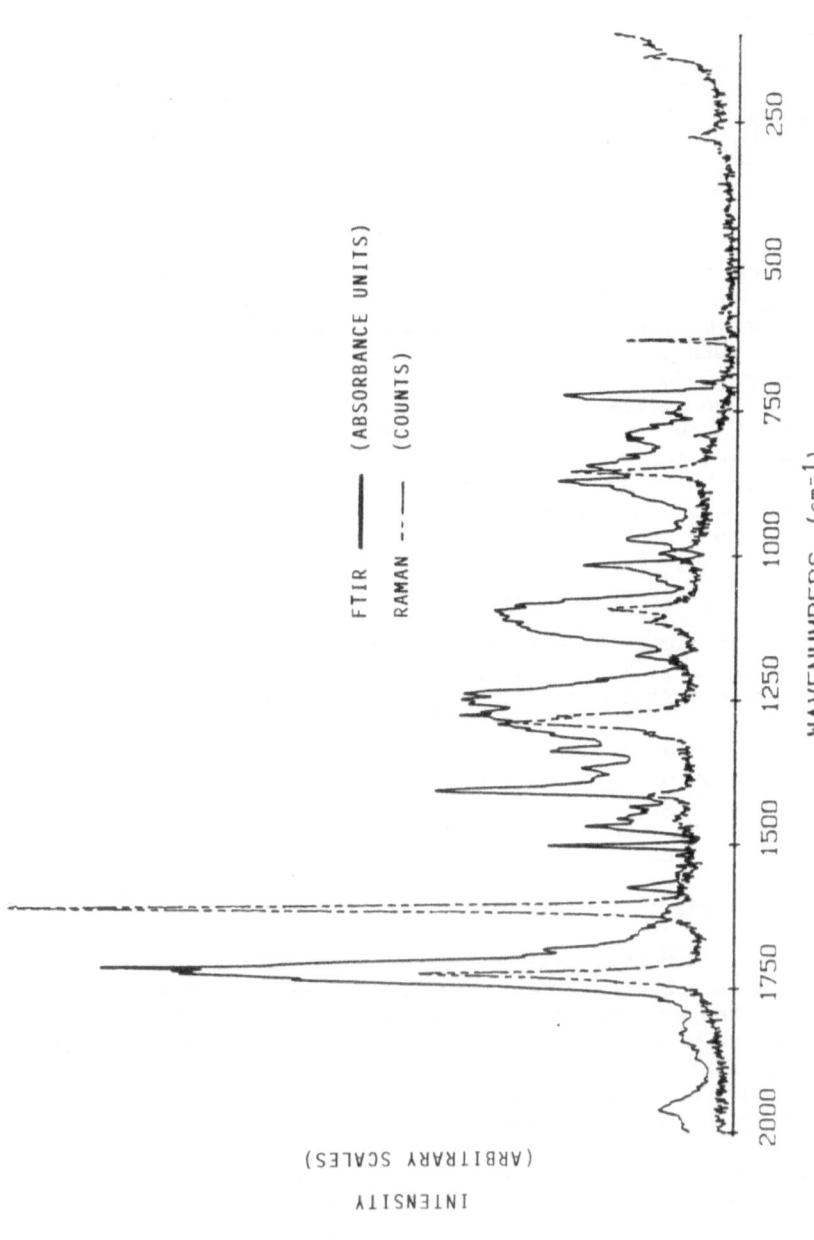

Figure 2. Infrared and Raman spectra of a 17 μm diameter polyester fiber showing the complementarity of the two techniques. Reprinted from Ref. (23), p. 329, by courtesy of Academic Press, Inc.

149

are used together, one can obtain a more complete picture of a molecule's functionalities. In addition, Raman spectra on compounds in a highly polar matrix, such as water, can be obtained without the use of special accessories or spectral subtraction, often needed to obtain usable infrared spectra.

An example of complementary information resulting from different instrumental limitations can also be observed in Figure 2. Infrared microscopes typically require the use of a highly sensitive detector such as a mercury-cadmium-telluride (MCT) detector. The cut-off for a narrow range MCT is at about 700 cm^{-1}. However, the Raman spectrum shown is recorded down to 100 cm^{-1}, thus providing low frequency information which is important for the identification of inorganics. In the PET spectrum, the band at 142 cm^{-1} is characteristic of titanium dioxide, a commonly used delusterant in textiles.

APPLICATIONS

The semiconductor industry recognized the capabilities of infrared microspectroscopy for the identification of contaminants on electronic devices well before the technique's recent popularity.[24-26] Their scientists were some of the few users of the dispersive microspectrometers. In fact, the industry's increasing need to identify smaller and smaller contaminants may have provided the impetus for the coupling of the accessory with the FTIR spectrometer.[27] The Raman microprobe was early used for semiconductor contamination analysis[28-31] as well. These earlier applications, in addition to later ones, showed that both techniques could be used to identify organic and inorganic contaminants on silicon wafers and on the larger, circuit board package. Raman microspectroscopy has also been used to characterize silicon, germanium, and ceramic surfaces.[32-37]

Two related applications performed at the Molecular Microspectroscopy Laboratory (MML) involved the identification of solder flux contaminants on printed circuit boards,[29] and the identification of dust particles.[38] An Analect AQS-20M system was used to obtain the 4 cm^{-1} resolution infrared spectra on microscopic samples. Raman spectra were obtained using an Instruments SA Ramanor U-1000 microspectrometer. The 514.5-nm line of an argon-ion laser was used as the excitation source. Experimental details of each experiment have been previously described.[29,38]

Solder Flux Contamination

One cause of electronic failure involves the contamination by flux residues left on the printed circuit board from the soldering process. Water-soluble fluxes and activated rosin fluxes contain ionic metal activators which serve to remove metal oxides from the surfaces to be soldered. These ionic residues are particularly corrosive to tin, lead and copper and may ultimately result in the formation of conductive filaments leading to short circuits.

In order to be able to identify flux contaminants on printed circuit boards via microspectroscopy, a spectral library of a variety of rosin and water-soluble fluxes was established to provide a reference base for comparison. It was hoped that the spectrum obtained from any residue removed from a failed board could be matched to one in the library. Infrared spectra and Raman spectra were obtained on the bulk fluxes as well as on the residues.

The subsequent analysis of some failed circuit boards showed that infrared microspectroscopy can be successfully used to identify flux residues. Using our established reference base, abietic acid[29] and a metal activator containing ammonium chloride[39] (Figure 3.) were identified from the spectra of particles removed from two failed boards.

However, the analysis of failed circuit boards for flux residues is not always so simple. The difficulty arises because the fluxes are mixtures often containing several components. In addition, the fluxes can react or decompose at the high temperatures needed for soldering. Consequently, the residues left behind are often mixtures. For example, in the spectra shown in Figures 4 and 5[40] one can observe bands due to some reaction products, copper abietate and copper sulfate. Bands from the conformal coating, polymethylmethacrylate, can also be seen.

A circuit board with a solder adhesion problem was also analyzed. The solder pads adhered so poorly to the metallic substrate that they could easily be pulled off with a piece of cellophane tape. After removing several solder pads, however, no residue was apparent with visual examination under the stereoscope. A grazing angle reflectance microscope accessory coupled to a Perkin-Elmer Model 1800 spectrometer was used to analyze these areas. Figure 6. compares the spectrum from a copper resinate reference with the spectrum obtained on the metallic substrate directly under the removed solder pad.[40] Several bands match in frequency, suggesting the presence of copper abietate as the cause of poor adhesion. Therefore, infrared microspectroscopy can be useful even in situations where a contaminant is not visually apparent even under microscopic examination.

Although in situ analysis using Raman microspectroscopy seemed to be an ideal method for the analysis of these failed boards, it was not feasible due to excessive fluorescence.

Dust Characterization

Contamination by dust particulates can frequently be the cause of product defect or failure especially in the manufacture of very large scale integrated circuits. However, unless a material has been manufactured, stored, transported and analyzed in a controlled environment, it is possible that it will also become contaminated by "insignificant" dust, dust that is not the cause of the material's defect. The presence of extraneous contaminants often makes it difficult for the analyst to identify the source of the problem.

Although microscopists have been long aware of the optical microscopic characteristics of dust constituents[41], we carried out a microspectroscopic study of particles in settled office and laboratory dust to become more familiar with the spectra of these contaminants.

Cellulose was the most prevalent constituent identified in the dust samples collected. Most of the cellulose could be identified as wood cellulose from paper due to the presence of lignin bands, or paper additives such as, kaolin, calcium carbonate, and/or polymethacrylates. The infrared and Raman spectra of a cellulose fiber are compared in Figure 7.[42] Acquisition times for the two spectra are two minutes for the infrared spectrum and four hours for the Raman spectrum.

Other paper-related contaminants identified included epoxy resin and photocopying toner. Several black dust particles, which yield

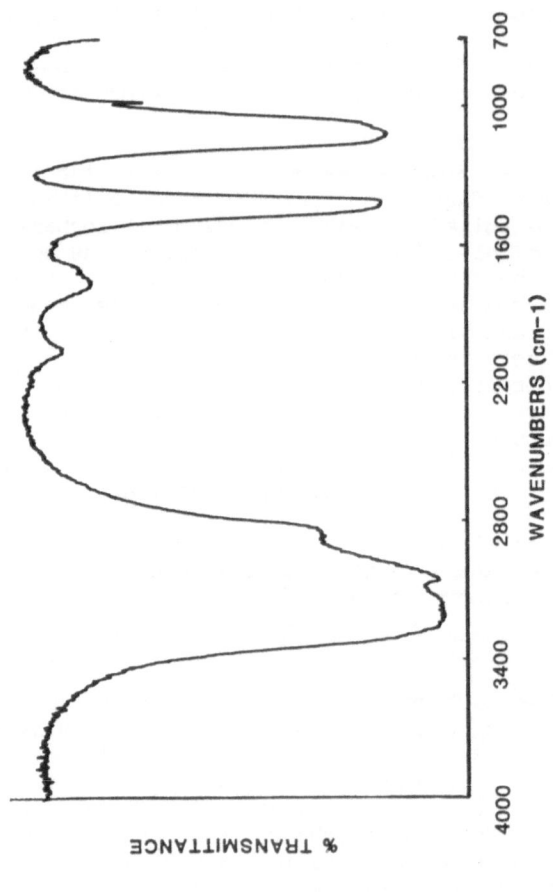

Figure 3. Spectrum of a residue removed from around a solder joint of a failed board. Reprinted from Ref. (39), p. 48 by courtesy of Marcel Dekker, Inc.

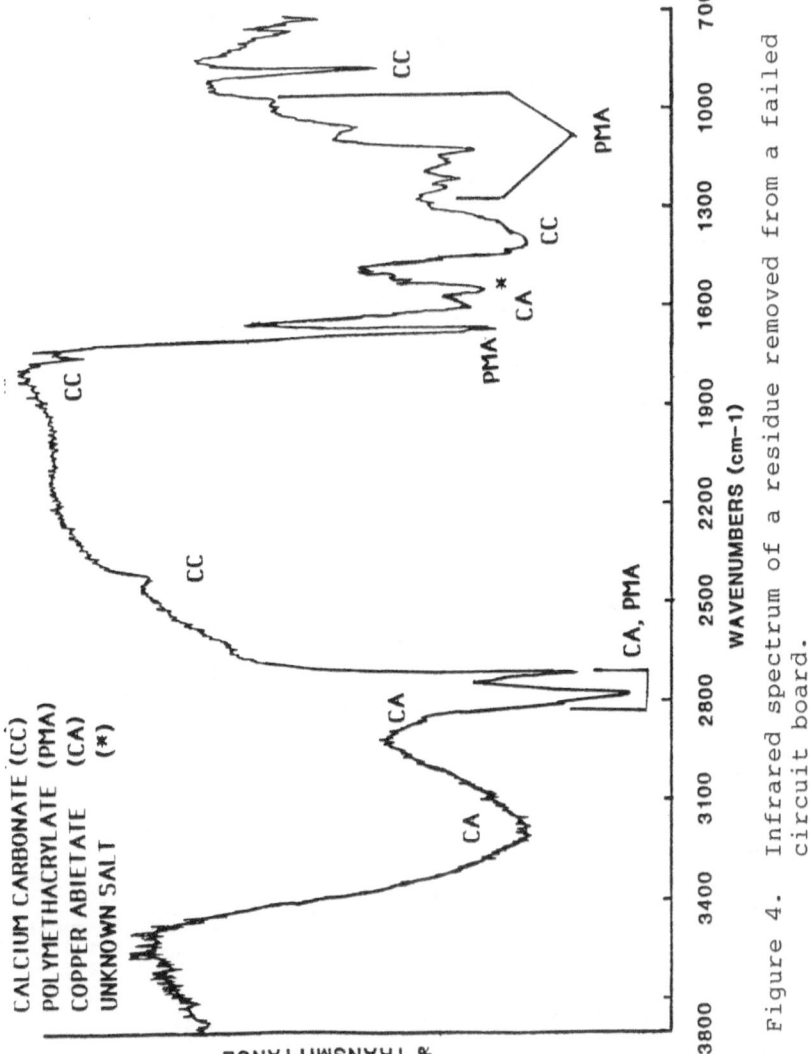

Figure 4. Infrared spectrum of a residue removed from a failed circuit board.

153

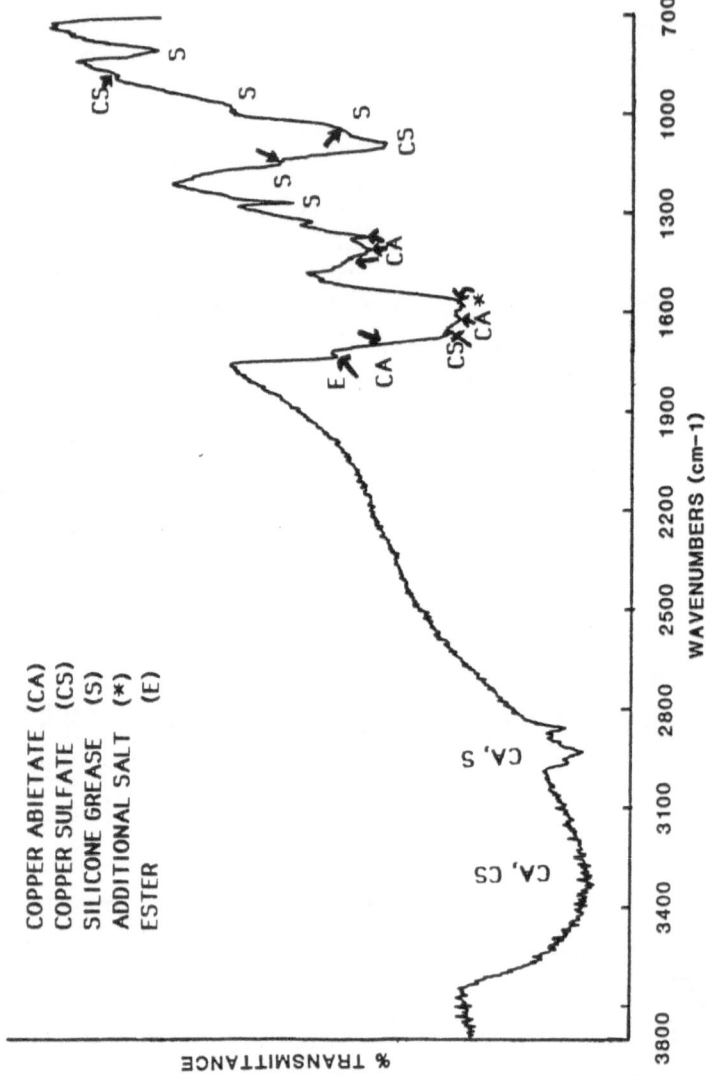

Figure 5. Infrared spectrum of a residue removed from a failed circuit board.

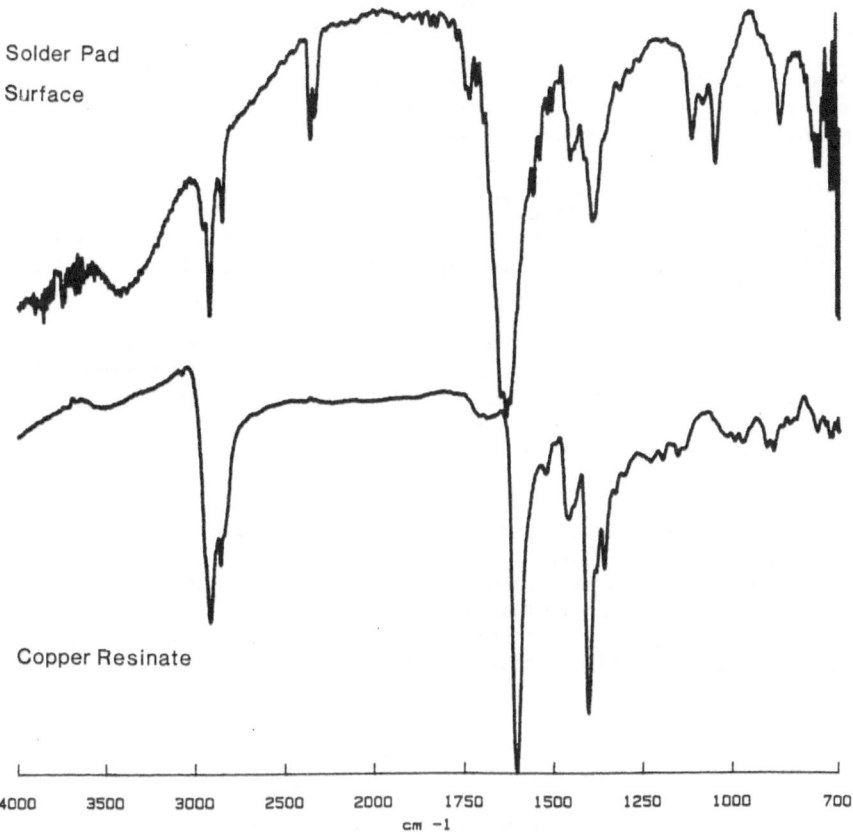

Figure 6.   Comparison of a reference spectrum of copper abietate with
            an infrared spectrum of the copper substrate under a poorly
            adhering solder pad.   A grazing angle reflectance microscope
            accessory was used.

virtually no infrared spectral features, could be identified from their Raman spectra as amorphous carbon. Since steam heat is used in the environments which were studied, it is thought that the source is the photocopying toner.

Many dust constituents identified are a result of the shedding of skin, hair, and clothing. In fact, following cellulose, polyamides were the most prevalent dust constituents. The overwhelming majority of polyamides were present as flat, flaky particles. Under microscopic examination (400X), these dust particles were recognized as epithelial cells, thereby confirming their skin origin. A representative infrared spectrum (two minutes collection time) and Raman spectrum (eight hour collection time) is shown in Figure 8.[42] It is clear from these spectra, as well as those in Figure 7, that infrared is the method of choice for identification of cellulose and protein. These two materials are very weak Raman scatterers but give strong infrared absorption.

Natural and synthetic polyamide fibers were occasionally observed, too. Scott and Ramsey[26] have previously recognized that natural polyamides are a possible source of contamination and also have pointed out some of their differing spectral characteristics relative to synthetic polyamides.

Other particles identified included talc, dried liquid make-up, calcium and sodium sulfate, polyvinyl chloride, amorphous silica, glass wool, and various textiles such as polyethylene terephthalate, polypropylene, cotton, and wool. The spectrum of a particle found in the dust, which is believed to originate from a liquid cosmetic, is shown in Figure 9.[38]

Aware of the possible extraneous contaminants, the microspectroscopic analyst can then take the necessary steps to insure that the reported spectra are relevant to the problem under investigation. These steps may include obtaining a detailed history of the material, sampling from several areas, or sampling from areas where contaminants are systematically arranged.

CONCLUSIONS

Infrared and Raman microspectroscopy are powerful tools for the identification of microscopic particulates. Because the methods provide molecular information, as opposed to information on elemental composition, the methods are generally conclusive. In addition, the specificity of the methods can usually be improved by utilizing both infrared and Raman microspectroscopy, thus taking advantage of their complementary nature. This complementarity has both a theoretical origin in the nature of the phenomena being studied and a practical origin in the experimental treatment of the samples and in the equipment used.

Since infrared and Raman microspectroscopy allow the acquisition of vibrational spectra from a sample size which is much smaller than heretofore possible, they provide a means of solving problems which have previously been intractable. In fact, the application of infrared and Raman microspectroscopy by industry for the identification of microscopic contaminants, which severely impact the performance or appearance of commercial products, has gained rapid acceptance.

However, the various kinds of dust particles which are generally found with samples can be easily sampled along with or instead of the

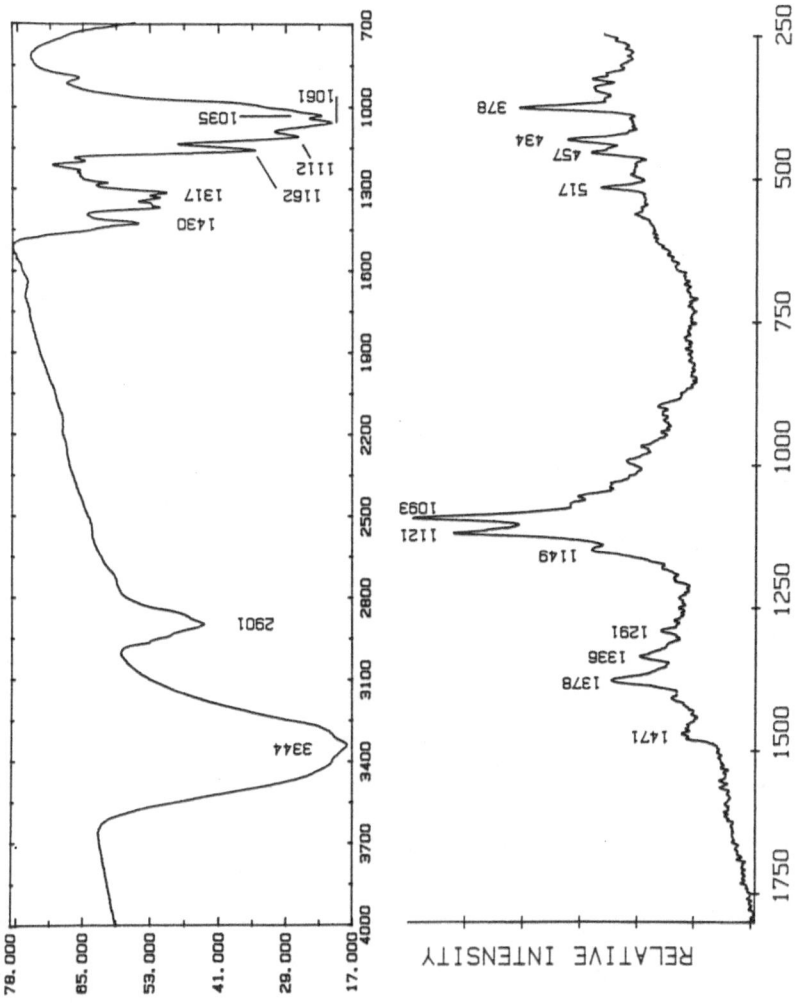

Figure 7. Infrared spectrum of cellulose (top, two minute collection time) and Raman spectrum of cellulose (bottom, two hour collection time). Reprinted from Ref. (42), p. 208, by courtesy of San Francisco Press, Inc.

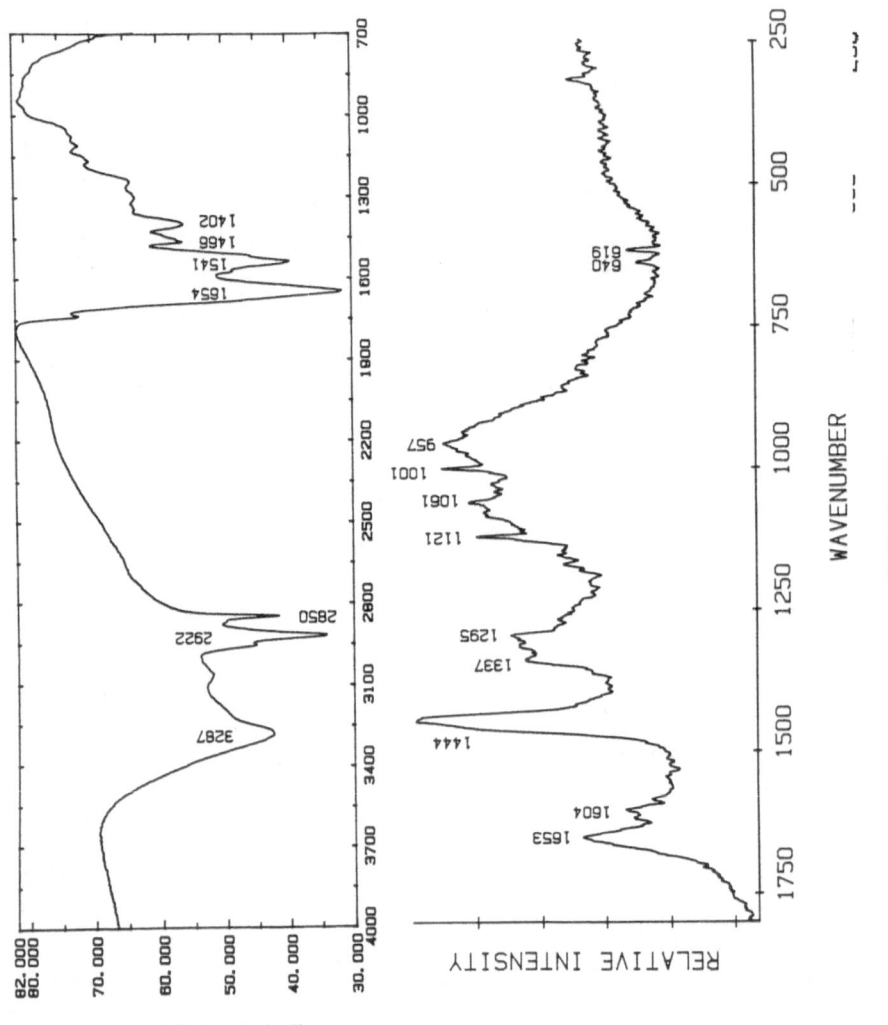

Figure 8. Infrared spectrum of protein (top, two minute collection time) and Raman spectrum of a protein (bottom, eight hour collection time). Reprinted from Ref. (42), p. 209, by courtesy of San Francisco Press, Inc.

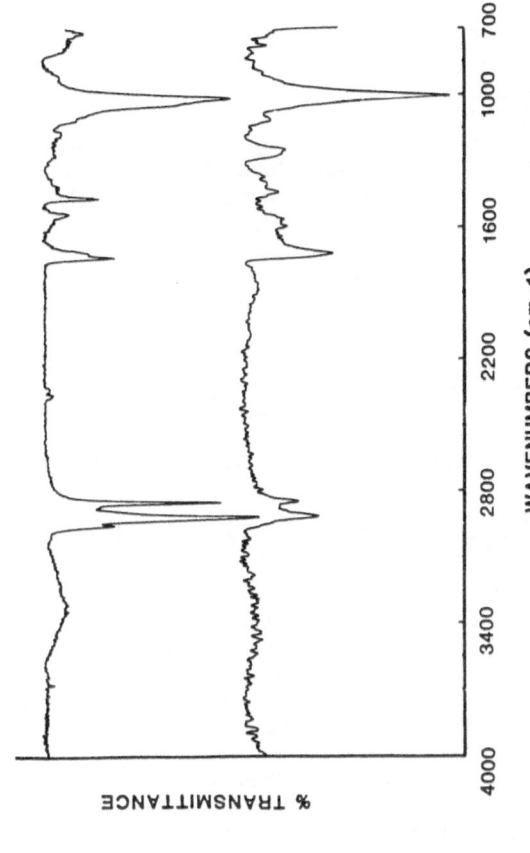

% TRANSMITTANCE

WAVENUMBERS (cm−1)

4000   3400   2800   2200   1600   1000   700

Figure 9. Comparison of the spectrum obtained from a dust particle (bottom-baseline flattened)
with that obtained from a dried liquid make-up base (top-193 scans, 15 point
Savitsky-Golay smoothing, baseline flattened).
Reprinted from Ref. (38), p. 316, by courtesy of The Society for Applied Spectroscopy.

problem-causing particle. These ubiquitous particles are primarily cellulose and protein and appear to be deposited at some time after the sample was first acquired. The analyst must always be aware that certain particles may simply be adventitious and therefore unrelated to the problem at hand. Nonetheless, infrared and Raman microspectroscopy provide rapid and accurate solutions to many modern manufacturing problems.

REFERENCES

1. J. E. Katon, G. E. Pacey, and J. F. O'Keefe, Anal. Chem., 58, 465A (1986).
2. C. R. Burch, Proc. Phys. Soc. (London), 59, 41 (1947).
3. K. P. Norris and W. E. Seeds and M. H. F. Wilkins, J. Opt. Soc. Am., 41, 111 (1951).
4. Vincent J. Coates, Abe Offner, E. H. Siegler, Jr. J. Opt. Soc. Am., 43, No. 11, 984 (1953).
5. Elkan R. Blout, George R. Bird, and David S. Grey, J. Opt. Soc. Am., 40, No.5, 304, (1950).
6. Peter R. Griffiths and James A. de Haseth, "Fourier Transform Infrared Spectrometry", John Wiley and Sons, New York, 1986.
7. David W. Schiering, in "Infrared Microspectroscopy", Robert G. Messerschmidt and Matthew A. Harthcock, Eds., Chap. 16, Marcel Dekker, New York, 1988.
8. Robert G. Messerschmidt, Ch.1, Ibid.
9. Application Note, "The Surface Scope," (Spectra-Tech, Inc., Stamford, CT).
10. P. Dhamelincourt, Analytica Chimica Acta, 195, 33 (1987).
11. T. Hirschfeld, Microbeam Analysis 1982, 247.
12. R. Grayzel, M. LeClercq, F. Adar, J. Lerner, M. Hutt, and M. Diem, Microbeam Analysis 1985, 19.
13. Fran Adar, Jeremy Lerner, and Yair Talmi, Microbeam Analysis 1987, 141.
14. G. J. Rosasco, Raman microprobe spectroscopy in "Advances in Infrared and Raman Spectroscopy", R. J. H. Clark and R. E. Hester, Eds., Vol. 7, p. 223, Heyden, London, 1980.
15. F. Adar, Microbeam Analysis 1981, 67.
16. R. G. Messerschmidt, Microbeam Analysis 1987, 169.
17. J. A. Lander, Anal. Proc., 23, 270 (July, 1986).
18. Bernard Guineau, J. Forensic Sci., 29, No. 2, 471 (1984).
19. K. L. Higgins and C. L. Stein, Microbeam Analysis 1986, 31.
20. Fran Adar, Michel LeClercq and Roy E. Grayzel, Amer. Lab., 52, (March 1982).
21. M. E. Andersen and R. Z. Muggli, Anal. Chem., 53, 1772 (1981).
22. David W. Schiering, personal communication 1988.
23. Patricia L. Lang, J. E. Katon, J. F. O'Keefe, and David W. Schiering, Microchem. J., 34, 319 (1986).
24. J. N. Ramsey, The application of small area molecular species analysis techniques to S/C device and package processing, Paper presented at the International Forum for the Production of Electronic Components, Rhein-Main-Halle Wiesbaden/F.R.G., 1982.
25. J. N. Ramsey and H. H. Hausdorff, Microbeam Analysis 1981, 91.
26. R. M. Scott and J. N. Ramsey, Microbeam Analysis 1982, 239.
27. Robert G. Messerschmidt and Matthew A. Harthcock, Eds., "Infrared Microspectroscopy", p. iii, Marcel Dekker, New York, 1988.
28. Fran Adar, in "Microelectronics Processing: Inorganic Materials Characterization", Lawrence A. Casper, Ed., ACS Symposium Series No. 295, Chp. 13, p. 230, 1986.

29. Patricia L. Lang and J. E. Katon, Microbeam Analysis 1986, 47.
30. Robert Z. Muggli and Mark E. Andersen, Solid State Tech., 28, 309 (April 1985).
31. Karen Madden and John Ramsey, Test Measurement World, 4, No. 2, 54-59 (February 1984)
32. Shinji Hayashi and Hiroya Abe, Jap. J. Appl. Phys. 23, No. 11, L824 (November 1984).
33. R. J. Nemanich, Mat. Res. Soc. Symp. Proc., 69 23 (1986).
34. D. R. Clarke and F. Adar, Raman microprobe spectroscopy of polyphase ceramics, in "Advances in Materials Characterization", David R. Rossington, Robert A. Condrate, and Robert L. Snyder, Eds., p. 199, Plenum Press, New York, 1983.
35. P. M. Fauchet, Scanning Electron Microscopy 1986, II, 425 (1986).
36. Shinji Hayashi, J. Phys. Soc. Japan, 56, No. 1, 243 (1987).
37. Philippe M. Fauchet and Ian H. Campbell, Raman microscopy of semi-conductor films, in "Modern Optical Characterization Techniques for Semiconductors and Semiconductor Devices", O. J. Glembochi, Fred H. Pollach, J. J. Song, Eds., (SPIE-The International Society for Optical Engineering, Washington, 1987).
38. Patricia L. Lang, J. E. Katon, Anthony S. Bonanno, Appl. Spec., 42, 313 (1988).
39. Patricia L. Lang, J. E. Katon, Anthony S. Bonanno, and G. E. Pacey, in "Infrared Microspectroscopy", Robert G. Messerschmidt and Matthew A. Harthcock, Eds., Chap. 3, Marcel Dekker, New York, 1988.
40. Patricia L. Lang, "Application of Infrared and Raman Microspectroscopy and the Vibrational Spectra, Structure, and Conformational Behavior of Dimethyl Dicarbonate," Ph.D. Dissertation, Miami University, University Microfilms (1987).
41. W. C. McCrone, R. G. Draftz and J. G. Delly, "The Particle Atlas", Ann Arbor Science Pub., Ann Arbor, MI 1967.
42. Andre J. Sommer and J. E. Katon, Microbeam Analysis 1988, 207.

PARTICLE IDENTIFICATION BY AUGER ELECTRON SPECTROSCOPY

Kenneth D. Bomben and William F. Stickle

The Perkin-Elmer Corporation, Physical Electronics Laboratory
2305 Bering Drive, San Jose, California 95035

The use of Auger Electron Spectroscopy (AES) as a method for particle characterization has grown in importance, particularly as semiconductor geometries have become smaller. The surface sensitivity and small analysis size of AES makes it the technique of choice for the characterization of sub-micron particles because X-ray techniques approach the limit of their spatial and depth resolution. AES provides particle identification with a minimum of interference from the surrounding matrix. The analysis and identification of particles found during the processing of electronic circuits is discussed. A comparison of surface and near-surface techniques for applicability to sub-micron particle analysis is also made.

INTRODUCTION

Particle identification and characterization is important in any technological field where particles are of the same size as the critical dimensions for the process. A number of well-known techniques are available for the detection of particles and these have found widespread use. However, the ability of "typical" techniques to unambiguously identify the composition of a particle is limited by the spatial and chemical resolution of the technique. For example, Energy Dispersive X-Ray (EDX) spectroscopy uses an electron beam to probe a sample and detects the characteristic X-rays of the elements that are present. It is limited to a depth resolution and spatial resolution of about one micrometer. (The resolution varies as a function of the primary beam voltage.) Furthermore, most EDX spectrometers have difficulty with elements with atomic masses lighter than sodium, primarily because of the low cross section for X-ray emission from these elements, although thin windows will allow the detection of low z elements[1]. For sub-micron particle analysis, EDX contains a back-

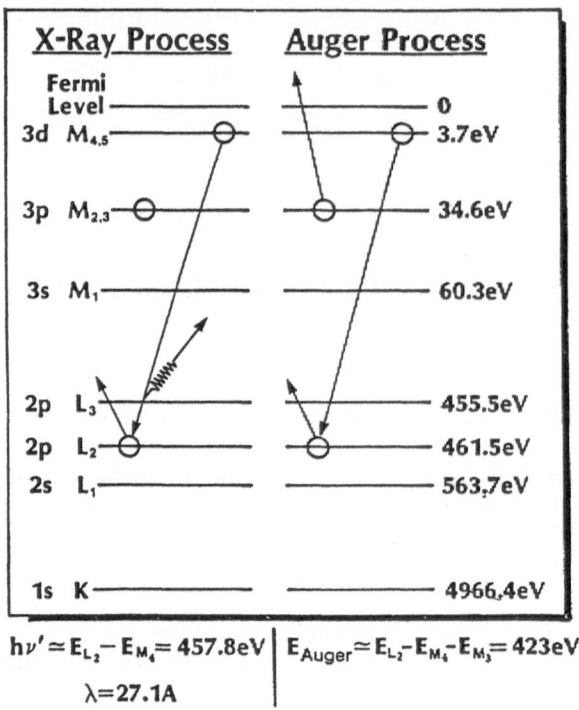

$$h\nu' \simeq E_{L_2} - E_{M_4} = 457.8eV \quad | \quad E_{Auger} \simeq E_{L_2} - E_{M_4} - E_{M_3} = 423eV$$

$$\lambda = 27.1A$$

Figure 1. Schematic representation of Auger electron emission and X-ray emission in titanium.

ground from the matrix surrounding the particle. As particle dimensions approach 300nm, for example, the particle volume contributes less than 3% to the total EDX analysis volume.

An alternative technique for particle analysis is high-resolution Scanning Auger Microscopy (SAM), based on Auger Electron Spectroscopy (AES)[2]. Like EDX, AES uses an electron beam to probe a sample, however, the detected particles are electrons characteristic of the elements that are present. The depth resolution of AES is limited to the escape depth of the Auger electron, the distance the electron will travel in the sample before it interacts with another particle and loses energy. The escape depth is dependent upon the composition of the sample but may be generalized to a few atomic layers, about 3-5nm. The spatial resolution of AES is primarily limited by the diameter of the electron beam and scattering of electrons in the sample. Typically 50-100nm is considered the limit of SAM spatial resolution. Hence, for a 300nm diameter particle, all of the AES signal originates from the particle and none from the surrounding matrix. Furthermore, all elements except hydrogen and helium can be detected. AES, then, is the preferred technique for sub-micron particle characterization.

An Auger electron is produced when an electron vacancy in a core level is filled by an electron from a higher-energy orbital along with the ejection of the Auger electron, as shown schematically in Figure 1. The figure also shows the competing X-ray production process typical for EDX. The core vacancy can be produced by an electron beam, an X-ray beam or an ion beam; the energy of the simultaneous two-electron Coulombic rearrangement, that is the Auger process, is independent of the excitation source.

164

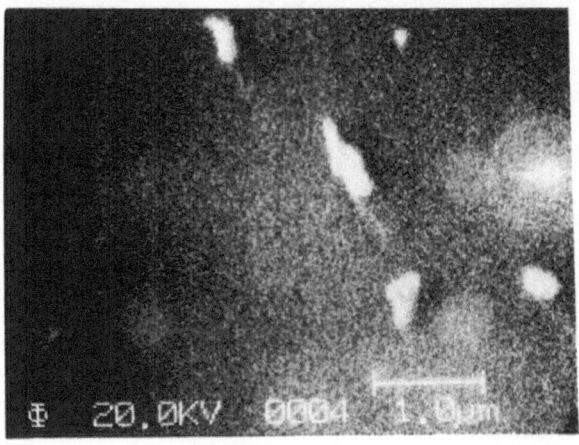

Figure 2. Secondary electron micrograph of particles on a laser mirror.

EXPERIMENTAL

All samples were examined in a Perkin-Elmer PHI Model 600 SAM system. Primary electron beam analysis conditions were usually 20kV and 100nA. Samples were mounted on conductive stubs and a grounding clip was attached to the sample to carry away any excess charge. The sample tilt was typically 60° to minimize the effect of the electron beam on insulating samples. The surfaces were examined as received, without the deposition of a conductive layer which would distort the analysis.

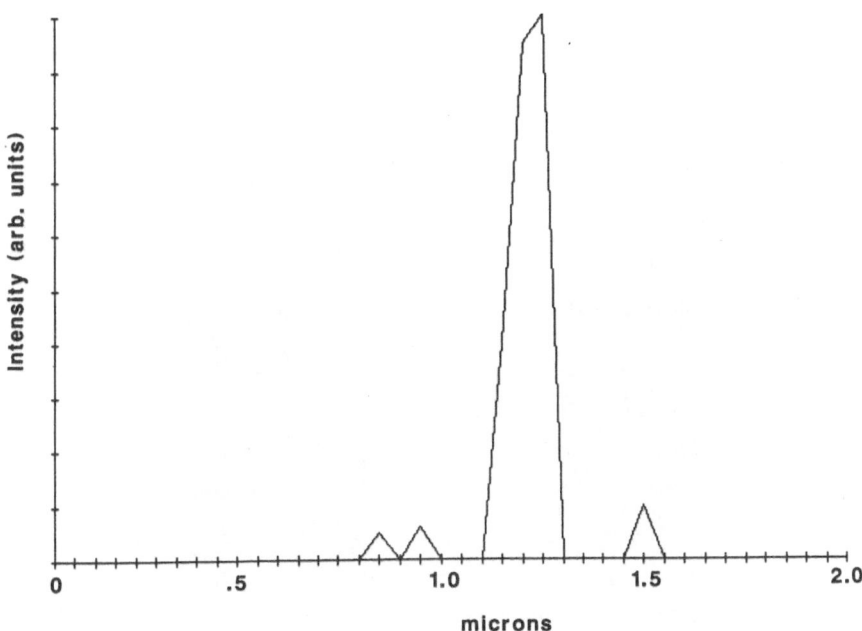

Figure 3. Scanning Auger line scan for silicon. The line scan is taken along a horizontal line through the center of Figure 2.

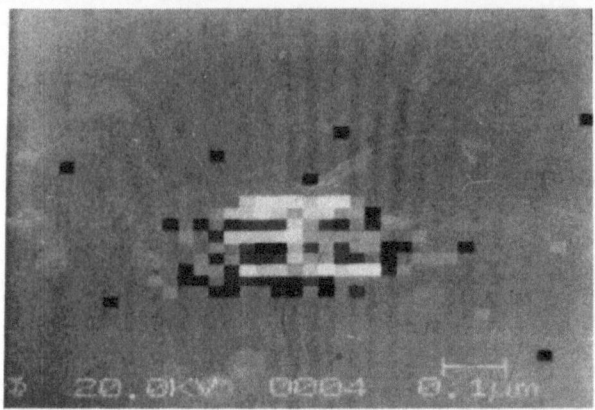

Figure 4. Scanning Auger map for silicon. The particle is at the right edge of Figure 2.

Particles on a Laser Mirror

Particles of unknown origin and composition were observed on the surface of an optical laser mirror. The multi-layer mirror with its anti-reflective coatings consisted of 1 μm thick ZnSe on 1.5 μm of ThF$_4$ on 0.3 μm of Th on 0.2 μm of Ag on a silicon substrate. Particles were between 0.1 and 0.6 μm in diameter and caused an unacceptable level of light scattering. Figure 2 is a secondary electron micrograph of the surface showing several typical particles. Attempts to characterize these particles using EDX failed. Using AES, the particles were found to be silicon. The outermost surface of the mirror was ZnSe.

Because of the insulating nature of this surface, addition spectra were taken on and off the particles to guarantee the absence of charging on either surface. In this case the high tilt angle between the incident electron beam and the sample surface was sufficient to insure undistorted spectra and maps. It is important, however, to recognize that special care must be taken in the investigation of insulating surfaces by AES because non-uniform charging can distort mapping of elemental distribution by reflecting charging points. All maps were checked by gathering data in the regions of high intensity and looking for evidence of charging. No charging artifacts were observed during the analysis.

The low energy silicon transition (92eV) was monitored as a function of position along a line which was chosen to pass through a small particle. The SAM line scan for Si is shown in Figure 3. This particle is 150 nm, edge to edge. No Ag, Th or F is seen on either the surface or in the particle. A SAM map of the surface for Si, Figure 4, done at a resolution of 25nm per pixel, clearly shows the particles to be Si on the

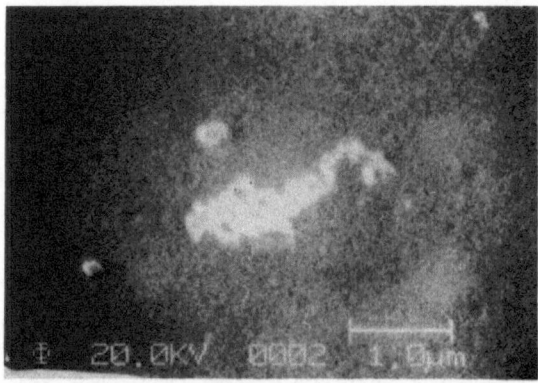

Figure 5. Secondary electron micrograph of particles on an unpatterned silicon wafer.

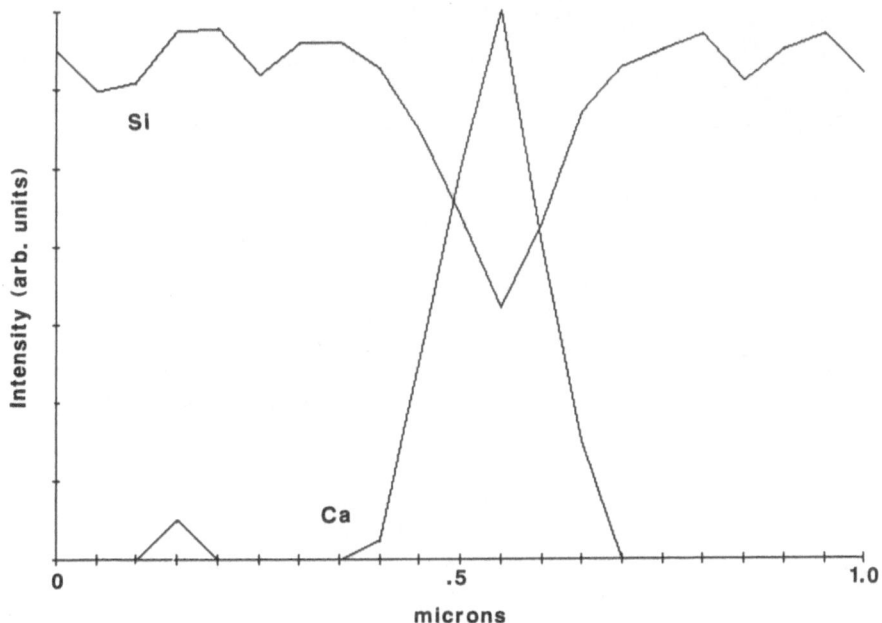

Figure 6. Scanning Auger line scan for silicon (top) and calcium (bottom). The line scan is a horizontal line through the particle at the lower left of Figure 5.

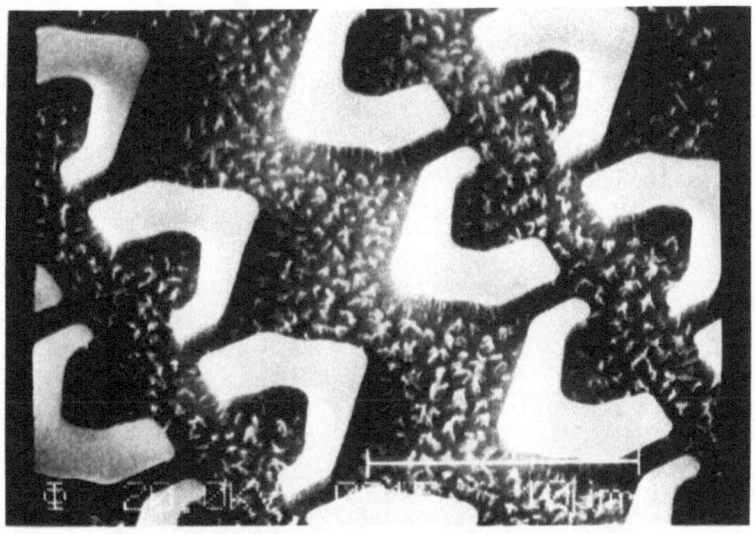

Figure 7. Secondary electron micrograph of residues on a patterned wafer following reactive ion etching

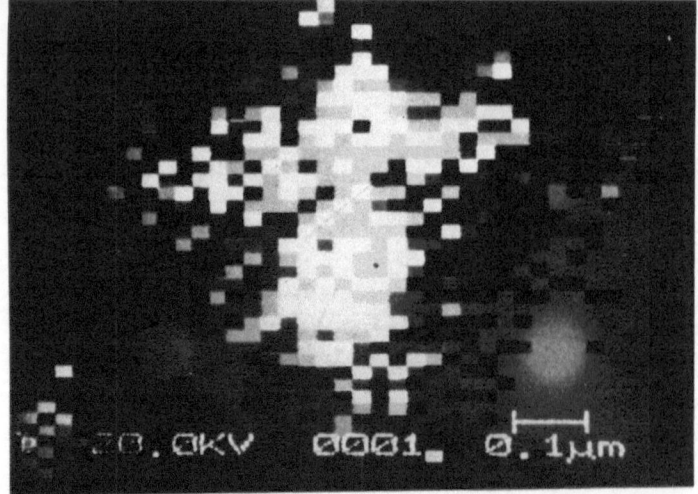

Figure 8. Secondary electron micrograph at 100,000x of the residue illustrated in Figure 7. This feature is in the center of Figure 7.

ZnSe substrate. The time required for complete characterization of the particles, including line scans and maps, was less than two hours.

The absence of Ag, Th and F in the particles suggests that the particles could not have come from the anti-reflective surface of the mirror or the Si substrate before deposition. They must have been deposited after film deposition and are probably the result of rough handling of the Si blanks or are from the bottom of other mirrors.

Particles on Unpatterned Wafers

Freshly-cleaned silicon wafers were found to be contaminated with particles, as illustrated in Figure 5. Using AES a point analysis of the surface showed only Si and O (from the native oxide) while the large particles showed a strong gold signal and a line scan was done through the particle. The smaller particle (lower left of Figure 5) did not show any Au but did show Ca, S and Cl. This particle is 250nm, edge to edge. A SAM line scan for Ca and Si, shown in Figure 6, clearly show the Ca to be associated with the particle.

Each particle is, therefore, from a different source. The most likely source of the Au is from the backplane of other wafers, probably deposited by scraping or rough handling. The composition of the 250nm particle is typical of human residues and is probably the result of carelessness.

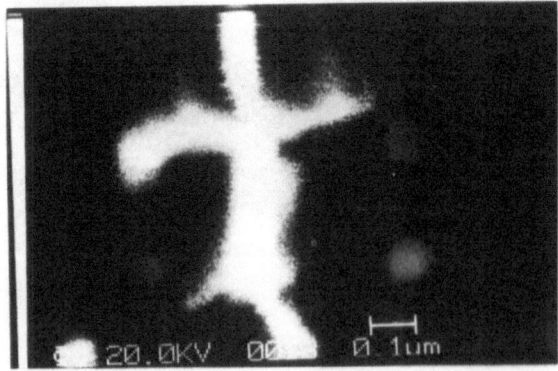

Figure 9. Scanning Auger map for tungsten. The map dimensions correspond to the dimensions of Figure 8. The residue dimensions are 50 to 100nm.

Particles on Patterned Wafers

A fiber-like residue, shown in the micrograph of Figure 7, was found between Al lines on a patterned wafer following reactive ion etching. The normal composition of this area of the wafer is 60nm of W on Si. A photoresist residue was suspected and analysis by EDX detected only Si and W As EDX is not sensitive to the light elements present in a photoresist, an analysis was done using AES. The micrograph of Figure 8 shows the residue at 100,000x. Point analysis shows that the strands contain W, O, C and Si; while the surface on which the fiber rests contains only W, O and C. The W map is shown in Figure 9; these features are 5 to 10nm, edge to edge. The composition of the residue suggests that the residue was formed during the reactive ion etching in which the silicon formed an unreactive compound with the photoresist.

SUMMARY

The surface sensitivity of SAM is able to unambiguously identify sub-micron particles and features in surfaces. The ability to examine a fine particle, in exclusion of its surrounding matrix, is necessary if the identification of the chemical composition is important. It is particularly important for the determination of the origin of contaminants. Furthermore, it is possible to completely characterize insulating particles or particles on insulating substrates if the proper percautions are observed during the course of the analysis.' Within the sub-micron geometries of semiconductor devices, or on optical surfaces, the ability to analyze small areas of contamination such as particles is only possible by a technique such as AES.

ACKNOWLEDGEMENTS

The authors wish to thank Dr. A. M. Turner, now of LTV Aerospace in Dallas, Texas for his work on the analyses of these particles.

REFERENCES

1. G. Aden and D. Isaacs, Res. and Develop., 29(8), 45-48 (1987)
2. P. A. Lindfors, R. W. Kee, and D. L. Jones, in "Microelectronics Processing," L. A. Casper, Editor, ACS Symposium Series No. 295, pp. 118-143, American Chemical Society, Washington, D. C., 1986

IDENTIFICATION AND CHARACTERIZATION OF NONMETALLIC PARTICULATE
CONTAMINATION REMOVED FROM AEROSPACE COMPONENTS

C.E. Wilson and D.A. Scheer

McDonnell Douglas Corporation
P.O. Box 516, Dept. 256
Bldg. 102, Post L315
St. Louis, MO 63166

The performance and reliability of today's
highly advanced aircraft and space hardware have
become increasingly dependent upon the elimination
or very tight control of particulate contamination.
The presence of particles on optical surfaces of
communication components can cause beam scattering,
as well as destruction of optical surface coatings,
thereby resulting in significant power losses.
Particulate contamination within critical tolerance
aircraft hardware components can degrade system
performance. These particles can originate from
either external sources during assembly or from
internal materials of construction. Identification and
characterization of particulate material is essential in
determining its source. Once the source is identified,
corrective steps can be initiated to reduce or eliminate
the particulate contamination. This paper will describe
Fourier Transform Infared Spectroscopy and Probe Mass
Spectrometry techniques which have been employed in the
analysis of particulate contamination to increase
performance and reliability of today's aerospace products.

INTRODUCTION

The analysis of particulate contamination has become extremely
important in the development, manufacturing and operation of today's
advanced aerospace hardware. Before modern technology, there was not as
much need for "fine" particulate contamination control as today, but
with current technology particulate contamination can have an adverse
effect on hardware systems. In order to eliminate or control fine
particles, frequently it is necessary to determine their identity and
source. Fortunately the same technology that is adversely affected by
fine particle contamination has also provided the analytical tools to
control this contamination.

Fourier transform infrared (FT-IR) spectroscopy and probe mass spectrometry have become important tools for the identification and characterization of this particulate contamination. The greater use of nonmetallic components in optical systems as well as a continuing increase in the use of polymeric materials in the aircraft industry have increased the role these two instruments play in particulate contamination identification. Although probe mass spectrometry is not used as frequently as FT-IR for polymer identification, it does offer a unique analytical technique for comparison of various polymers, as well as an identification capability for droplet type contaminants. The combination of FT-IR and probe mass spectrometry provides a powerful analytical capability for the analysis of nonmetallic particulate contamination.

EXPERIMENTAL

One of the first steps in chemical analysis is sample preparation, or in this case, sample retrieval. A solvent flush with hexane or petroleum ether is one of the more expedient ways of collecting particles. This technique, however, is limited to the surface of a material that is compatible with the solvent and to a surface that can be positioned in such a manner that the hexane and particles can be retrieved.

An Adhesive tape, SPV-224 made by Nitto Demco, provides an excellent means for removing particles from surfaces which have coatings that are too delicate for use of a pick and forceps or for surfaces which are not compatible with solvents. Another positive feature of this technique is that the tape leaves no significant adhesive film on the surface which makes it particularly good for removing particles from optics. This was shown experimentally by analyzing the hexane rinse from a glass surface where strips of the tape had been placed and then removed. Similarly, it leaves no significant interfering adhesive on the particles to be analyzed.

The third particle retrieval technique employs the use of a fine pointed probe and/or fine tipped forceps. Frequently the most appropriate way to collect particulate contamination for analysis is to transfer particles from the contaminated surface to the analysis matrix with a fine-tipped probe or forceps. Even in cases when particles are collected by the solvent flush or adhesive tape techniques described above, it is often necessary to isolate specific particles or particle types under the microscope using this technique.

Once the particles are isolated, one method of analysis is FT-IR. To prepare the particles for FT-IR analysis they are placed in a case-hardened stainless steel mortar and ground into finer particles. Potassium bromide (KBr) is then added and ground with the sample. Potassium bromide is used because it is transparent to infrared energy for the wavelength of interest. The sample, along with KBr, is placed in a die and pressed under approximately 7×10^8 N/m^2 of pressure. A 4 mm KBr pellet is then obtained and placed in the infrared beam of the spectrometer. The collected data are then processed and provide a unique fingerprint or spectrum of the particles.

The FTIR system includes an interferometer, a 20 bit computer to process the data and a 160 megabyte storage module where the spectral libraries are stored. There are about 20,000 spectra divided into 6 different libraries. The library most frequently used is our user-prepared library which has about 500 spectra. This library is most often used in identifying unknown compounds because it is composed of materials most frequently used in the aerospace industry and, more specifically, our company.

Figure 1 shows a spectrum of particles removed from a Unified Fuel Control filter of a fighter aircraft. This fuel filter is crucial to the maximum performance of the aircraft. It is a 35 μm filter that filters fuel which is used to hydraulically control the engine throttle. Although this filter was supposed to be self washing, fine particles were restricting the flow of fuel resulting in a spongy and inadequate throttle of the engine.

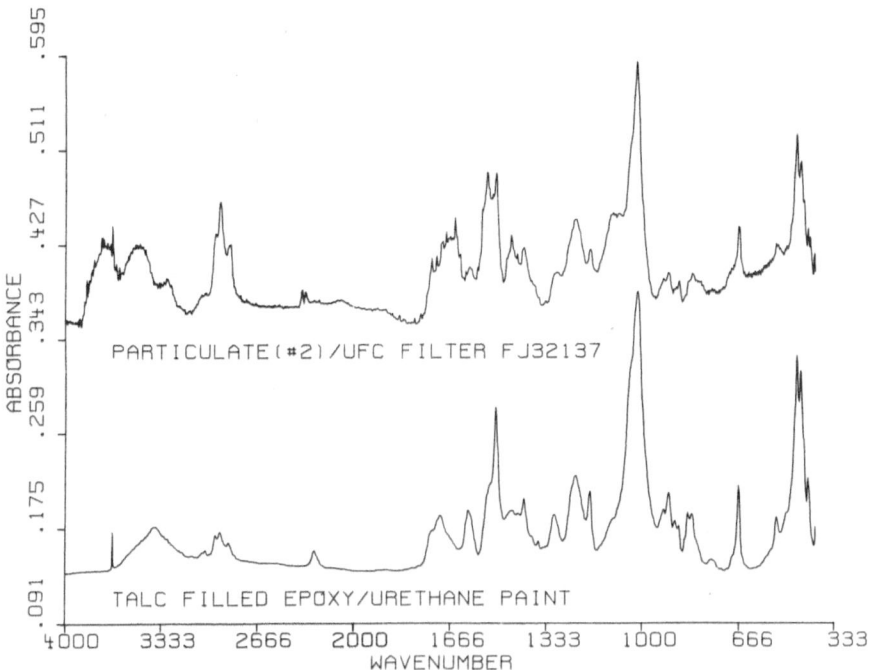

Figure 1. FT-IR spectrum of particles removed from a Unified Fuel Control filter and a library reference spectrum of talc filled Epoxy/Urethane Paint.

A spectral search of the particles identified them as primarily talc with some epoxy and urethane as shown by the reference spectrum in Figure 1. It was determined that the source of these particles was a talc-filled polyurethane/epoxy paint. The particles resulted from paint overspray in the fuel tank during manufacturing.

Once the particles and source were identified, the problem was corrected by modifying the filter and manufacturing process.

A spectrum of a particle removed from a space laser communication hardware component is shown in Figure 2. Particles left on this hardware prior to launch can migrate to optically critical locations resulting in dramatic loss in laser power.

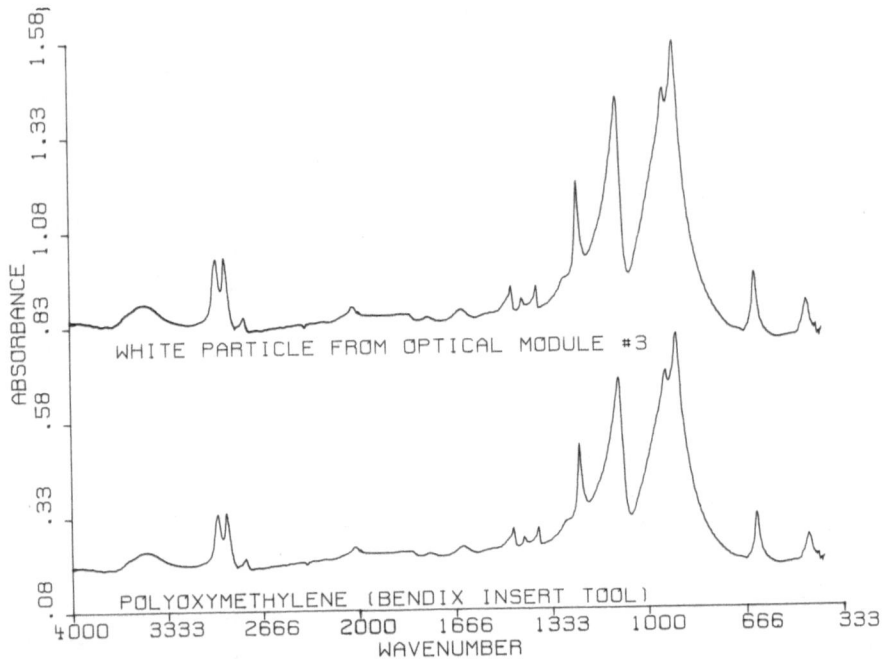

Figure 2. FT-IR spectrum of a particle removed from a space laser communication hardware component and a library reference spectrum of polyoxymethylene.

A spectral search identified the particles as polyoxymethylene as shown by the reference spectrum in Figure 2. The source of these particles was removed by modifying the use of the installation tool.

In the two examples of particle analysis just described, the samples were analyzed using the conventional FT-IR/KBr pellet technique. With this technique it was desirable to have a particle of 100 μm in size or a composite of several smaller size particles to provide sufficient sample for a good quality spectrum. With the development of the infrared microscope, it is no longer necessary to have a 100 μm size particle. A good quality spectrum can be obtained with a 10 μm particle. An added benefit is that potassium bromide sample preparation is not required on most particles.

Our laboratory has recently acquired a Spectra-Tech, Inc IR-PLAN II infrared microscope accessory, and attached it to our existing FT-IR Spectrometer. Figure 3 shows a spectrum of a 30 µm foreign particle removed from the Dacron braiding of a multi-conductor wire cable. This spectrum was obtained with the use of the microscope attachment. A spectral search identified the particle as polycarbonate as shown by the reference spectrum in Figure 3.

An additional feature of the infrared microscope is that the infrared beam is focused on the same particles through the microscope that the spectroscopist views. This not only increases the capability of identifying smaller particles but also reduces analysis time because of less sample preparation.

Since most nonmetallic particulate contamination is relatively non-volatile, FT-IR is the method of choice for most analyses. However, probe mass spectrometry offers a unique analytical technique for comparison of organic polymers. The technique of probe mass spectrometry involves placement of unknown particulate material into a glass or quartz tube which is then inserted into the tip of the probe. For particulate contamination which is of sufficient size, fine pointed forceps are used to place the particles in the probe tip. For very fine particulate contamination, the particles can be placed in a solvent and a small amount of the solvent is then allowed to evaporate in the probe

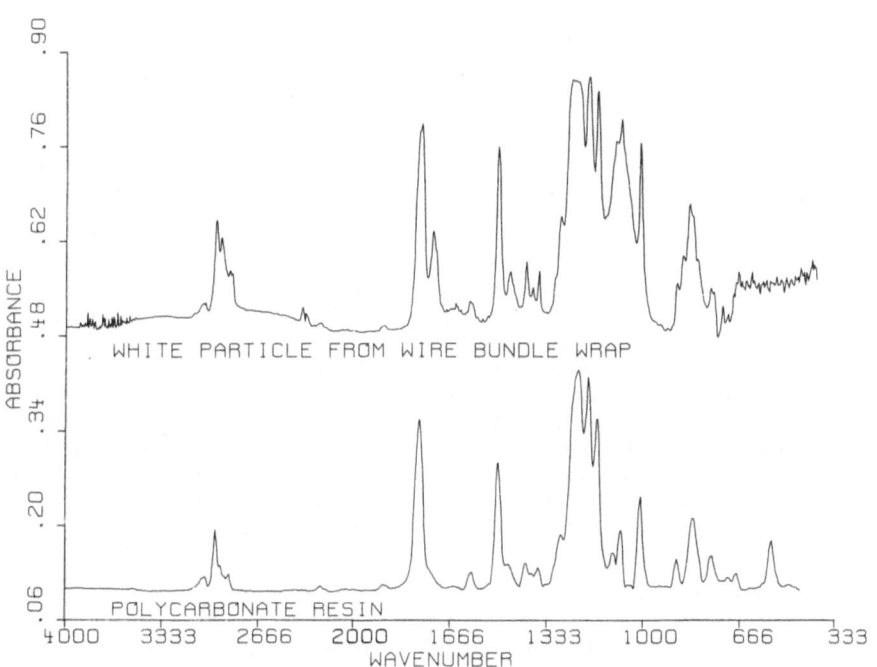

Figure 3. FT-IR spectrum of a 30 µm particle removed from an electrical wire bundle braiding and a library reference spectrum of polycarbonate resin.

tip. This technique is especially useful for fine particles which have been filtered from a liquid solution.

Following sample preparation the probe is inserted into the mass spectrometer through a vacuum lock. The probe is then heated to a defined temperature (350°C maximum) at a programmed rate.

During this heating cycle the mass spectrometer is set for continuous scanning over a defined mass range. Probe analyses of organic polymers produce spectra showing various organic components such as residual monomers, additives, plasticizers, and residual processing aids. In addition, depending on the polymer and the heat employed, various breakdown components may also be observed. This type of analysis produces a unique fingerprint pattern.

Figure 4 shows a probe mass spectrum of particles filtered from fuel taken from an aircraft that was experiencing problems with clogging fuel filters. Initial analysis by FT-IR indicated spectral similarity between the particles and some gaskets and O-rings used in the fuel system. Because of the high carbon content of the particles, which blocks infrared energy, a specific identification of the particles could not be made by FT-IR. It was important to determine if the suspect gasket and O-ring materials were showing some incompatibility or degradation with the fuel system. Each suspect material was analyzed by probe mass spectrometry under the same conditions as the contaminant particles. Each of the suspect materials produced uniquely different fingerprint spectra when analyzed by probe mass spectrometry showing the particles did not come from the gasket or O-ring material. An example of this spectral difference is shown in Figure 5. In this example the source of the particles was not determined. However, it was most essential to determine that the particles were not originating from the degradation of the gasket or O-ring material. Although particulate contamination is generally considered to be solid material, the presence of liquid material in the form of microdroplets can also be classified as particulate contamination. An example of this type of contamination is the condensation of a vapor on cooled optics which have been placed under vacuum. Various components present in an optic system may release volatiles when exposed to vacuum conditions. The presence of cooled optics in the system results in a condensation of these materials on the optic surface. Because the optics are frequently mounted with special adhesives, the use of solvents to remove and analyze the contamination is sometimes undesirable. Probe mass spectrometry has aided in the analysis of this type of contamination. The tip of the probe is gently touched to the optic surface or an adjacent cooled area. A small amount of the contamination is transferred to the probe for analysis, thus avoiding the use of solvents which may have an adverse effect on the adhesives. Figure 6 shows a spectrum of a trace amount of hydrocarbon removed from an optical surface. This technique is substantially more sensitive than infrared for this type of contaminant.

RESULTS AND DISCUSSION

Analytical instrumentation has made dramatic advancements in recent years. FT-IR with its speed and signal/noise advantage over dispersive

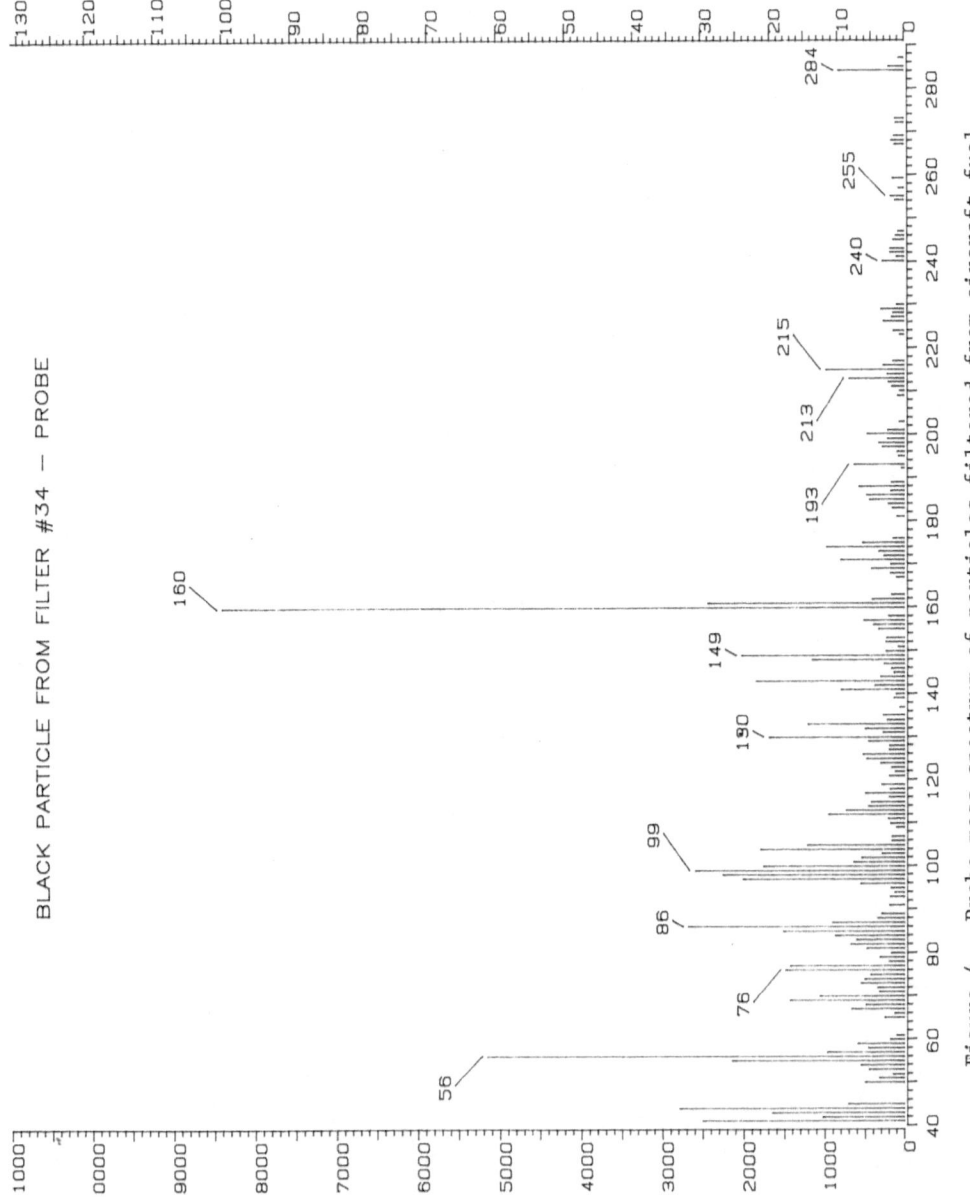

Figure 4. Probe mass spectrum of particles filtered from aircraft fuel.

177

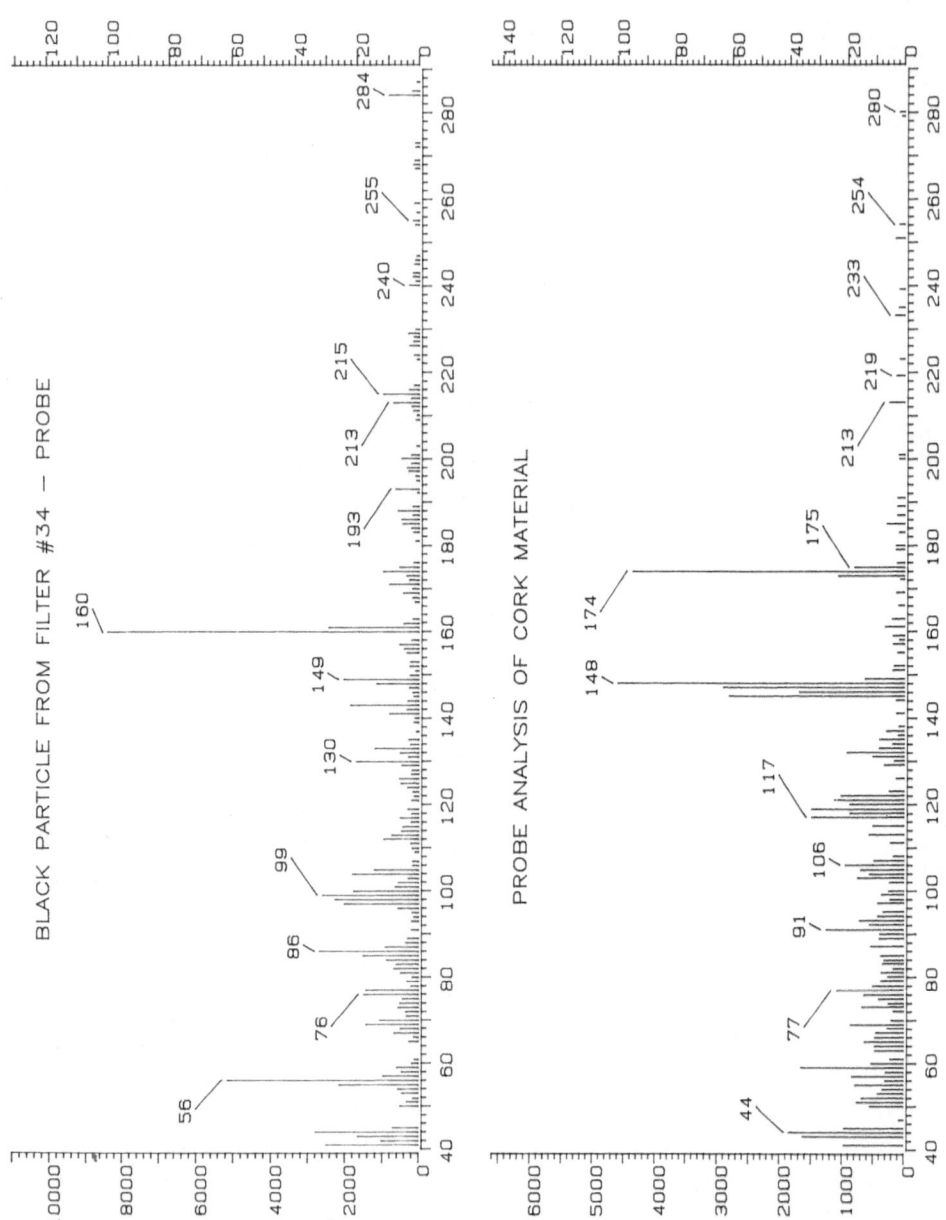

Figure 5. Probe mass spectrum of particles filtered from aircraft fuel and reference mass spectrum of O-ring material.

OPTICAL LENS — PROBE ANALYSIS

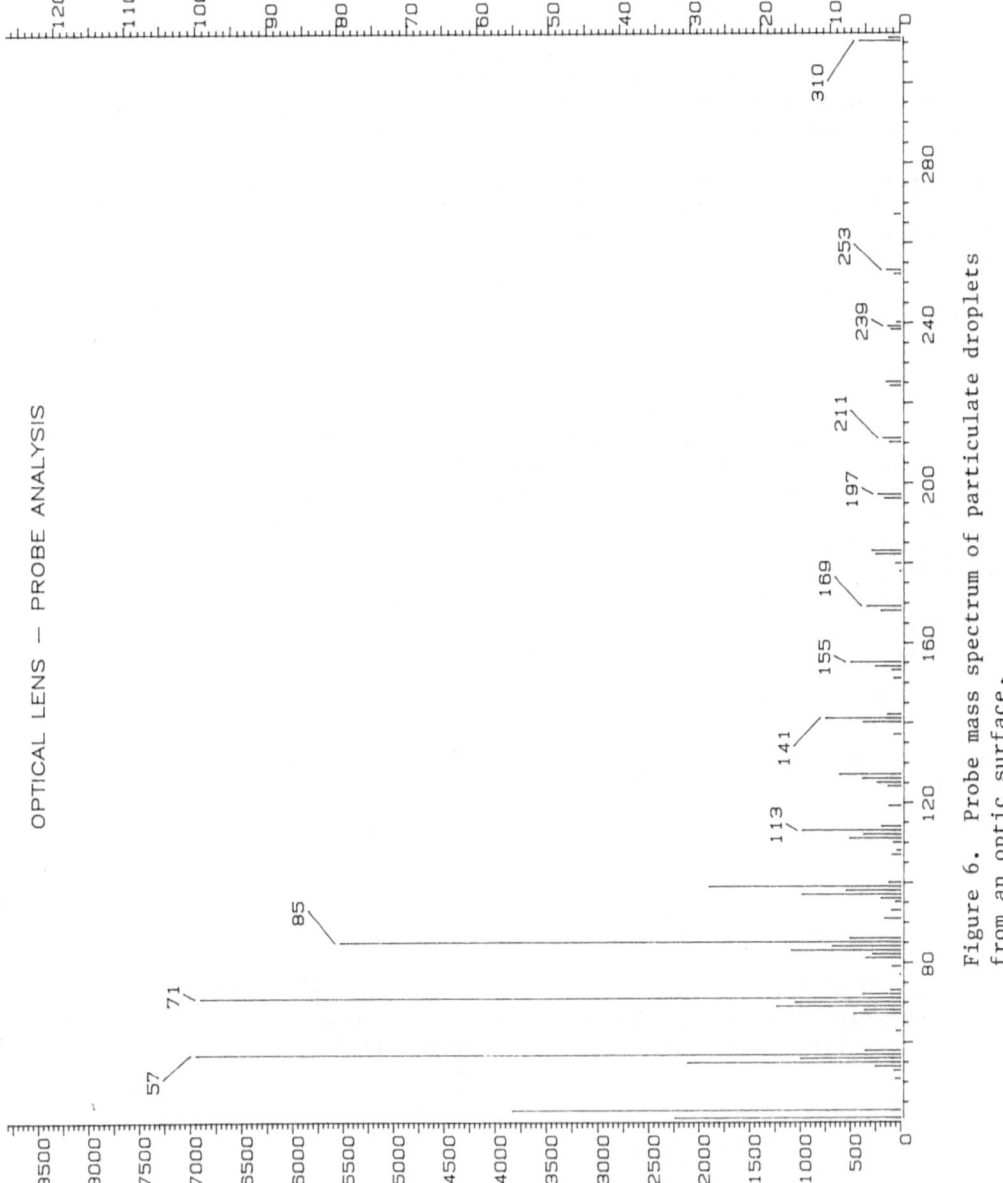

Figure 6. Probe mass spectrum of particulate droplets from an optic surface.

infrared has improved the chemist's ability to identify nonmetallic particulate contamination. The addition of the infrared microscope has enhanced the chemist's analytical ability even further, to the extent of identifying single particles with nanogram quantities of material. With the advances in capability, however, certain analytical techniques and precautions have become more important, such as using the highest quality solvents or reagents and constant awareness of possible background contamination.

The work described in this paper has shown how these advanced capabilities can be used as an analytical tool for solving problems related to nonmetallic particulate contamination. This work has also shown the use of probe mass spectrometry as an additional tool for "fingerprinting" particulate contamination. The combination of these analytical tools greatly enhances one's capabilities to identify and characterize particulate contamination. Although these techniques may not always provide specific identification, comparison between known materials and the particulate contamination can be vitally important in helping to solve manufacturing and operating problems.

SUMMARY

As long as we live in a world of matter and materials, we will have to cope with particulate contamination and the problems it causes. In most cases particulate contamination can not be completely eliminated but it can be controlled if it and its source are identified. The ultimate goal is to eliminate all particulate contamination, but with current technology that is essentially impossible or economically impractical. A more attainable goal is to eliminate intolerable particulate contamination and control that which cannot be eliminated. Infrared spectroscopy and probe mass spectrometry are two of the techniques available which can be used to achieve that goal.

REFERENCES

1. J.T. Vondeber, Chairman; D.G. Anderson, J.K. Duffer, J.M. Julian, R.W. Scott, T.M. Sutcliff and M.J. Vaickus, "An Infrared Spectroscopy Atlas for the Coating Industry", pp. 17-18, Federation of Societies for Coatings Technology, Philadelphia, Pennsylvania, 1980.

2. J. Throck Watson, "Introduction To Mass Spectrometry", 2nd ed., pp. 140-143, Raven Press, New York, 1985.

3. D.O. Hummel and F. Scholl, "Atlas of Polymer and Plastics Analysis" 2nd ed. pp. 13, 139-140, Verlag Chemie International, Deerfield Beach, Florida, 1981.

SURFACE PARTICLE INSPECTION PLANS IN SEMICONDUCTOR MANUFACTURING

Zinovy Fichtenholz

Siliconix Corporation
2201 Laurelwood Road
Santa Clara, California 95054

Leon L. Pesotchinsky

Department of Mathematics and Computer Science
San Jose State University
One Washington Square
San Jose, California 95192-0103

The review of two existing inspection policies for estimation of particle densities has shown that the policy which reduces the number of inspected wafers leads to a biased density estimator. In the present paper, a class of sequential inspection policies with unbiased density estimators is proposed. These policies reduce the number of inspected wafers. In many practical situations the variances of density estimators for these sequential policies are smaller than the variances of the estimators for the standard existing policies.

INTRODUCTION

In many problems involving surface inspection the standard operation policy calls for the inspection of a fixed (e.g., unit) surface area "collected" from a number of sampled units. One of the common examples is wafer surface inspection in semiconductor manufacturing, which is the most important step in determining spatial particle density. In most schemes an operator inspects a total of one unit of surface area accumulated over a few wafers from a batch of 20 to 50 wafers processed simultaneously, and uses the cumulative number of particles as an estimate of particle density. Inspection wafers and locations are chosen according to some blocking arrangement in order to eliminate (or at least to reduce) the influence of the order of processing and of the surface spatial patterns on the particle counts on individual wafers. The location "blocks" may specify center, top, right, left and bottom of the wafer, and processing "blocks" may prescribe selection of inspection wafers from different parts, say, of a wafer boat.

However, the inspection itself is one of the major particle sources, and in contemporary "super-clean" manufacturing environment it may indeed be responsible for more than doubling the number of defects on a wafer.

It is also known that any increase in the wafer fabrication cycle time has a detrimental effect on yield[1,3] (both these phenomena elude accurate quantitative description in large part because any review/inspection step will only increase the number of defects). This explains why the major semiconductor houses, especially in Japan, put such a great emphasis on shorter cycle times and on "no-to-low" wafer inspection. On the other hand, many of the processes in the semiconductor industry are not yet established and automated at levels which may allow for a "no inspection" manufacturing. As an example, one may consider a traditional develop inspection which, as one of the process steps, could prevent a destruction of several wafer lots caused by a .5µm misalignment of a critical layer (e.g., due to a temporary equipment malfunction). Thus, the "no inspection" processes are highly desirable but will become economically feasible only when the process cleanliness and equipment can guarantee extremely low misprocessing levels (or if the material and labor costs become smaller than the cost of inspection steps). There have been significant achievements (cleaner and more automated fabs, better engineered processes, etc.) along the way to the "no inspection" goal, but for the time being the inspection cannot be totally abolished and should remain as a process monitoring tool and as an indicator of how close the manufacturing is to this goal. Quite naturally, the wafer inspection should get less and less extensive as the process yields improve, which leads to the problem of developing economical and reliable inspection methods and techniques.

In particular, the common sense suggests that the number of inspected wafers should decrease along with the decrease in the particle density (although the total inspected area will remain the same). Unfortunately, the processes employed in the fabrication of integrated circuits are very sensitive and frequently demonstrate an excessive batch-to-batch and wafer-to-wafer variability. This, along with rigid internal quality requirements, makes choice of a small fixed number of inspection wafers impractical and rather risky.

Two inspection policies have been widely accepted in manufacturing: one relies upon inspection of equal areas on a fixed large number W of wafers, and the other is based upon a sequential approach with the decision to continue inspection on the same wafer (or to shift to another) depending on the order of inspected wafer and on the location of a found defect.[3] In both policies only N specified die locations (numbered 1,2,...,N) with the total area of one unit are inspected. Under the first (so-called fixed) policy, N/W locations are inspected on each of W wafers. Under the second (variable) policy[2,3], inspection shifts from the particle location $x(i)$ on the ith wafer to the location $x(i)+1$ on the next i+1st wafer and continues until a particle is found at a location $x(i+1)$. Then inspection shifts to the location $x(i+1)+1$ on the next i+2nd wafer, and so on. If no particles are found on the i+1st wafer then the inspection is terminated on this wafer at the last location N (and the cumulative particle count is i); on the last Wth wafer the inspection continues through all the remaining locations. Thus, under the variable policy the small particle counts reduce the number of inspected wafers, whereas, in general, this number is bounded by the same value W as under the first policy.

Particle density estimates obtained through both inspection policies have been virtually identical for the similar processes in older fabrication areas (fabs). However, in the modern "super-clean" fabs the estimates have not been in good agreement. The explanation lies in the fact

that unlike the fixed policy, the variable inspection policy yields an unbiased particle density estimator only for a Poisson particle distribution. The bias for any non-Poisson distribution is negligible when the
particle density, μ, is large compared to the number of inspected wafers
(the situation in the older fabs); and thus the bias of the estimator
could not have been detected even when the Poisson distribution model was
not valid. There is a good reason to believe that in contemporary semiconductor manufacturing environment (limited operator participation,
total laminar flow, etc.) the behavior of a few remaining particles is
"structured" more than allowed under Poisson spatial distribution, thus
introducing a noticeable bias under the variable inspection policy.[2]

We propose a probabilistic particle distribution model based upon
arrival times in a counting process. This allows us to use some known
results on the distribution of arrival times for the quantitative analysis
of the inspection procedures and for the development of new classes of
inspection policies.

Probabilistic Model of Particle Surface Distribution

Without much loss of generality, we assume that N's are sufficiently
large (which is always true in practice) so that there may be no more
than one particle at each die (inspection) location, and that k particles
at the locations $x(1) < x(2) < \ldots < x(k)$ on the same wafer can be viewed
as an ordered sample $X(1) < X(2) < \ldots < X(k)$ from a continuous distribution on $(0,1)$, $X(i) \approx x(i)/N$. (If k=0 we agree to set $X(0)=1$.) The
same sequence of $X(i)$'s can also be treated as arrival (waiting) times of
a counting process. In particular, under a commonly assumed Poisson spatial particle distribution the values $X(1), \ldots, X(k)$ become arrival times
in a Poisson process with the rate equal to the particle density μ (so
that the intervals between arrival times have negative exponential distribution with parameter μ). It is important to recall that given the
number of arrivals, k, the Poisson arrival times have the same joint distribution as the order statistics corresponding to k independent random
variables uniformly distributed on the interval $(0,1)$.

SEQUENTIAL INSPECTION POLICIES WITH UNBIASED ESTIMATORS

Let Y_i denote the number of arrivals in the ith subinterval of
length $1/W$ under the fixed policy (Y_i = number of particles on the inspected dies of the ith wafer when N/W dies are inspected on each of W
wafers), and let M_i be as before the number of arrivals in the ith interval $(0,1)$, so that given $M_i=m$, Y_i is binomial $B(m, p=1/W)$, $i \leq W$. It
is easy to show that $E\{Y_i\} = \mu/W$ and

$$\text{Var} Y_i = (\mu/W)(1-1/W) + \sigma^2/W^2 = \mu/W + (\sigma^2-\mu)/W^2 \qquad (1)$$

where μ and σ^2 are, respectively, mean and variance of M_i. We also introduce a variable $S(T)$ which corresponds to an inspection of a total area
of 1-a square unit over a random number T of wafers with $(1-a)N/T$ dies
inspected on each of them. Here a is a constant which shows the area not
yet inspected, $1 > a \geq 0$, and T is an integer random variable which shows
how many wafers have been inspected. It is convenient to introduce variables Z_i, which stand for the number of particles on the inspected area of
the ith wafer, so that the conditional distribution of Z_i given $M_i = m$
and $T = t$ is binomial $B(m,(1-a)/t)$. Variables Z_i are independent,
$1 \leq i \leq t$, and it is clear that $S(T)$ is the sum of them:

$$S(T \mid T=t) = \Sigma_{i \leq t} Z_i.$$

In the following we may consider the fixed policy estimator as a special case of $S(T)$ when $P\{T=W\} = 1$ and $a=0$, and use notation $S(W)$. One can easily see that

$$E\{S(T)\} = (1-a)\mu;$$
$$\text{Var} S(T) = (1-a)\ \mu(1-(1-a)\ E\{1/T\}) + (1-a)^2\sigma^2 E\{1/T\} \qquad (2)$$
$$= (1-a)\mu + (1-a)^2\ (\sigma^2-\mu)E\{1/T\}$$

We may notice from (1) and (2) that the number of inspected wafers should be small when $\sigma^2 < \mu$ and large when $\sigma^2 > \mu$. Quite appropriately for Poisson particle distribution ($\sigma^2=\mu$) the variance of the estimator does not depend on the number of inspected wafers.

Since the parameters and type of the particle distribution are frequently not known, we may consider a class of inspection plans in which at the first stage N/W dies are inspected on each of r ($r \geq 1$) wafers, and the number $T \geq 1$ of additional wafers (as a function of the observed particle counts $Y_1,\ldots,Y_r$) for the second stage inspection is determined. To keep the total inspected area equal to one unit, $(1-r/W)N/T$ dies are inspected on each wafer at the second stage (it should be mentioned that with T given the counts Z_j's from the second stage are independent of counts Y_i's from the first stage). The cumulative particle count serves as an estimator for μ. Let $S(r,W)$ denote the cumulative particle count at the first stage; we have $E\{S(r,W)\} = \mu r/W$, and $\text{Var} S(r,W) = r\text{Var} Y_i$. The total cumulative count $C = S(r,W) + S(T)$, and it is easy to see that $E\{C\} = \mu$ and $E\{S(r,w)S(T)\} = E\{S(r,w)\}E\{S(T)\}$ (although $S(r,W)$ and $S(T)$ are dependent). Thus,

$$\text{Var} C = \mu(1-r/W^2) + \sigma^2 r/W^2 + (1-r/W)^2\ (\sigma^2-\mu)E\{1/T\} \qquad (3)$$

It is useful to notice that $\text{Var} C \geq \min\{\sigma^2,\mu\}$;
$\min \text{Var} C = \mu + (\sigma^2-\mu)/W$ when $\sigma^2 > \mu$ and $r+T = W$ and
$\min \text{Var} C = \sigma^2$ when $\sigma^2 < \mu$ and $r=0$, $T=1$.

Let $\Omega(r,W)$ be the class of the inspection policies (that is, variables T) defined as above. We may introduce in $\Omega(r,W)$ an optimal T as the one which minimizes $\text{Var} C$ over all T from $\Omega(r,W)$. This makes sense when variable T is bounded so that $P\{T+r \leq W\} = 1$, or at least when $E\{T\}+r \leq W$. The first class (bounded T) seems to be more realistic because from practical considerations (throughput, etc.) the total number of inspected wafers cannot exceed certain predetermined value. We assume that either of these restrictions on T holds for the policies from $\Omega(r,W)$. One can see from (9) that for $\sigma^2 \geq \mu$ the minimum of $\text{Var} C$ corresponds to the minimum of $E\{1/T\}$ which is delivered by constant $T = W-r$ (this obviously defines the fixed policy). For $\sigma^2 \leq \mu$ the minimum of $\text{Var} C$ corresponds to the smallest T so that $P\{T=1\} = 1$. We may also consider a class of bounded variables T ($t_1 \leq T \leq t_2$) in which case the minimum to the $\text{Var} C$ will be delivered by $T=t_2$ when $\sigma^2 \geq \mu$ and by T taking on both extreme values when $\sigma^2 < \mu$. Thus, in the class of sequential procedures $\Omega(r,W)$ the cumulative particle count $C = \Sigma\ Y_i + \Sigma\ Z_k$ is an unbiased estimator of particle density μ. If the variance σ^2 of the distribution $r(x)$ of the number of particles M on a unit surface area is larger than its mean μ, then the fixed policy is optimal in $\Omega(r,W)$; if $\sigma^2 \leq \mu$ and $w=W$, then the optimal policy minimizes the number of wafers inspected at the second stage by choosing $T=1$.

It is clear that in most cases we do not know which of the two values σ^2 or μ is larger; thus the choice of r and W for the first stage of the inspection should address this question in addition to helping us to keep the number of inspected wafers low when μ is small. The latter consideration is the only one we can take into account when $W \leq 3$ because to compare estimates of σ^2 and μ we need at least r=2, and since it is assumed that $T \geq 1$ all three wafers would have to be inspected in either case. In the first of the following two examples of sequential policies, we consider W=3 thus limiting the choice of r and w to r=1 and w=2 or 3. In the second example, we review the policy which tests whether $\sigma^2 \leq \mu$ on the basis of the inspection of the first two wafers (r=2), and then depending on the result of the test either calls for the inspection of all of the remaining W-2 wafers(W $\geq$ 4), or for the inspection of one additional wafer. This latter policy can be easily generalized for r > 2.

Policy With No More Than Three Inspected Wafers

Table I contains expected values of inspected wafers E{T} + 1 and variances of both estimators S(W) and C for the Poisson P[μ], and uniform U[μ-1, μ+1], U[0,2μ] and U[μ] particle distributions for small values of μ. The type of sequential policy used in the table is defined by parameters W=3, w=2 or 3, r=1, and by the threshold value y which directs the choice of one or two wafers for the second stage:

$$
T(Y_1) = \begin{cases} 1 & \text{if } Y_1 = 0,\ 1,\ \ldots,\ y \\ 2 & \text{if } Y_1 \geq y + 1 \end{cases}
\tag{4}
$$

One can notice that for large μ this policy will almost certainly coincide with the fixed policy because $P\{T(Y_1) = 2\} = P\{Y_1 \geq y+1\} \to 1$ as $\mu \to \infty$. In the table we use value y=1; increase in the threshold value will make it more adaptable to larger values of μ but will not change the asymptotics.

Table I. Comparison of Unbiased Estimators for a Sequential and Fixed Inspection Policies, W=3

| | | Sequential | | | | Fixed |
| | | w=2 | | w=3 | | |
f	μ	E{T}+1	VarC	E{T}+1	VarC	VarS(W)
U[0.2μ] uniform on	1	2.083	.837	2.037	.818	.889
0, 1, ..., 2μ	2	2.288	2.000	2.156	2.000	2.000
$\sigma^2=(\mu^2+\mu)/3$	3	2.449	3.444	2.281	3.493	3.333
	4	2.560	5.147	2.390	5.251	4.889
	5	2.638	7.102	2.478	7.247	6.667
	6	2.693	9.307	2.549	9.469	8.667
	∞	3.000	1.000*	3.000	1.000*	
U[μ-1, μ+1] uniform on	1	2.083	.837	2.037	.818	.889
μ-1, μ, μ+1	2	2.250	1.375	2.123	1.296	1.556
$\sigma^2 = 2/3$	3	2.479	1.973	2.259	1.838	2.222
	4	2.667	2.611	2.402	2.446	2.889
	5	2.797	3.265	2.532	3.105	3.556
	6	2.880	3.927	2.642	3.797	4.222
	∞	3.000	1.000*	3.000	1.000*	
U[μ]	1	2.000	.750	2.000	.444	.667
f(μ) = 1	2	2.250	1.438	2.111	1.136	1.333
$\sigma^2 = 0$	3	2.500	2.063	2.259	1.835	2.000
	4	2.688	2.656	2.407	2.535	2.667
	5	2.813	3.244	2.539	3.231	3.333
	6	2.891	3.832	2.649	3.922	4.000
	∞	3.000	1.000*	3.000	1.000*	
P[μ]	1	2.090	1.000	2.045	1.000	1.000
Poisson	2	2.264	2.000	2.144	2.000	2.000
$\sigma^2 = \mu$	3	2.442	3.000	2.264	3.000	3.000
	4	2.594	4.000	2.385	4.000	4.000
	5	2.713	5.000	2.496	5.000	5.000
	6	2.801	6.000	2.594	6.000	6.000
	∞	3.000	1.000*	3.000	1.000*	

*This value is the limit of ratio VarC/VarS(W) as $\mu \to \infty$

Policies With Three or More Inspected Wafers

Inspection of two or more wafers at the first stage ($r \geq 2$) provides some information about the relationship between mean and variance of the particle distribution (we assume that the variance exists). This may be used to decide on the number T of additional wafers inspected at the second stage. One of the simplest policies may require testing hypothesis $H_0 : \sigma^2 \leq \mu$ against alternative $H_1 : \sigma^2 \geq \mu$ and choosing $T = W-r$ (that is, the maximum possible number of wafers) when H_0 is rejected and $T=1$ otherwise.

Let us consider $r=2$ and introduce statistics

$$U = (Y_1-Y_2)/\sqrt{(Y_1+Y_2)}, \text{ and } L = U^2 = (Y_1-Y_2)^2/(Y_1+Y_2),$$

where variables Y_1, Y_2, given $M_1 = m_1$, $M_2 = m_2$, $p = 1/W$, are, respectively, binomial $B(m_1,p)$ and $B(m_2,p)$. Since $E(Y_1+Y_2) = 2\mu p$ and $E\{(Y_1-Y_2)^2\} = 2\mu p + 2(\sigma^2-\mu)p^2$, we may expect that when H_0 is correct (that is, when $\sigma^2 \leq \mu$), the quantity L is somewhat close to 1. More precisely, if M_i's have Poisson $P(\mu)$ distribution, then Y_i's are Poisson $P(\mu p)$ variables, and as $\mu \to \infty$ $(Y_1-Y_2)/\sqrt{2\mu p}$ converges in distribution to a standard normal variable and $(Y_1+Y_2)/2\mu p$ converges to 1 in probability. Thus, U is asymptotically standard normal and L is Chi-square with one degree of freedom (the approximation is quite good as long as $W=4$ ($p=.25$) and $\mu \geq 10$). This may be used to test H_0 for larger (≥ 10) values of μ; whereas, for small μ's the distribution of U (or L) can be computed.

In general case, $(Y_1+Y_2)/2\mu p$ still converges to 1 and, even though numerator and demoninator in U are dependent, the asymptotic distribution of U coincides with that of $(Y_1-Y_2)/\sqrt{2\mu p}$ if the latter exists. We can write

$$(Y_1+Y_2)/\sqrt{2\mu p} = (M_1-M_2)p/\sqrt{2\mu p} + (Y_1-M_1 p)/\sqrt{2\mu p} - (Y_2-M_2 p)/\sqrt{2\mu p} \qquad (5)$$
$$= \{(M_1-M_2)/\sqrt{2\mu}\}\sqrt{p} + V_1 - V_2$$

Variables $V_i = (Y_i-M_i p)/\sqrt{2\mu p}$ are independent and, given $M_i=m_i$, asymptotically normal with means $m_i\sqrt{p}/\sqrt{2\mu}$ and variances $m_i q/2\mu$. But M_i/μ converges to 1 as $\mu \to \infty$, and with $\sigma^2 = \mu$ we may show that V_1-V_2 is asymptotically normal with mean 0 and variance q: $V_1-V_2 \approx Z\sqrt{q}$, where Z denotes a standard normal variable. Under the same conditions M_1-M_2 and V_1-V_2 are asymptotically independent so that if $(M_1-M_2)/\sqrt{2\mu}$ converges in distribution to, say, variable $K(E\{K\} = 0, Var\{K\} = 1)$, then U converges in distribution to $K\sqrt{p} + Z\sqrt{q}$, and for large μ

$$U \approx K\sqrt{p} + Z\sqrt{q} \qquad (6)$$

If K is normal, then $U \approx Z$ (in distribution); however, this is not the case for some of the models considered earlier. For instance, a uniform integer distribution with $\mu \approx \sigma^2$ is concentrated on interval $[\mu-\sqrt{3\mu}, \mu+\sqrt{3\mu}]$, and limiting distribution for $(M_1-M_2)/\sqrt{2\mu}$ is triangular on interval $[-\sqrt{6},\sqrt{6}]$.

We can use (6) to computer power function $P\{U>u_\mu\}$ of a test of H_0 based on the critical region $U>u_\mu$ for various distributions of M_i and fixed values of σ^2,μ. Of course we would like the power (i.e., the probability that $U>u_\mu$) to be "small" when $\sigma^2 \leq \mu$ no matter what is the real distribution of defects, and we would like the power to be "large" if $\sigma^2 \geq \mu$. This would make the choice of small number of inspected wafers more likely when $\sigma^2 \leq \mu$, and vice versa. For instance, we may notice that if $(M_i-\mu)/\sqrt{\mu}$ converges in probability to zero (as in case of uniform $U[\mu]$ and $U[\mu-1,\mu+1]$ distributions), then $U \approx Z\sqrt{q}$ and the tail probabilities of U can be computed using the standard normal cumulative distribution function (cdf) $\Phi(x)$:

$$P\{U > u_\mu\} \approx 2(1-\phi^{-1}(u_\mu/\sqrt{q})) \qquad (7)$$

For a uniform on interval $[0,2\mu]$ distribution and for large μ

$$P\{U > u_\mu\} \approx (1-u_\mu/\sqrt{2\mu})^2 \qquad (7$$

and for a uniform on interval $[\mu-\sigma\sqrt{3},\ \mu+\sigma\sqrt{3}]$ with variance σ^2

$$P\{U > u_\mu\} \approx (1-u_\mu(\sqrt{\mu}/\sigma)/\sqrt{6})^2 \qquad (7")$$

In Table II we compare VarC, VarS(W) and expected number of inspected wafers for the same values of μ and for most of the distributions used in previous examples, under the policy in which W=4, r=2, and

$$T(Y_1,Y_2) = \begin{cases} 1 & \text{if } U > u_\mu \\ 2 & \text{if } U > u_\mu \end{cases} \qquad (8)$$

In other words, under this policy we test $H_o:\sigma^2 \leq \mu$ and choose T based on whether we reject or do not reject H_o. Values of u_μ are chosen in such a way that the size (i.e., the probability) of the critical region of the nonrandomized test defined by (8) under Poisson distribution $P[\mu]$ is as close to 5% as possible. (It can be estimated from (5) that for large μ and an arbitrary distribution in H_o the size of the critical region does not exceed $1/(u_\mu)^2 \approx 26\%$.)

Table II. Comparison of Unbiased Estimators for a Sequential and Fixed Inspection Policies, W=4

f	μ	u_μ	$P\{U>u_\mu\}$	Sequential E{T}+2	VarC	Fixed VarS(W)
U[0.2μ] uniform on	1	1.42	3.21%	3.03	.876	.917
0, 1, ..., 2μ	2	1.51	.46	3.00	2.000	2.000
$\sigma^2 = (\mu^2+\mu)/3$	3	1.64	4.80	3.05	3.369	3.250
	4	1.74	9.10	3.09	4.970	4.667
	5	1.74	11.74	3.12	6.866	6.250
	6	1.89	6.33	3.06	8.937	8.000
	∞	1.96	100.00	4.00	1.000*	
U[μ-1, μ+1] uniform on	1	1.42	1.61%	3.02	.876	.917
μ-1, μ, μ+1	2	1.51	.30	3.00	1.501	1.667
$\sigma^2 = 2/3$	3	1.64	.96	3.01	2.128	2.417
	4	1.74	1.84	3.02	2.758	3.167
	5	1.74	2.64	3.03	3.389	3.917
	6	1.89	.77	3.01	4.005	4.667
	∞	1.96	2.40	3.02	.837*	
U[μ]	1	1.42	.00%	3.00	.625	.750
f(μ) = 1	2	1.51	.00	3.00	1.250	1.500
$\sigma^2 = 0$	3	1.64	1.32	3.01	1.875	2.250
	4	1.74	3.21	3.03	2.516	3.000
	5	1.74	4.91	3.05	3.156	3.750
	6	1.89	1.35	3.01	3.750	4.500
	∞	1.96	2.40	3.02	.837*	
P[μ]	1	1.42	4.13%	3.04	1.000	1.000
Poisson	2	1.51	1.74	3.02	2.000	2.000
$\sigma^2 = \mu$	3	1.64	3.90	3.04	3.000	3.000
	4	1.74	5.95	3.06	4.000	4.000
	5	1.74	7.74	3.08	5.000	5.000
	6	1.89	3.24	3.03	6.000	6.000
	∞	1.96	5.00	3.05	1.000*	
V[0, 2μ]	1	1.42	4.88%	3.05	1.000	1.000
f(0) = f(2μ) = $\frac{1}{2}$	2	1.51	3.34	3.03	2.242	2.500
$\sigma^2 = \mu^2$	3	1.64	10.08	3.10	5.241	4.500
	4	1.74	18.74	3.19	8.359	7.000
	5	1.74	25.95	3.26	12.176	10.000
	6	1.89	18.33	3.18	17.078	13.250
	∞	1.96	51.20	3.51	1.244*	

*This value is limit of ratio VarC/VarS(W) as $\mu \to \infty$

One can see from the table that under $V[0,2\mu]$ distribution in H_1 the power does not converge to 100% as $\sigma^2/\mu \to \infty$. This, of course, is due to the fact that as $\sigma \to \infty$ $\lim P\{[(M_1-M_2)/\sigma] < c\} > 0$ for any constant c. It is easy to notice that the power will approach 100% if this limit equals zero because then for any constant c' with $\sigma^2/\mu \to \infty$

$$P\{K > c'\} = P\{[M_1-M_2)/\sigma] > c'\sqrt{2\mu}/\sigma\} \to 1.$$

Thus, the policy in (8) will, for large μ and for most of the particle distributions, reject H_o when $\sigma^2 > \mu$. This will increase the number of inspected wafers and, therefore, minimize the variance of the density estimator. On the contrary, a policy of a type used in Table I (see (4)) always tends to increase the number of inspected wafers when μ is large which, in turn, increases the variance of the density estimator when $\sigma^2 < \mu$. This suggests that for large μ the policies which utilize tests of $H_o : \sigma^2 \leq \mu$ are better than simpler policies in which either only one wafer is inspected in the first stage or no tests of H_o are conducted.

The policy defined by (8) can be generalized to accommodate any $r \geq 2$ by introducing statistic

$$L = \sum_{i=1}^{r} (Y_i - \overline{Y})^2/\overline{Y}$$

which in the Poisson case has an asymptotic Chi-square distribution with $r-1$ degrees of freedom.

CONCLUSIONS

All procedures based on an inspection of an accumulated unit surface area lead to unbiased density estimators with constant variance if and only if the underlying spatial partical distribution is Poisson. This popular assumption has to be challenged in view of the fact that some of the common inspection procedures invented to reduce the expected number of inspected wafers led to biased density estimators. The bias depends on the distribution type but tends to diminish with the increase in the defect density, which most likely explains why this effect has not been noticed in manufacturing environment of the recent past, and came to light only with the introduction of state-of-the-art "super-clean" fabs.

The proposed class of sequential inspection procedures yields unbiased density estimator for all particle distributions and reduces the expected number of inspected wafers. The variance of this estimator depends on the first two moments of the particle distribution, and when the distribution type (or at least the ratio of distribution mean to variance) is known, an optimal procedure (which minimizes the estimator's variance) can always be chosen. It is also possible to choose procedures in this case which reduce the variance of the density estimator even when the distribution type is unknown.

REFERENCES

1. B. Hardegen and A.P. Lane, Solid State Technol., 28, 189 (1985).
2. L.L. Pesotchinsky and Z. Fichtenholz, IEEE Trans. Semicond. Manuf., 1, 15 (1988).
3. Thomas Group Inc., "Defect Density," Techn. Rep., Ethel, LA (1982).

PARTICULATE GENERATION AND DETECTION ON SURFACES

Marvin Fein

System Technology Division
IBM Corporation
Endicott, New York 13760

This paper describes a tool that was developed to
evaluate particulate generation of materials. The tool
provides realtime data of both airborne and settling
particulates in a consistent and repeatable manner by
inputting a continuous and controlled stress on the
material in a controlled test chamber environment. Sample
test results on several cleanroom materials are presented
to demonstrate the use of the tool.

INTRODUCTION

All materials generate particulates. When dealing with cleanrooms
and contamination sensitive products and processes, these particulate
generators must be minimized. Materials and commodities used in
cleanrooms (gloves, garments, wipes, etc.) are potential fiber and
particle generators. It is, therefore, essential that optimum materials
be used in all commodities to minimize the contamination levels.

To effectively evaluate materials, a tool was developed that
provides a unique method that is both consistent and repeatable and
provides both airborne and settling data simultaneously.

The tester can accommodate a large variety of materials from
garments, wipes and gloves to paper substitutes and impregnated cloths.
In addition to testing materials for particulate generation, the tester
can also be used to determine the permeability to known contaminants by
backloading standards behind the sample.

The purpose of this paper is to describe the operation of the tester
and to present the results of an evaluation on several cleanroom wipes.

DESCRIPTION AND OPERATION

The tester is designed to characterize contamination levels on
sample materials, refer to Figure 1. The sample is contained within a
clean chamber and repeatedly flexed by a variable flow of compressed gas
(filtered nitrogen was used in this test but compressed air can be

substituted) which is injected into the deflection core to drive against
the sample. (A membrane can be used behind the material if simulation of
flexing, i.e., an elbow of a garment, is desired.)

Debris from the material is immediately entrained in the airstream,
refer to Figure 2. The airstream is generated by a rheostat controlled
blower to maintain the desired air flow velocity.

As the debris is moved in the airstream, particulates too large to
be maintained in the airstream gradually settle onto a "witness plate"
surface in the settling chamber of the tester. The witness plate is a

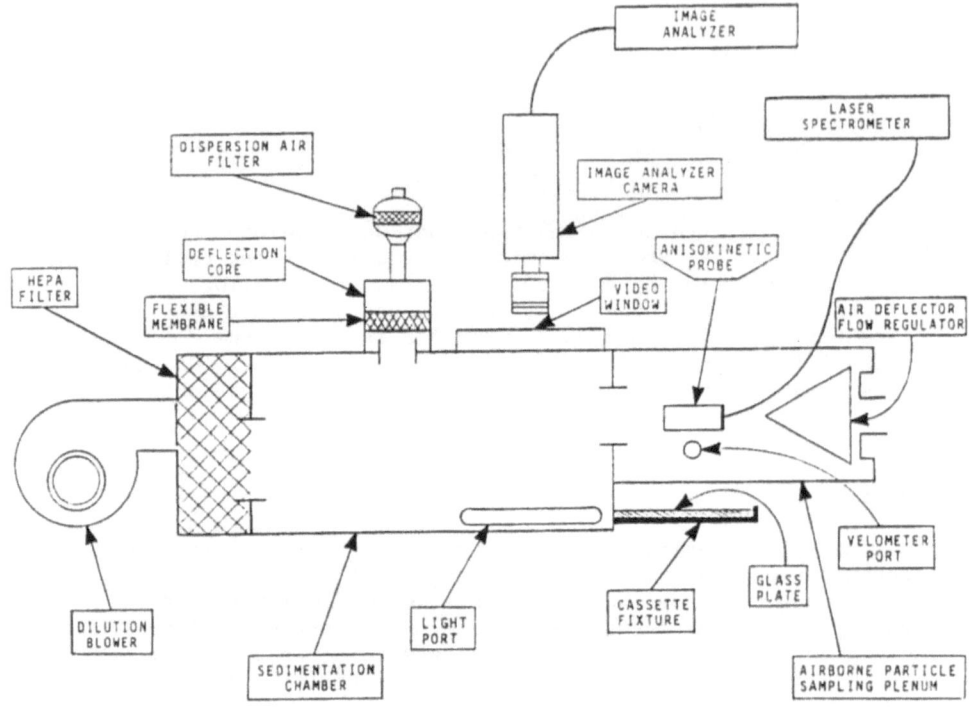

Figure 1. Tester Schematic.

highly polished piece of black glass which is held on a slide tray
cassette for easy removal for cleaning.

The witness plate is 9 in^2 but the effective settling area can be
decreased if desired.

The trajectories these particulates follow are dependent on the flow
velocity and their size (mass) and can be readily calculated by using
Stokes Law[1]. Once settled, the particles are illuminated by a fiber
optics source and reflect light toward an optical port or video window.
An image analysis video camera scans these particles to characterize by
size and quantity and presents the data in a programmed format.

Particles smaller than the critical diameter (approximately 10 μm)
are maintained by the airstream and carried into the airborne particle
plenum.

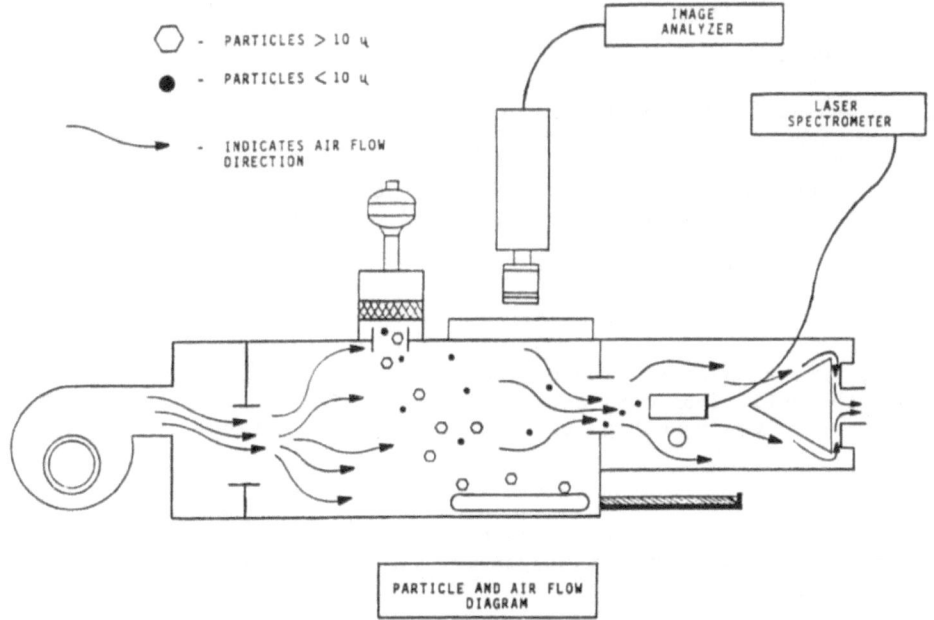

PARTICLES > 10 μ

PARTICLES < 10 μ

INDICATES AIR FLOW DIRECTION

IMAGE ANALYZER

LASER SPECTROMETER

PARTICLE AND AIR FLOW DIAGRAM

Figure 2. Particle and Air Flow Diagram.

The airborne particle plenum is designed to ensure that no settling of small particles occurs before the parcel of air is sampled by the anisokinetic probe within the plenum[2]. Here, any larger (greater than 10 μm) particles that have escaped settling within the settling chamber are eliminated by the restrictive effects of the small diameter, high velocity probe. Only the small diameter particles are able to follow the sudden constriction of flow and are efficiently sampled, sized and counted by a laser spectrometer.

Controlling the flow of air within the plenum and the entire unit is a triangular air deflector located at the end of the plenum. This deflector is positioned downstream of all sampling and is shaped so that minimal airflow disturbance is caused as the air leaves the unit. The air deflector may be moved relative to the terminal slotted plate in order to change the vent area and consequently the flow through the unit. The air deflector is also designed to prevent light from entering through the vent area.

The air supply is passed through a Hepa filter to prevent background contamination and positive pressure is maintained within the system to prevent incursion of outside air. The system is also light blind, and the interior surfaces are treated to be non-reflective and anti-static.

The modular design of the system allows for any particle counter or image system to be substituted for those currently in use. This could provide a means for comparing different units under strictly controlled conditions.

The use of an optical port and an image analyzing system allows continuous realtime monitoring of large particulate activity as the material wears. Wear amplitude and stress can be varied to simulate actual material motions and wear.

MATERIAL EVALUATION

Materials Used

An evaluation was performed to indicate the results that can be obtained from the tester in comparing materials. The materials selected were three types of cleanroom wipes from a particular manufacturer. The wipes are supplied in varying weights and weave openings within the same packaging and are classified for Class 10K use.

Type 1 -- nylon
Type 2 -- polyester
Type 3 -- polyester/nylon blend

The test samples require no special preparation. The material is held in the deflection core as is, with a 2 in^2 section of the material being flexed. Consequently, the test is non-destructive.

Sample Size

Twenty-five samples from each type were used to provide a statistically significant sample of the population from each of the three lots.

Test Procedure

All the materials were set-up in the same manner. A "new" sample was used on each run except when testing for second and third cycle generation. The center of the wipe was used to prevent edge particulates from confounding the results. The physical size of the sample is 2 in^2 (controlled by the opening of the deflection core). The tester parameters were maintained constant throughout.

1. Core pressure -- 25 psi
2. Air flow -- 20 fpm
3. Number of cycles -- 19/minute
4. Airborne particle size measured -- 0.5 to 10 μm
5. Settling particle size measured -- 10 μm and greater

RESULTS AND DISCUSSION

Table I. Average and Standard Deviation

	Type 1	Type 2	Type 3
Settling Particles (particles/in^2) average ($\bar{x}$)	14.88	13.91	7.86
Standard Deviation ($\bar{\sigma}$)	11.13	6.25	5.95
Airborne Particles (particles/ft^3) average ($\bar{x}$)	15.31	10.81	5.48
Standard Deviation ($\bar{\sigma}$)	16.58	10.61	4.73

Significance Tests

A Students T and F Test[3] was performed on the data to determine if there were any significant differences between the means of the samples. The results are presented in Table II. The results indicated that all three types of materials were from different populations at the 95% confidence level. This was true for both settling and airborne particles.

After the initial one minute cycle, two additional cycles on each sample were run. This provided information as to whether the material would continue to generate debris and deteriorate or whether the initial generation of particles was an adequate "burn in" test to remove loose debris and particles from the manufacturing process.

Table III shows a significant decrease in particle generation during the second cycle and then a levelling off during the third cycle. This suggests that each material has residual and/or loose particles that can be purged after several stress cycles. Examination of the material after the third cycle did not indicate any noticeable wear.

Table II. Significant Differences at 95% Confidence Level

	F Test	T Test
Settling Particles		
type 1 vs. type 2	no	yes
type 1 vs. type 3	no	yes
type 2 vs. type 3	no	yes
Airborne Particles		
type 1 vs. type 2	no	yes
type 1 vs. type 3	yes	yes
type 2 vs. type 3	no	yes

Table III. Comparison of Extended Cycles vs. Particle Generation

	Cycle I ($\bar{x}$)	Cycle II ($\bar{x}$)	Delta (%)	Cycle III ($\bar{x}$)	Delta (%)
Settling Particles					
type 1	14.88	4.60	-69.1	3.70	-19.6
type 2	13.91	4.31	-69.0	4.00	-7.1
type 3	7.86	4.98	-36.7	4.87	-2.2
Airborne Particles					
type 1	15.31	2.22	-85.5	1.24	-44.4
type 2	10.81	8.40	-22.3	3.48	-58.6
type 3	5.48	4.79	-12.6	4.43	-7.5

Table IV.   Regression Analysis

	Correlation Coefficient (R)	R^2 x 100 (%)
Type 1	0.413	17.02
Type 2	0.268	7.18
Type 3	0.056	0.31

Airborne vs. Settling Particles

The data collected from this evaluation provided an ideal vehicle to test the hypothesis of the correlation between airborne and settling particles and whether the settling rate can be predicted from airborne readings in similar environments.

A regression analysis at the 95% confidence level was performed and Table IV presents the correlation coefficients between each of the types of materials in relation to airborne versus settling particles.

The square of the coefficient is the percent of the total variation of the dependent variable (airborne) that can be attributed to the relationship with the independent variable (settling).   The data suggest that there is no significant correlation between the two variables.

CONCLUSIONS

The test results indicate that the subject tester can provide important information on material evaluation of cleanroom commodities in a consistent and repeatable manner in a realtime format.

REFERENCES

1.   W. C. Hinds, "Aerosol Technology," Wiley and Sons, New York, 1982.

2.   B. Y. H. Liu, "Fine Particles - Aerosol Generation, Measurement, Sampling and Analysis," Academic Press, New York, 1976.

3.   J. E. Freund, "Modern Elementary Statistics," Prentice Hall, New Jersey, 1973.

IDENTIFICATION OF SMALL PARTICLES ON A SILICON WAFER

N. Fujino, S. Miyazaki, M. Takeshita, H. Horie, and S. Sumita

Kyushu Electronic Metal Co.
Kohoku-cho, Saga, Japan 849-05

T. Shiraiwa

Osaka Titanium Co.
Amagasaki, Hyougo, Japan 660

An SEM with EDX is usually used to analyze particles on silicon wafer surfaces, but since the particles are so few and so small, much time must be spent searching for them. In the present report, a combined method is described for particle detection and analysis. First, a surface particle counter (Tencor Surfscan 4500) is used to locate particle positions. The coordinates of each particle are then converted to SEM coordinates, and each particle observed and analysed. Using this new method, the time spent for full particle analysis of a wafer is greatly reduced.

INTRODUCTION

Silicon wafers are an important material for LSI and VLSI. The line width for these devices is now smaller than 1 micron, and particles with diameters 20% of line width must be controlled and eliminated.[1,2] To do so, particles must be analyzed and the origin of the particles located, allowing process modifications to eliminate groups of particles.

A scanning electron microscope (SEM) with energy-dispersive X-ray analysis (EDX) capability is usually used to analyze particles on silicon wafer surfaces, but since the particles are so few, so widespread and so small, much time must be spent searching for each particle. In the present work, a quicker method is presented, using a combination of SEM/EDX and surface particle counter (SPC).

EXPERIMENTAL

Equipment

Table I lists the equipment used in this work. The SPC can detect particles and supply their coordinates by analyzing light scattered from particles on the wafer surface while such surface is illuminated by a scanning laser beam.

Table I. Methodology for Particle Detection and Identification

Detection
     Surface Particle Counter (SPC)
     TENCOR SURFSCAN 4500
     Detection Limit   : 0.173 μm dia.(latex)
     Position Repeatability : 50 μm
     Wafer Size   : 50 to 150 mm
Observation
     Scanning Electron Microscope (SEM)
     Akashi DS-130
     Resolution   :50Å
     Wafer Size   :100 to 150 mm
Identification
     Energy Dispersive X-ray Analysis (EDX)
     PHILIPS ECON III
     Detectable Element   : B$\leq$

Particle Location and Analysis Method

Fig.1 illustrates the method used to convert coordinates from the SPC's axes to the SEM's axes.
(1) Particles are detected on a wafer by the SPC.
(2) To obtain the new coordinates: The wafer is placed on orthogonal axes, one axis is fixed to the orientation flat and the second axis is set tangent to the wafer. These calculations are carried out by a microcomputer.
3) The computer-supplied coordinates are used by the SEM operator to locate each particle for analysis.

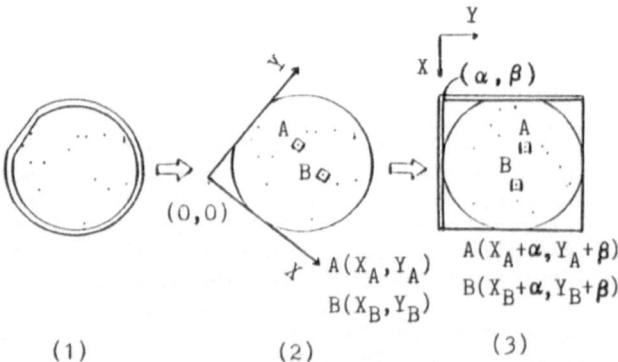

Figure 1. The method used to convert coordinates from the SPC's axes to the SEM's axes. (1) Particle detected on a wafer by the SPC. (2) To obtain the new coordinates: The wafer is placed on orthogonal axes, one axis is fixed to the orientation flat and the second axis is set tangent to the wafer. These calculations are carried out by a microcomputer. (3) The computer supplied coordinates are used by the SEM operator to locate each particle for analysis.

RESULTS AND DISCUSSION

Position Repeatabilities of the SPC and SEM

The position repeatabilities were found to be 50 um for the SPC and 80 μm for the SEM, a total of 130 μm, so by using this method, a particle can be located in a 260 μm x 260 μm area. When searching for a 0.1 μm particle on the SEM, we usually use x1000 magnification, making the scanned area about 100 μm x 80 μm . The 260 μm x 260 μm thus corresponds to only fifteen frames of the SEM. Thus quick location and observation of even small particles can be accompalished.

Detection Limit of Visual Inspection

The inspection of particles on silicon wafers has historically been done by human eye. We investigated the detection limit of human vision; Fig.2 shows the results. The points marked by • were detected by the SPC while the points marked by ○ were detected by visual inspection. The particles found by visual inspection were about 0.2 μm or larger, thus it is thought that this is the detection limit of visual inspection.

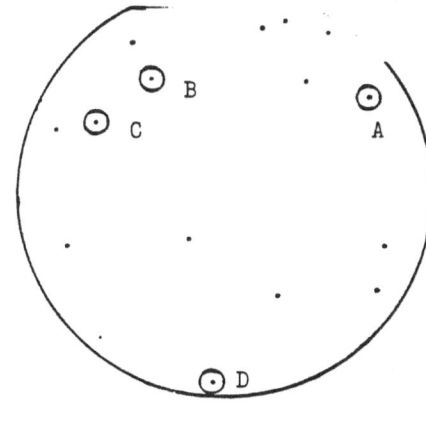

	Scattering Cross Section (μm^2)	Latex dia. (μm$\emptyset$)
A	>0.255	>0.32
B	>0.255	>0.32
C	0.040	0.24
D	0.010	0.19
E	>0.255	>0.32
F	0.022	0.21
G	0.021	0.21
H	>0.255	>0.32

Figure 2. Comparison of particles on a wafer found by visual inspection with those found by SPC. The points marked by • are detected by SPC and those marked by ○ are detected by visual inspection.

Relationship Between Scattering Cross Section and Particle size

Fig.3 shows the relationship between scattering cross section by SPC and particle size by SEM. Particles remaining on a wafer after cleaning were classified as spherical or non-spherical. Spherical particles show a

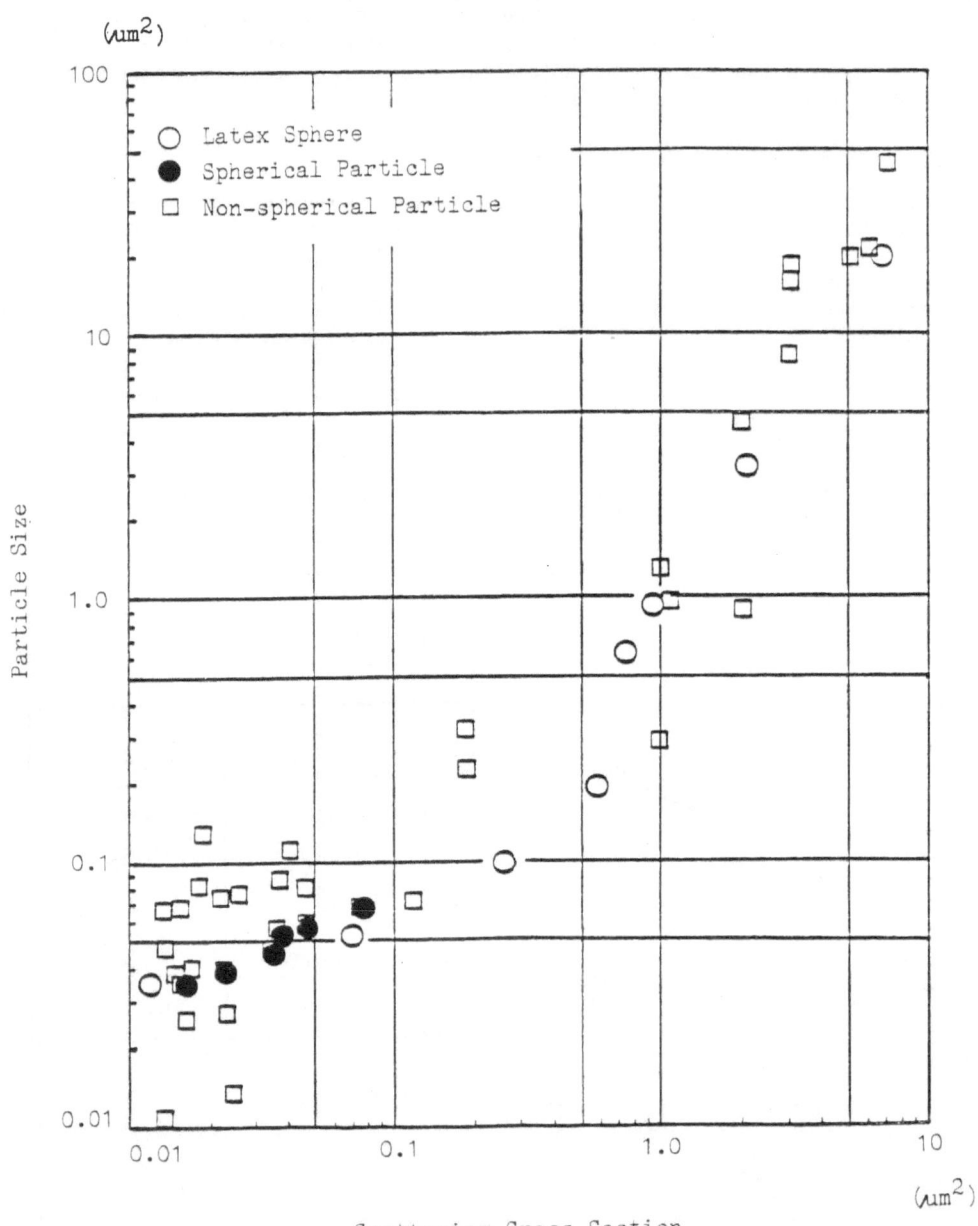

Figure 3. The relationship between scattering cross section by SPC and particle size by SEM. The spherical particles(●) show relation similar to the relation of polystyrene latex sphere(○). But, non-spherical particles (□) are not similar to latex sphere.

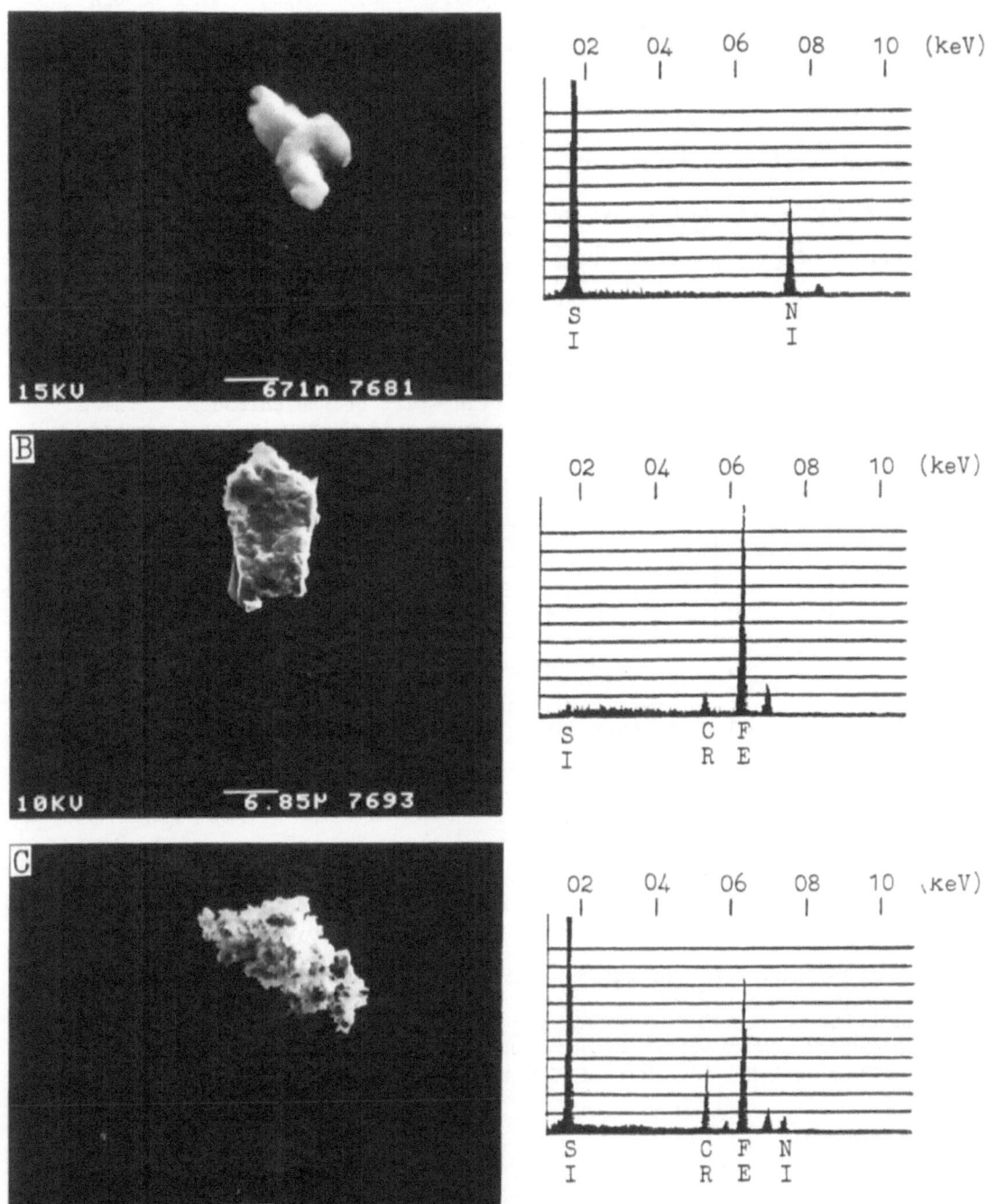

Photo.1. Photographs and analytical results of particles generated from the flatness tester (see figure 4.).

similar relation to polystylene latex spheres, but non-spherical particles show no such relation. Since non-spherical particles have various shapes, their surface areas vary greatly with their shape. It is thought that the estimation of particle size by scattering cross section is very difficult when the particles are so small.

APPLICATION

Analysis of Particles Generated from a Flatness Tester

Many kinds of measuring equipment are used in the production of silicon wafers; it is important to learn the particle generation from such equipment. Fig.4 shows an example of particles generated by a flatness tester. In this tester, the loading unit is Ni-coated and the unloading unit is made of stainless steel. Monitor wafers were kept at the sampling points shown in Figure 4 for two hours, and the particles thus captured were analysed by this method. The element appearing in particle A in Photo.1 is Ni, the material from the loader. Particles B and C show Fe, Cr, and Ni, the elements of the moving part of the unloader. These results indicate that the origin of particles can be determined, making particle elimination easier.

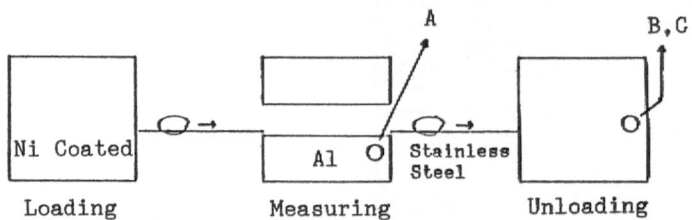

Figure 4. An example of particle generated from a flatness tester. Sampling points are marked O. A,B,C were typical particle-generated from these points. Photo. 1. shows the photographs and analytical results.

CONCLUSION

A quick method for observation and identification of small particles on a silicon wafer is established by means of correlation of coordinates of an SPC(Surface Particle Counter) and an SEM(Scanning Electron Microscope).

The detection limit of small particles on a silicon wafer by visual inspection is about 0.2 μm .

Using this method, the generating source of particles in the manufacturing process can be determined, making it easier to eliminate the source, thus decreasing the total wafer particle count.

ACKNOWLEDGMENT

The authors would like to express their sincer thanks to Mr. B. Johnston of OTC Semiconductor Corp. who provided assistance in this work.

REFERENCES

1. I. K. Bansal J. Environ. Sci., 26-4, 20(1983).
2. E. M. Juleff, W. J. McLeod, E. A. Hulse, and S. Fawlett. The advances in contamination control of processing chemical in VLSI. Solid State Technol, 25, 82(September).

THE EVALUATION OF PWA AND SMA CLEANLINESS LEVELS FOR "IN-LINE" DEFLUXING BY HIGH PERFORMANCE LIQUID CHROMATOGRAPHY

R.F. Klima and J.K. Bonner

ALLIED-SIGNAL INC.
Genesolv®/Baron-Blakeslee
Applications Laboratory
Melrose Park, IL 60160

Recently surface mount technology (SMT) has gained popularity. Surface mount assemblies (SMA's) are more electrically efficient, sophisticated in function and optimized with respect to board real estate. Boards containing SMD's (surface mount devices) require solder paste screening, surface component attachment by infrared or vapor phase reflow processing as well as conventional processing if they contain both SMD's and through hole components (THC's).

Solder paste defluxing is not well understood. Most conventional defluxing solvents will not dissolve solder paste. The rheological properties of solder paste are quite different from those of solder flux. The chemical composition of solder paste is also different than conventional solder flux. Solder paste will not dissolve in isopropanol and water; therefore, the conventional ionic contamination tests are questionable at best. Other cleanliness criteria are based on "eyeballing" which is a subjective, non-quantitative evaluation. In this study a quantitative method has been developed for solder paste. This method was then applied to a paste defluxing operation. Cleanliness was determined and expressed in absolute and relative (to initial paste application) terms. The effect of defluxing equipment parameters on cleanliness was also investigated. The paste data was then compared to data obtained from defluxing a conventional RMA solder flux in an in-line defluxer.

INTRODUCTION

There are two major classes of analytical chemical techniques that can be applied to printed wiring assembly (PWA) contamination analysis: direct surface analysis techniques most commonly employed are: ESCA (Electron Spectroscopy and Chemical Analysis), SAM (Scanning Auger Microprobe), and SIMS (Secondary Ion Mass Spectrometry)[1]. These methods are very costly and extremely time consuming.

The indirect extractive approach consists of extracting a surface residue into a suitable solvent and analyzing the solvent by an analytic technique. The most common techniques are FT-IR (Fourier Transform Infrared Spectroscopy)[2], GC (Gas Chromotography)[3] and HPLC (High Performance Liquid Chromatography)[4]. The extractive techniques are moderately priced, fast and accurate.

The principles of cleaning solder flux are fairly well understood. After manufacturing printed wiring assemblies, which consists of bare board processing, electrical component placement and attachment, the finished circuit board must be cleaned. Cleaning, or defluxing, entails removing ionic flux components (activators) and organic (rosin) components. Inefficient cleaning can lead to electrical short circuits, electrical opens, corrosion, and properties, resistance, etc.)

Defluxing solvents are bipolar in nature. The less polar material, which is usually the major component, will solubulize the flux rosin, while the polar solvent constituents dissolve the ionic activators.

Recently surface mount technology (SMT) has gained popularity. Surface mount assemblies (SMA's) are more electrically efficient, sophisticated in function and optimized with respect to board real estate. Boards containing SMD's (surface mount devices) require solder paste screening, surface component attachment by infrared or vapor phase reflow processing as well as conventional processing if they contain both SMD's and through hold components (THC's).

Solder paste defluxing is not as well understood. Most conventional defluxing solvents will not dissolve solder paste. The rheological properties of solder paste are quite different from those of solder rflux. The chemical composition of solder paste is also different than conventional solder flux. Solder paste will not dissolve in isopropanol and water; therefore, the conventional ionic contamination tests are questionable at best. Other cleanliness criteria are based on "eyeballing" which is a subjective, non-quantitative evaluation. In this study a quantitative method has been developed for solder paste. This method was then applied to a paste defluxing operation. Cleanliness was determined and expressed in absolute and relative (to initial paste application) terms. The effect of defluxing equipment parameters on cleanliness were also investigated. The paste data were then compared to data obtained from defluxing a conventional RMA solder flux in an in-line defluxer.

EXPERIMENTAL METHOD

Circuit boards containing SMD's with a 1.27×10^{-4} meter standoff were vapor phase reflow soldered using a RMA solder paste and cleaned with Genesolv DFX in a Baron-Blakeslee in-line defluxer at various conveyor speed. Conventional boards were wave soldered using a conventional RMA solder flux and defluxed in the same equipment. The objective of the experiment was to determine the effect of conveyor speed on cleanliness of the circuit boards.

To evaluate cleanliness quantitatively, an analytical method was developed. This method consisted of extracting the paste or flux residue from circuit board surfaces after reflow or wave soldering into an organic solvent (such as Genesolv DTA, tetrahydrofuran, acetonitrile, 1,1,1-trichlorethate or and FC-112 based solvent) and then analyzing this solution by high performance liquid chromatography (HPLC). A schematic diagram of a typical HPLC is shown in Figure 1. The analytical method is outlined in Figure 2.

Best resolution for paste components was achieved on a reverse phase column. The mobile phases were acetonitrile and water. Components separation was accomplished using a 15 minute, 20% to 80% acetonitrile gradient. Sample peaks were monitored by a diode array detector which is part of the Hewlett Packard 1090 a high performance liquid chromatography system.

Surface residues were extracted by soaking PWA's for a fixed amount of time (paste: 24 hrs.; flux: ½, 2, 6, 24 hrs.) Before the actual analysis, the HPLC output (peak area) had to be correlated to the amount of flux or paste residue extracted from a circuit board. This was done by a calibration curve. Solutions of different flux or paste concentrations were made. Since the paste consists mainly of metal solder balls (90%), the actual paste concentration is

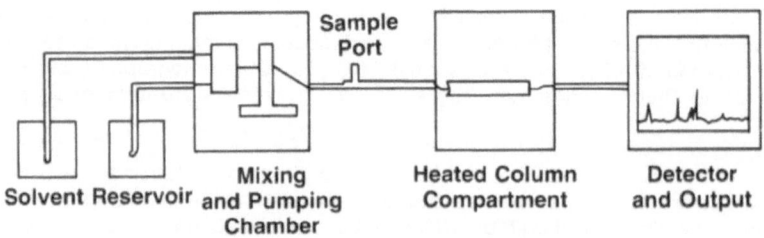

Sample Port

Solvent Reservoir   Mixing and Pumping Chamber   Heated Column Compartment   Detector and Output

Figure 1. High Performance Liquid Chromatography Apparatus

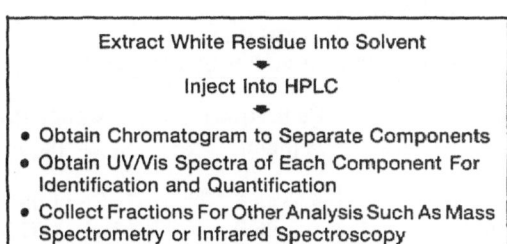

Figure 2. Analysis Technique

only 10% of the total material used. The standard solution concentrations (kilograms per cubic meter) can be expressed in terms of weight per unit area of circuit board (10^{-6} cubic meters extraction solvent used for every 6.45×10^{-4} square meter of surface area). The calibration curves are shown in Figure 3 and 4. The numerical data are displayed in Tables I and II.

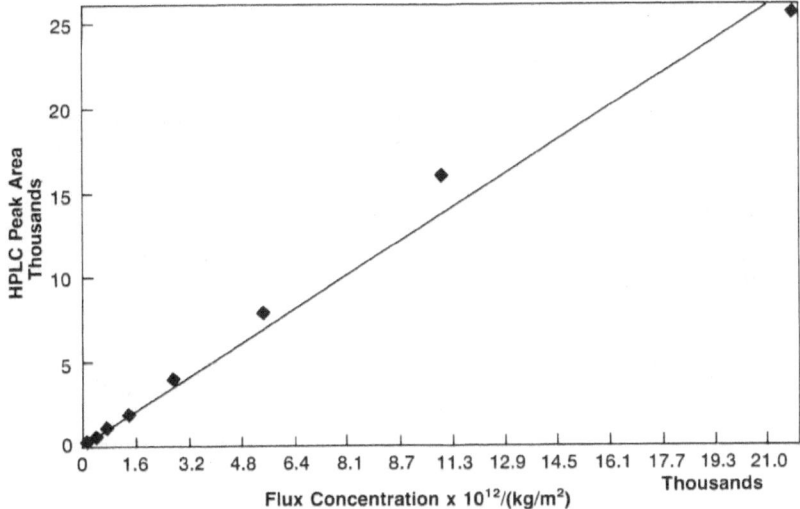

Figure 3. HPLC Calibration Curve for Flux

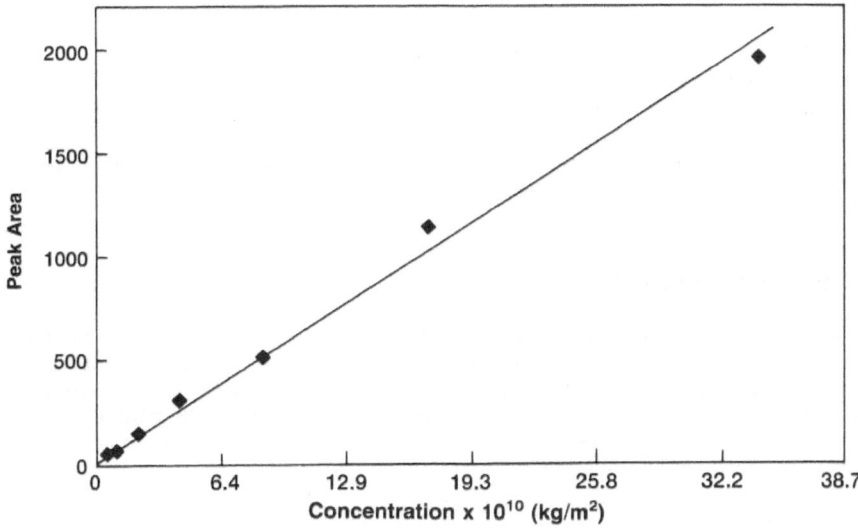

TABLE I
Calibration Curve For Solder Paste

Concentration (kg/m^2)	Peak Area (mau)	Concentration/Area $(kg/m^2 \cdot mau)$
0	0	
5.31×10^{-11}	48.7	1.09×10^{-12}
10.64×10^{-11}	66.7	1.59×10^{-12}
21.27×10^{-11}	156.5	1.36×10^{-12}
42.54×10^{-11}	308.2	1.36×10^{-12}
85.09×10^{-11}	598.6	1.42×10^{-12}
170.17×10^{-11}	1152.3	1.48×10^{-12}
340.35×10^{-11}	1968.3	1.73×10^{-12}

Average 1.43
Std. Dev. 0.20

TABLE II
Calibration Curve For Solder Flux

Concentration (kg/m^2)	Peak Area (mau)	Concentration/Area $(kg/m^2 \cdot mau)$
0	0	
4.12×10^{-11}	74	0.57×10^{-12}
8.51×10^{-11}	140	0.61×10^{-12}
34.04×10^{-11}	557	0.61×10^{-12}
68.08×10^{-11}	1100	0.62×10^{-12}
272.33×10^{-11}	3984	0.68×10^{-12}
544.67×10^{-11}	7861	0.69×10^{-12}
1089.34×10^{-11}	16066	0.68×10^{-12}
2178.63×10^{-11}	25691	0.85×10^{-12}

Average 0.66
Std. Dev. 0.08

Absolute response factors were calculated for known amounts of paste and flux. The following calibration equation was used to determine all other paste concentration:

Conc (kg/sq.m.) = 1.43×10^{-12} X Total Area (mau).

For solder flux the following equation was used:

Conc (kg/sq.m.) = 0.66×10^{-12} X Total Area (mau).

The precision of the method was evaluated by repeat injection of a fixed volume into the HPLC. HPLC precision for this experiment is shown in Figure 7. For a 6.806 X 10 kg/sq.m. sample the standard deviation was ±10.4% of the average value.

RESULTS AND DISCUSSION

Solder flux is composed of rosin, activator and carrier solvent. The carrier solvent is usually vaporized during wave soldering leaving rosin and activator on the PWA surface. The rosin material is composed of a series of organic cyclic monocarboxylic acids[5]. The major acid components in rosin are abietic acid, dehydroabietic acid and neobietic acid; however, during wave soldering neoabietic acid is converted to abietic acid[6]. The liquid chromatograms of the rosin acids are shown in Figure 6 and 7.

---

* mau = milliabsorbance units

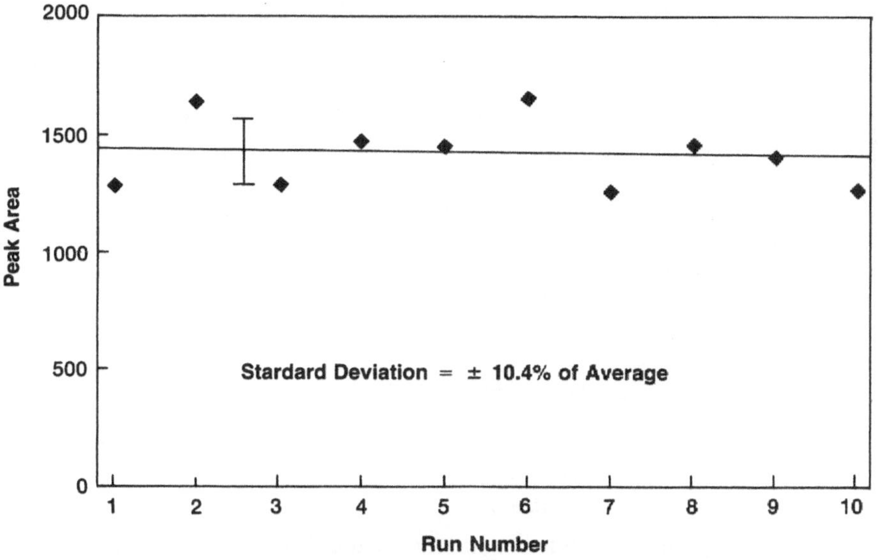

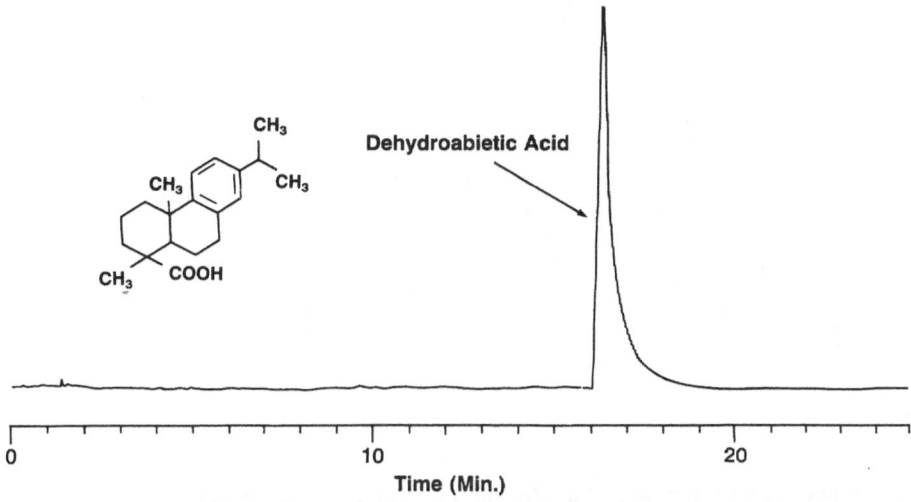

2000

1500

1000

500

0

Peak Area

1   2   3   4   5   6   7   8   9   10
Run Number

Stardard Deviation = ± 10.4% of Average

Figure 5. HPLC Precision

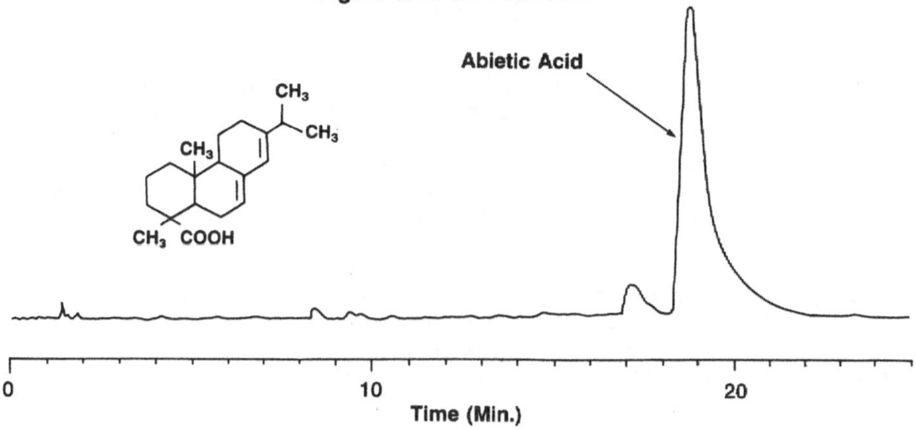

Abietic Acid

CH₃
CH₃
CH₃
CH₃ COOH

0   10   20
Time (Min.)

Figure 6. The Chromatogram of Abietic Acid

Dehydroabietic Acid

CH₃
CH₃
CH₃
CH₃ COOH

0   10   20
Time (Min.)

Figure 7. The Chromatogram of Dehydroabietic Acid

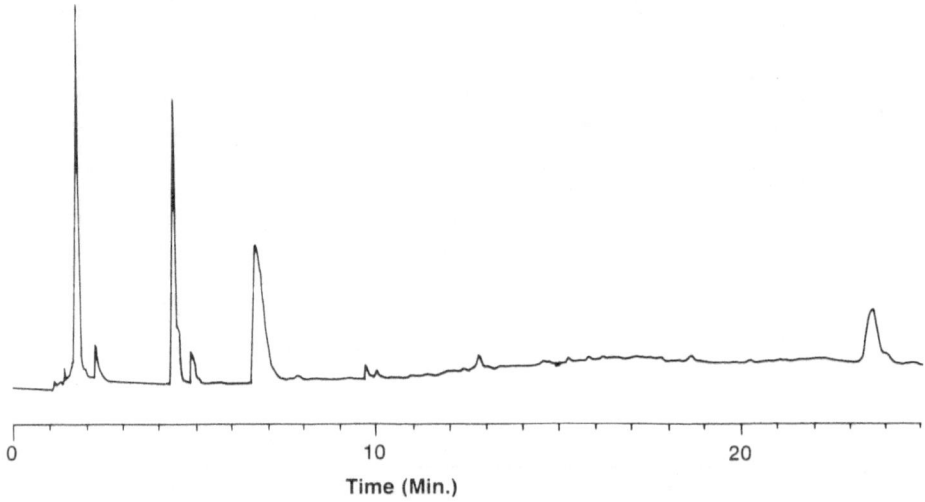

Figure 8. The Chromatogram of a Solder Paste Thickening Additive (Organic Derivative of Castor Oil)

Solder paste is composed of flux rosin (abietic acid and dehydroabietic acid – major components), activator, a thickening agent (rheological modifying additive) and a carrier liquid such as butyl cellosolve. The RMA paste used in the study contained Thixatrol ST, a rheological thickening agent (NL Industries). The liquid chromatogram of this additive is shown in Figure 8. This reological modifier is an organic derivative of castor oil. The thickening agent functions as a structural support to hold the solder powder in place after screening or stencilling.

Superimposing the chromatograms of rosin and thickening agent on top of the paste chromatogram confirms the components identity, Figure 9. Activator and carrier solvent peak identification is still under investigation. However, if total area under the chromatagram is integrated, the total amount of paste is accounted for regardless of where the peak is located.

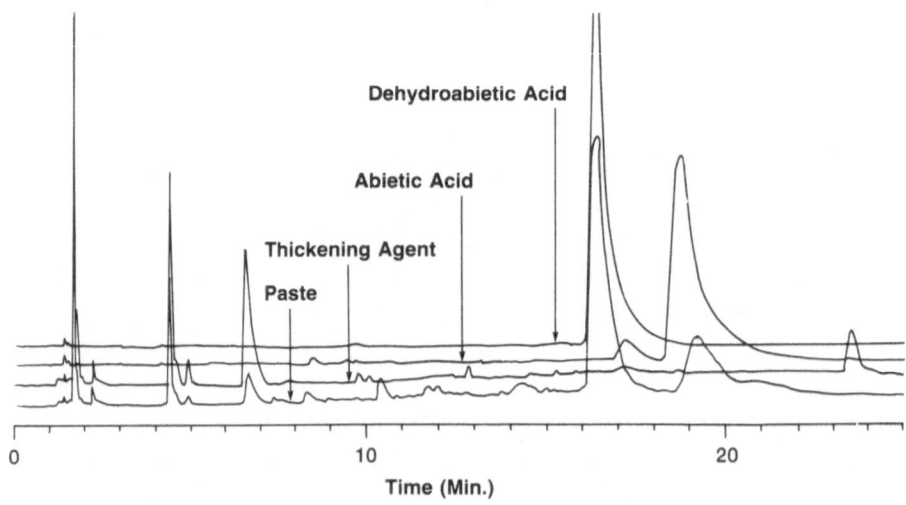

Figure 9. Solder Paste Component Identification By Computer Aided Superimposition of Components

The amount of organic contamination remaining on the PWA surface after soldering and defluxing at various "in-line", conveyor speeds is shown in Table III for an RMA solder flux. These values are compared to clean PWA's and contaminated PWA's which were waved soldered and not defluxed. Similar comparisons are given for various extraction times.

TABLE III
Flux Rosin Residue Extract From PWS's After Defluxing

Extraction Conc. Time (hr:min)	Conveyor Speed (ft/min)	Peak Area (mau)	kg/m^2
:30	1.25	39	25.7×10^{-12}
	2.50	68	44.9×10^{-12}
	5.00	518	341.9×10^{-12}
	10.00	1142	753.7×10^{-12}
	Unclean	9521	6283.9×10^{-12}
2:00	1.25	27	17.8×10^{-12}
	2.50	124	81.8×10^{-12}
	5.00	645	425.7×10^{-12}
	10.00	1617	1067.2×10^{-12}
	Unclean	11334	7480.4×10^{-12}
6:00	1.25	68	44.8×10^{-12}
	2.50	208	137.2×10^{-12}
	5.00	1059	698.9×10^{-12}
	10.00	1422	928.5×10^{-12}
	Unclean	11689	7714.7×10^{-12}
24:00	1.25	486	320.8×10^{-12}
	2.50	883	582.8×10^{-12}
	5.00	–	–
	10.00	3343	2206.4×10^{-12}
	Unclean	26243	17320.4×10^{-12}

The extraction time is the time a PWA is immersed in an extraction solvent (acetonitrile or tetrahydrofuran, etc.) prior to analysis of the solution by HPLC. Figures 10 through 13 depict the effect of extraction time on the amount of contamination leached out of the PWA into solution for different conveyor speeds. All the curves have similiar shapes indicating similiar desorption kinetics. Any extraction time can be used in future experiments as long as the same extraction time is used for each experiment in the cleaning evaluation comparison.

Figures 14 through 17 show the effect of conveyor speed on cleanliness levels of PWA's after defluxing. These curves all have similar shapes for different extraction times. As the conveyor speed increases, cleaning performance decreases. Cleaning performance for conventional leaded boards decreases rapidly above a conveyor speed of 2.5 ft/min.

Tables IV and V display the cleaning performance data for a five mil standoff RMA solder paste, defluxed from a SMD containing board.

Solder paste cleanliness levels are given various "in-line" conveyor speeds. Cleaning efficiencies are also presented (% removal of initial load). These data are displayed in the form of a bar graph in Figure 18.

Cleanliness levels at various conveyor speeds are compared to clean SMA's and "dirty" SMA's that were reflowed but not defluxed.

Figure 19 compares cleaning at various conveyor speeds on a linear scale. Again as conveyor speed increases cleanliness decreases. Above 2.5 ft/min cleaning deteriorates rapidly.

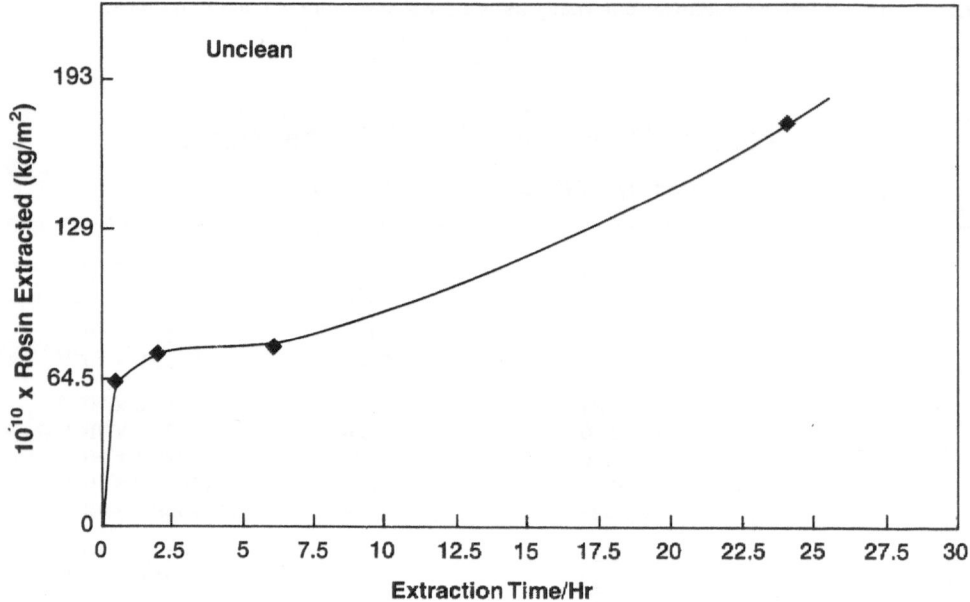

Figure 10. Rosin Extracted vs. Extraction Time for Unclean Boards

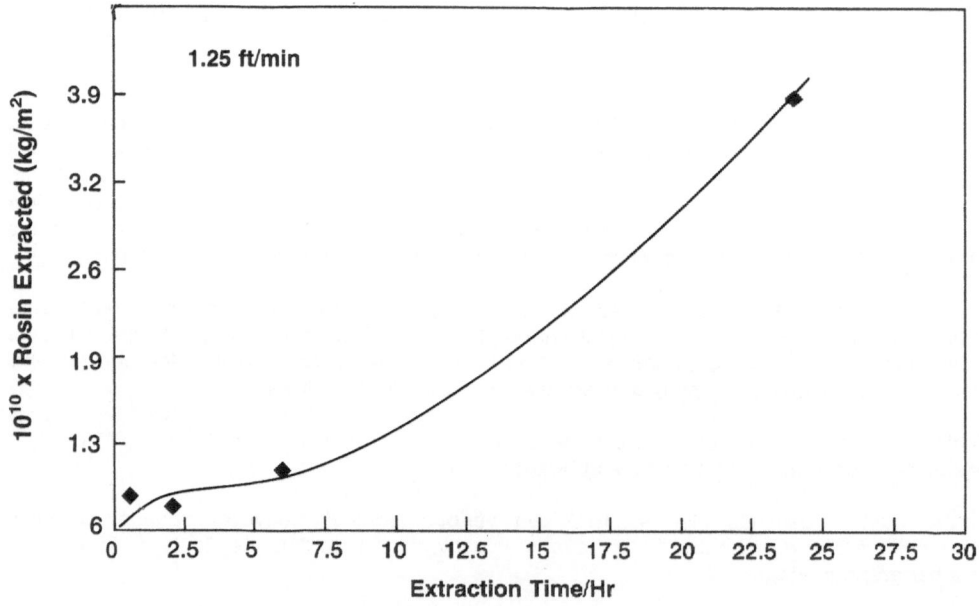

Figure 11. Rosin Extracted vs. Extraction Time for Boards Cleaned at 1.25 ft/min.

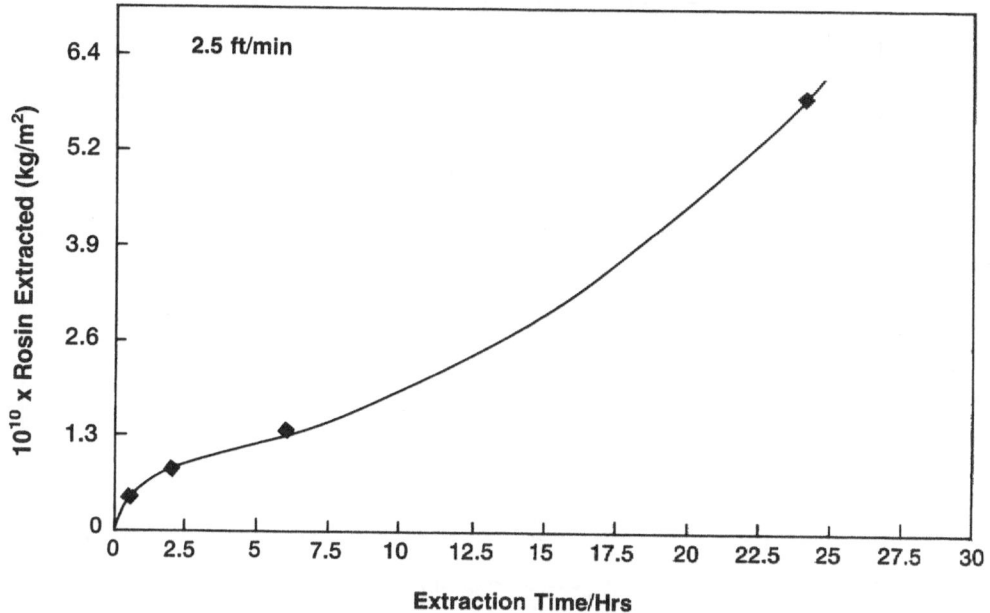

Figure 12. Rosin Extracted vs. Extraction Time for Boards Cleaned at 2.5 ft/min.

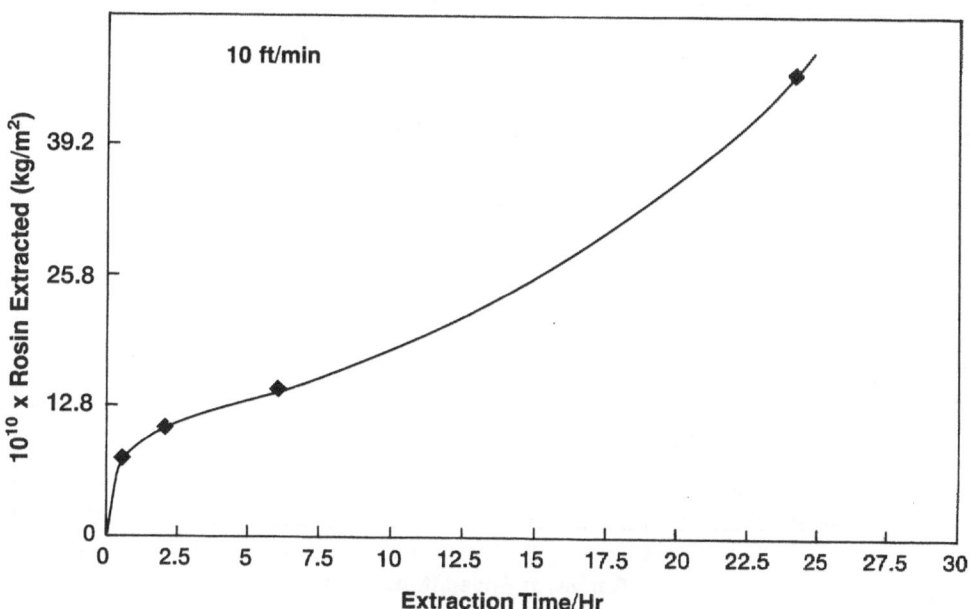

Figure 13. Rosin Extracted vs. Extraction Time for Boards Cleaned at 10 ft/min.

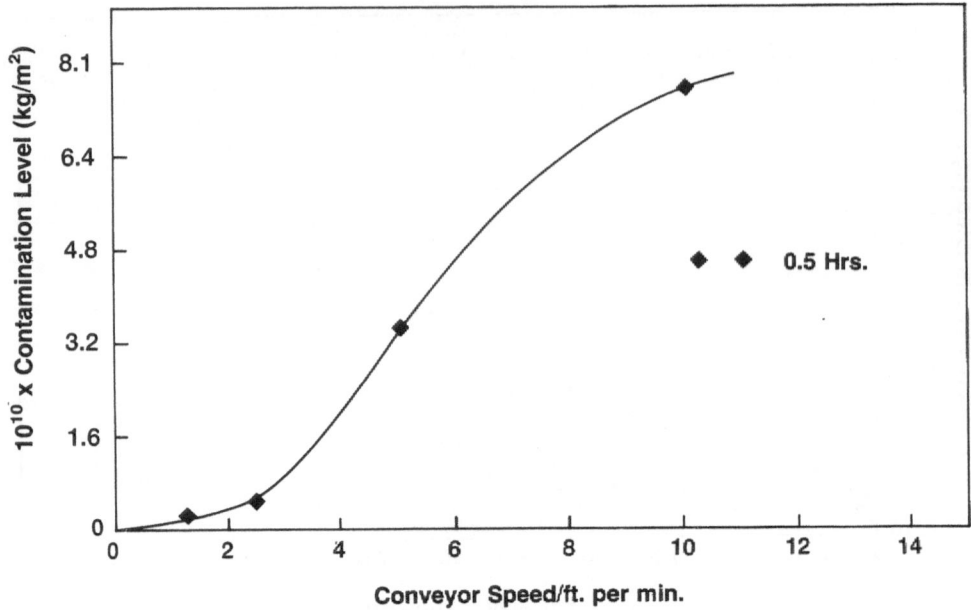

Figure 14. Contamination vs. Conveyor Speed for Boards Extracted ½ Hr.

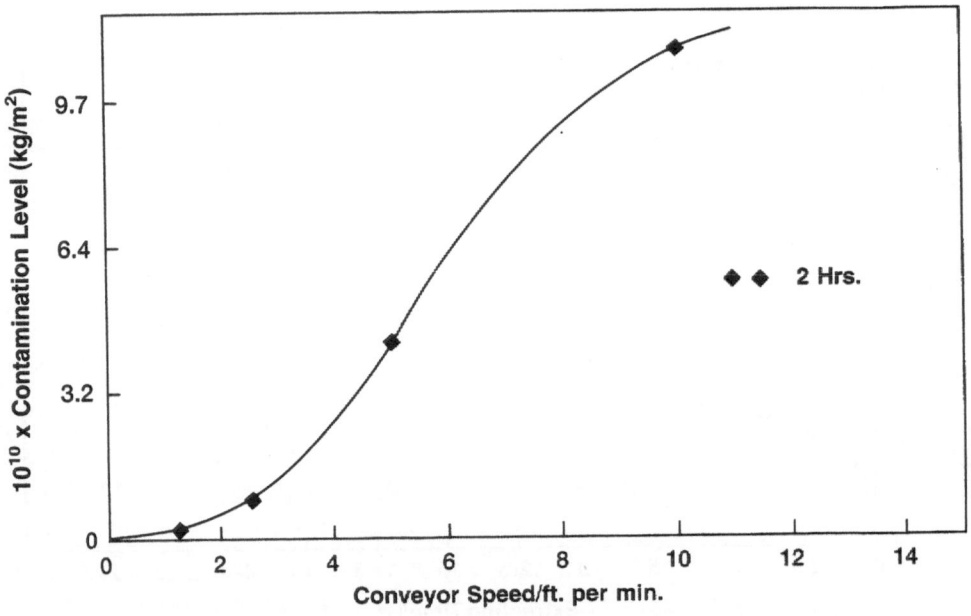

Figure 15. Contamination vs. Conveyor Speed for Boards Extracted 2 Hours

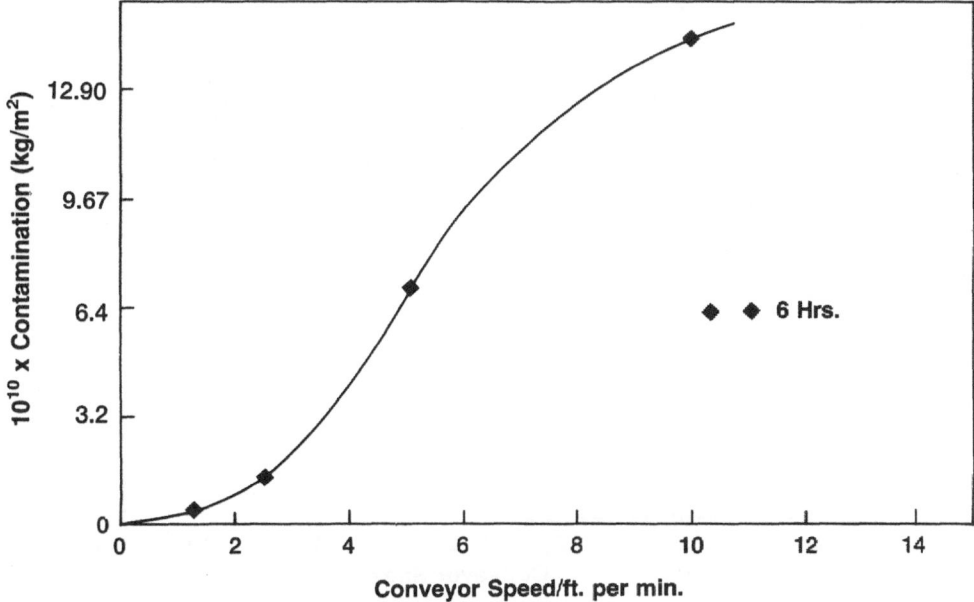

Figure 16. Contamination vs. Conveyor Speed for Boards Extracted 6 Hours

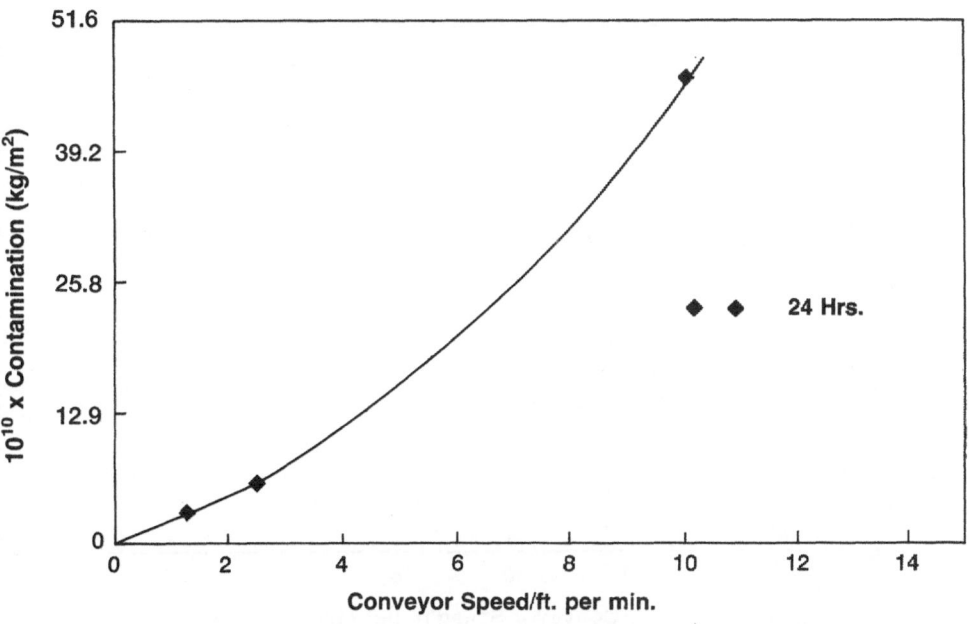

Figure 17. Contamination vs. Conveyor Speed for Boards Extracted 24 Hours

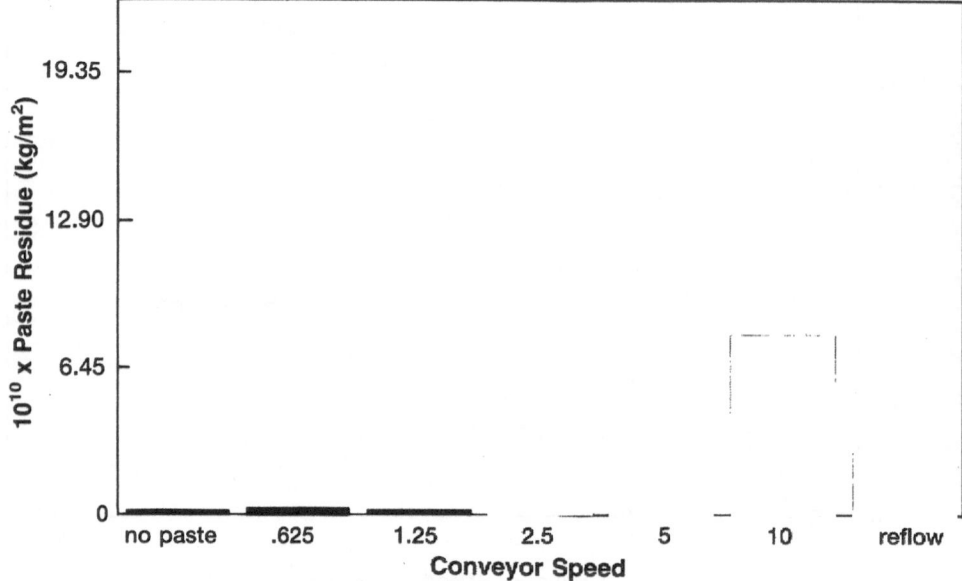

Figure 18. Paste Contamination Level vs. Conveyor Speed

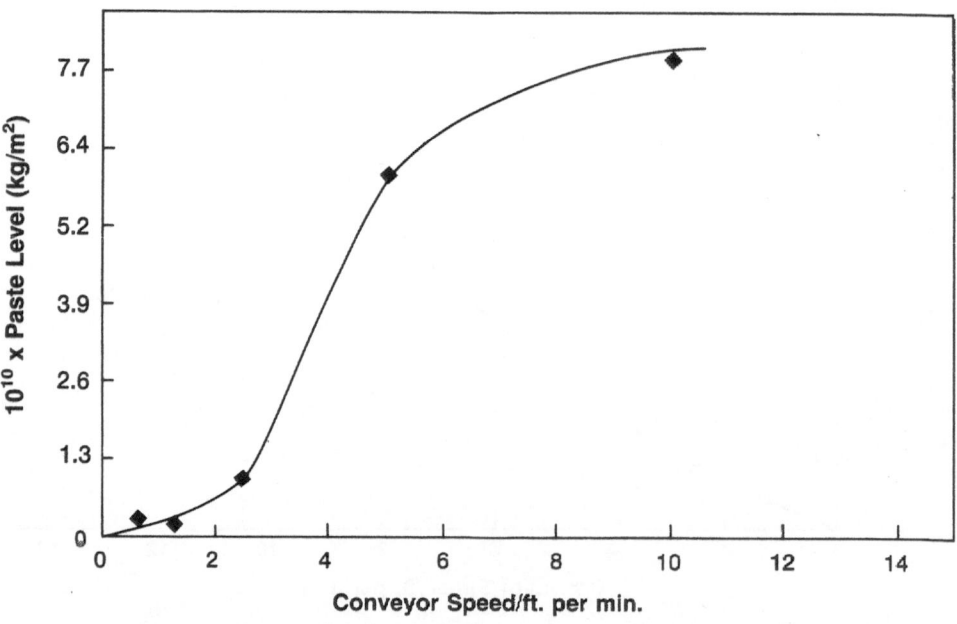

Figure 19. Contamination vs. Conveyor Speed for Paste Residue Extracts at 24 Hours

TABLE IV
Paste Residue Extract From SMA's After Defluxing

Conveyor Speed (ft/min)	Concentration kg/m^2
No Paste	15.0×10^{-12}
0.625	27.9×10^{-12}
1.25	16.9×10^{-12}
2.50	93.6×10^{-12}
5.00	603.2×10^{-12}
10.00	796.6×10^{-12}
Reflow (no cleaning)	20762.5×10^{-12}

TABLE V
Paste Residue Extracted From SMA's

Conveyor Speed	% Removal (Cleaning Efficiency)
No Paste	100
0.625 ft/min	99.3
1.25 ft/min	99.9
2.5 ft/min	96.2
5.0 ft/min	71.6
10.0 ft/min	62.3

CONCLUSIONS

A new method has been developed to identify and quantify paste components extracted from PWA's after defluxing. This method can be used to determine cleanliness levels at various process parameter settings with respect to solder flux, paste and white residue.

In general as conveyor speed increases cleanliness decreases. Cleaning performance for conventional RMA flux or 0.125 cm standoff paste deteriorates rapidly for "in-line" conveyor speeds greater than 2.5 ft/min.

REFERENCES

1. K. Bonner, A comparison of direct surface analysis techniques with solvent extraction contaminate profiling techniques for ascertaining production cleanliness of printed wiring, IPC-TP-323, IPC Annual Meeting, April, 1980

2. R.J. Kaier, A simple method for the detection and identification of organic non-ionic contaminants on PWB and PWA surfaces, IPC-TP-468, IPC 26th Annual Meeting, April, 1983

3. W.L. Archer and T.D. Cabelka, Behavior of rosin fluxes and solder paste during soldering operations, ISHM Conference, October, 1986

4. R.F. Klima and H. Magid, Analysis of fluxes and extracts of printed wiring by high performance liquid chromotography, Proc. NEPCON West 1986

5. R.F. Klima and J.K. Bonner, Analysis of white residue by high performance liquid chromatography, IPC-TP-713, March 1988 Meeting, Hollywood, FL

6. L.F. Fieser and M. Fieser, "Natural Products Related to Phenanthrene", 3rd Ed, Reinhold Publishing Corp., New York, 1949

PART III.   PARTICLE PREVENTION AND IMPLICATIONS

WEAR RESISTANT COATINGS REDUCE PARTICULATE CONTAMINATION IN A MAGNETIC DISK DRIVE

Wendy Jones, John McDowell, Walter Prater, and Garvin Stone

IBM Corporation
General Products Division
San Jose, California 95193

A test was performed to evaluate the particulate contamination coming from components in a magnetic disk drive. The 300-series stainless steel components undergo unlubricated sliding motion and are susceptible to adhesive wear and fretting.

To reduce the wear debris, the components were coated with six different hard coatings: electroless nickel, chromium, Nedox®, titanium nitride, amorphous diamond, and iron boride. Hardware was tested in a specially made chamber that permitted collection of the airborne particles on nucleopore filters as well as real-time particle counts. Particle size distribution and elemental composition were provided by automated image analysis using SEM and EDX.

Ceramic coatings tend to have a much lower particle generation rate than the metallic coatings. The most suitable coatings are titanium nitride because of its large particle size and low particle generation rate and amorphous diamond because of its extremely low particle generation rate.

INTRODUCTION

Particles generated during the operation of a magnetic disk drive can have adverse effects on the performance and reliability of the drive. With the shrinking of bit densities, track densities, and head flying heights (0.3 μm and less), particles one μm and smaller in diameter are becoming increasingly troublesome in the industry from a contamination perspective. Information about the particle composition, size distribution, quantity and morphology is fundamental for their control. This type of knowledge is invaluable to the designer of magnetic disk drives in which airborne contamination must be minimized.[1,2]

Magnetic disk drives cannot use conventional lubricants because of the strict cleanliness required for the head-disk interface. With the exception of the head, disk, and ball bearings, a sliding or rolling material pair must be selected where the interfacing materials have the required wear resistance or self-lubricating properties. If the base materials do not possess the needed tribological properties, then suitable surface coating may be applied. Often the substrate has specific physical properties needed to satisfy functional requirements, but it does not meet tribological requirements. Wear-resistant coatings are an efficient and cost-effective means to satisfy these conflicting requirements.

This paper differs from traditional wear studies, where the volume of material lost is measured with little attention paid to wear particle size. In those cases wear is gross and the consequences are typically a catastrophic loss of the product. This paper quantifies the particles themselves, coming from one part of the disk drive and attempts to understand how they are generated and how they can be prevented. There is a lack of quantitative information in wear-particle morphology and size distribution of the submicron to 5 μm size range of concern. Early studies dealt with large metal particles greater than 40 μm and the relationship between surface energy and size.[3] As yet, wear research has not investigated submicron size particles, particularly of ceramics.[4-6]

Recent advances in vacuum-deposition technology, evolving from the semiconductor industry, has not only found applications in the tool and die industry, but is now commonly available for many novel applications.[7,8] Specifically, disk drive components made of common metal alloys can be coated with exotic materials having unique material properties that reduce contamination.

In this magnetic disk drive a cam follower, sliding against a cam shaft under unlubricated, high-contact stress conditions (on the order of 345 M pascals) causes submicron and micron size wear debris. This particulate contamination may become airborne, transported to the read/write head vicinity, and then get trapped within the head-disk interface. This can result in an instantaneous head crash and a permanent loss of stored information. The most detrimental particles are those that are of a size equivalent to the head flying height and less. Another concern is that large particles are readily broken up by the impact of the head. The only 'good' particle is one that is too massive to become airborne or too solid to be crushed by the head and is swept away.

The cam shaft of the disk drive is made of 302-stainless steel and the follower is made of 301-stainless steel. Both parts are susceptible to adhesive wear and fretting. Reducing the metallic wear debris is achieved by applying hard and durable metallic and ceramic coatings to both of the contact surfaces. The coatings tested, which were applied over a base material of 300-series stainless steel, were: electroless-nickel, chromium, Nedox®, titanium-nitride, amorphous diamond and iron boride.

This paper discusses a test methodology developed to evaluate (during product operating conditions) the wear regime, particle generation rate, particle size distribution, and particle morphology for these six wear-resistant coatings.

COATING EVALUATION TEST METHOD

A special test chamber was developed for the purpose of duplicating the rolling-sliding motion of the two surfaces and for counting and collecting the airborne wear particles. This method utilizes the product cam shaft and follower.

The test chamber holds one cam shaft supported on each end by Teflon® bushings and the follower mounted by a screw to a plate as illustrated in Figure 1. Materials used in construction of the test chamber are aluminum and Plexiglass. To give the cam shaft a reciprocal motion at a rate of 4.7 cycles per minute, a stepper motor was mounted to the bottom to simulate the actual function. A mechanical stop, located on top of the chamber, limits the cam shaft's arc to a total of 86.5 degrees of which the cam shaft and follower are engaged for only 16 degrees. A contact sensor, used to count cycles, is located on one of the stops.

Class 10 air is supplied at 22.5 air changes per minute to the chamber which is sufficient to aerosolize the particles. A tube, comprising the air inlet, is located near the cam-follower interface. The air velocity impinging on the interface removes particles from the inter-

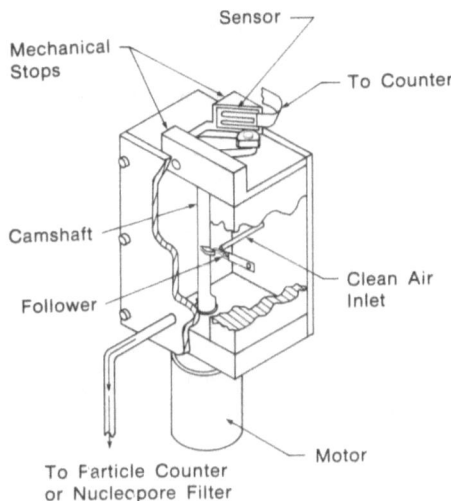

Figure 1.   Test chamber

face and mixes them throughout the chamber volume. This simulates the airflow forces encountered in the disk drive. Particles are sampled from the chamber via a plastic tube to either a Climet model CI-226M white-light particle-counter or to a Nucleopore® 0.4 μm filter. The Climet particle-counter draws air at the rate of 0.25 cubic feet per minute.

Testing begins after all parts are assembled, the temperature is stabilized at 60°C, and the chamber is purged of assembly debris. An IBM PC-XT controls the test and collects the data by interfacing with the particle counter monitor and the motor controller. See the schematic diagram in Figure 2. The data collected are the number of cam shaft cycles and the count of particles per cubic foot that are equal in size or are greater than 0.3 μm.

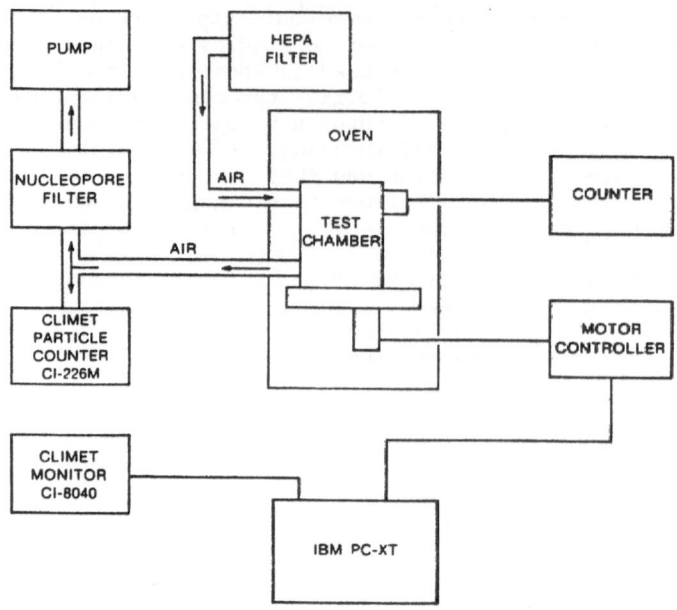

Figure 2.   Schematic of tester

The PC-XT program controlling the test was written to test the parts
to a minimum of 2500 cycles and to terminate the test at 27,000 cycles.
Based on known product sensitivities, if the particle count exceeds the
goal of 49 particles per cubic foot then the PC stops the motor; the
hardware is considered to have failed.

After failure, particles are collected for automatic image analysis.
In-line Gelman filter holders, containing 0.4 μm polyester filters, are
connected to the tube that feeds the Climet particle counter. The
Nucleopore® filter is removed from the holder and carbon coated. The
particles on the filter are analyzed by an automated image analysis
system.[9,10] In accordance with the traditional analytical techniques, each
particle is categorized by size using Scanning Electron Microscopy (SEM)
and by its elemental composition using Energy Dispersive X-rays (EDX).[11] A
histogram of the particle size distribution for the different particle
types is constructed, as well. SEM micrographs are taken of representative particles.

After the test has been completed, the cam shaft and follower are
removed from the chamber to inspect the wear scar by SEM. When required,
the debris and wear zone are analyzed for elemental composition using
EDX. The cam shafts are cross-sectioned to enable measurements to be
made of the coating hardness and the coating thickness.

RESULTS AND DISCUSSION

When selecting a coating to improve cam shaft and follower wear
resistance, several requirements must be met to be successful. The material must have high fracture toughness and hardness to be able to resist
high contact stresses. A low coefficient of friction is needed to reduce
both localized shear stresses and heating. Oxidation resistance is necessary to prevent fretting corrosion at the interface. The temperature
of the coating process must be low enough to prevent annealing the spring
temper of the follower and to avoid distorting the parts. (See Table I.)

Table I. Coating Properties.

Material	Fretting Resistance	Micro-hardness kg/mm²	Thickness, μm	Coefficient of Friction
301 & 302 Stainless Steel	Moderate	230 - 260	(Base Material)	0.09
Electroless-Nickel	High	500 - 1000	10 - 23	0.08
Chromium	High	450 - 1100	4 - 8	0.05
Nedox-SF2®	High	750 - 1000	19 - 25	0.10
Titanium-Nitride	High	1200 - 2000	2 - 4	0.19
Amorphous Diamond	High	5000 - 6000	2 - 3	0.20
Iron Boride	Poor	1700 - 1800	2 - 3	0.44

300-Series Stainless Steel

A basis for comparing all other coatings is provided by the test results of the five uncoated 300-series stainless steel cam shafts and followers. The cumulative airborne particle-count, with respect to the number of load/unload cycles is graphically illustrated in Figure 3. The average is 33,100 particles at 2500 cycles. Note that four of the shafts behaved similarly while the fifth shaft generated over 400% more parti- cles. Its failure mode was distinctly different than the rest. All of the graphs indicate a constant generation rate because the curves are straight lines. The failure points, when the particle count exceeded 49 particles per cubic foot, are denoted by an X.

The airborne stainless steel particle-diameter distribution is shown in Figure 4. While the mean particle-diameter is 1.31 μm in length, sizes can range up to 4.50 μm. The elemental composition provided by EDX analysis shows these particles to be mixtures of iron, nickel, and chromium, which are the alloying elements of 300-series stainless steel.

The SEM micrograph of a stainless steel particle on a nucleopore filter is shown on Figure 5a. The chromium-nickel-iron particle captured on the filter is very flat and is primarily one whole particle. It provides a good example of adhesive wear debris.

The wear scar, shown in Figure 5b provides an unmistakable image of adhesive wear, also known as galling. Adhesive wear occurs when two smooth surfaces under high contact stresses slide over each other, causing fragments to be pulled off one surface and to adhere to the other. Later, these fragments may smear or transfer back to the original surface or become loose. This action produced the irregular flat parti- cles within and along the margin of the wear zone.

Auger Electron Spectroscopy (AES), capable of light element detection, was performed on the area in and around the wear zone on a follower. A reddish-brown color was characteristic of some wear debris and was suspected to be fretting corrosion products of the stainless steel. After 10 minutes of Argon sputtering to clean the surface of the typical adsorbed layers of oxygen and water, the unworn area showed a very small oxygen peak. The freshly exposed metal in the wear scar has about 700% more oxygen while the wear debris has about 300% more oxygen than the unworn metal. This confirmed the tendency for the 300-series stainless steel to undergo fretting corrosion. Fretting corrosion arises when the two contacting surfaces, under load, are subjected to small amplitude vibration and slip. Fretting is a special case of erosion- corrosion that typically occurs in the air rather than in liquid media. In this case, stainless steel wear was caused by a three-step process, beginning with adhesive wear which generates particles and exposes the steel which then forms hard oxides. The fretting corrosion then initiates more aggressive abrasive wear of the stainless steel, thus generating more particles. Four out of five cam shafts showed adhesive wear while the fifth showed fretting corrosion.

Electroless-Nickel

Electroless-nickel plating readily generated contamination as shown on the cumulative airborne particle-count graph in Figure 6. The average number of particles at 2500 cycles is 204,000, which is five times that of stainless steel. Note that the particle generation rate is not linear but steadily increases. The failure points are denoted by an X.

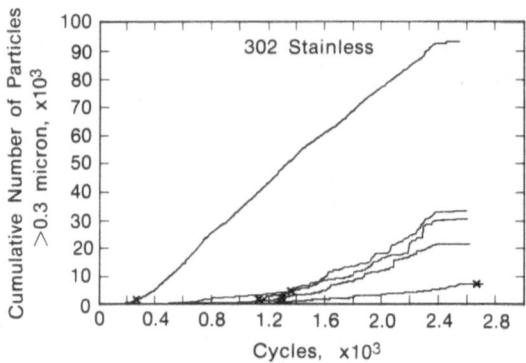

Figure 3.   Cumulative particle counts of 300-series stainless steel.

The airborne nickel particle-diameter distribution is shown in Figure 7. Because the nickel wore through, exposing the stainless steel substrate, the elemental composition of the particles now indicates two distinct populations. Electroless-nickel particles are composed of nickel and phosphorous, while stainless steel particles are made of iron, chromium and nickel. The mean diameter of the nickel particles is 1.20 μm but some may range up to 4.50 μm, which verifies the observation that the fine powder adheres to the surface and typically forms agglomerates that break loose to then become airborne. The stainless steel particles are fewer in number and larger in size than the nickel particles with a mean diameter of 1.4 μm.

An SEM micrograph of a nickel particle on a Nucleopore filter is seen at a magnification of 12,000X in Figure 8a. Also shown below is the fine submicron nickel powder on the cam lobe at a magnification of 500X. The single large particle is an agglomeration of 15 to 20 smaller particles of nickel. Some of the submicron particles in the powder are compacted by the rolling and rubbing action of the cam against the lever and are later ejected into the airstream. In Figure 8b signs of abrasive wear are indicated by the deep wear and the distinct straight lines in the wear scar and the presence of very fine debris, both in and out of the wear zone.

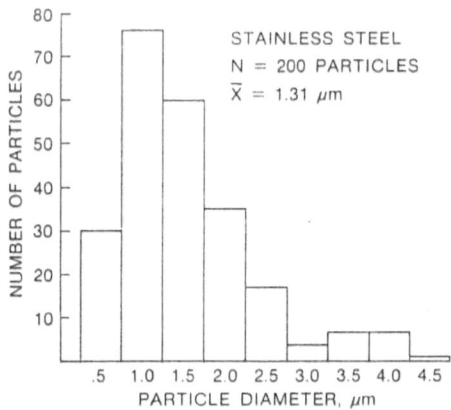

Figure 4.   Airborne stainless steel particle-diameter distribution.

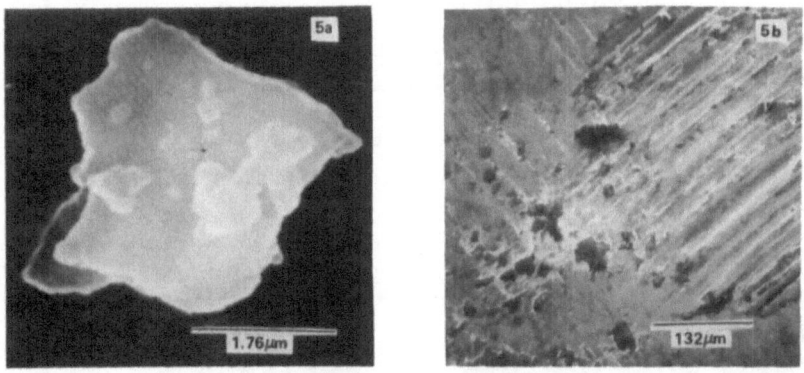

Figure 5.   a) Single stainless steel particle on filter;
             b) 302 stainless steel cam shaft adhesive wear scar.

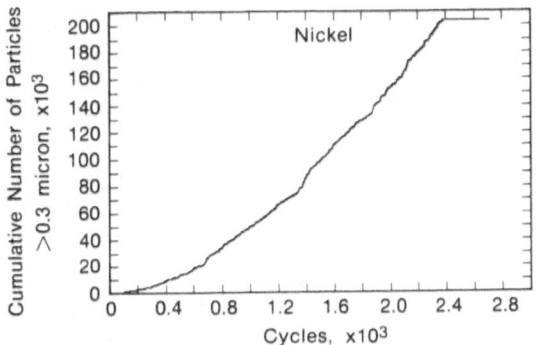

Figure 6.   Cumulative particle-counts of electroless nickel.

Chromium plating readily generates contamination as shown on the cumulative airborne particle-count graph in Figure 9. The average number of particles at 2500 cycles is 41,700, which is a little more than stainless steel. In one case the particle generation rate was nonlinear and was possibly caused by a combination of surface fatigue wear and abrasive wear. The failure points are denoted by an X.

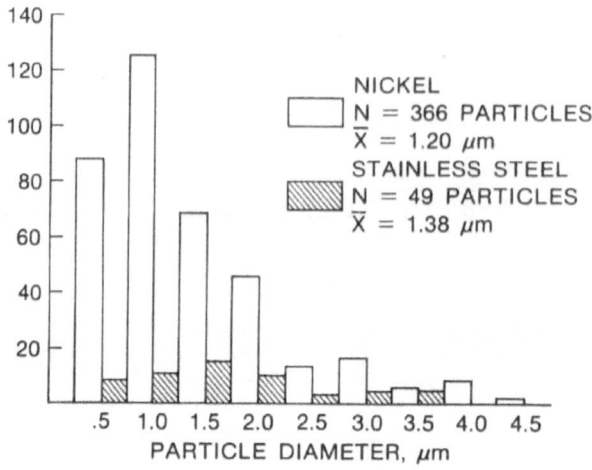

Figure 7.   Airborne nickel particle-diameter distribution.

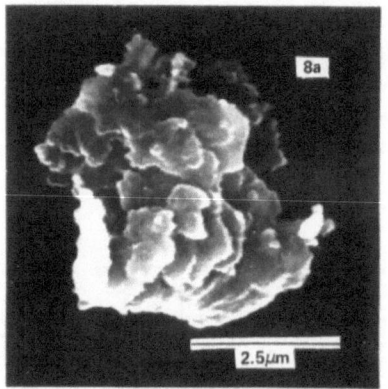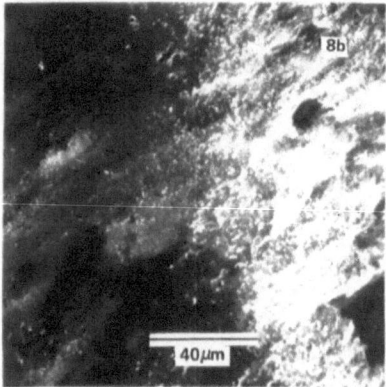

Figure 8.   a) Agglomerate of nickel particles on filter.
             b) Fine abrasive wear particles in the cam shaft wear scar.

The airborne chromium particle-diameter distribution is shown in Figure 10. All particles contain chromium; there was no wear through the chromium into the stainless steel. While the mean particle diameter is 1.12 μm, some may range up to 3.50 μm, which verifies that the fine powder adheres to the surface and usually only agglomerates become airborne.

224

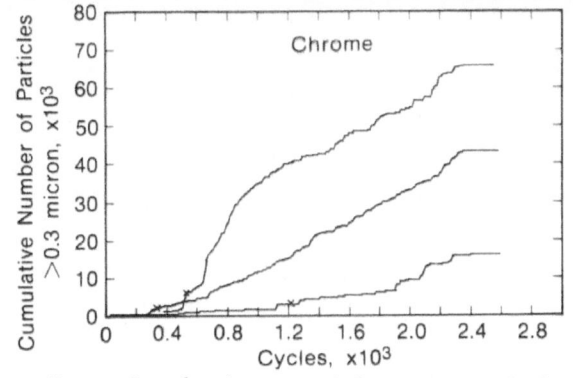

Figure 9.   Cumulative particle-counts of chromium.

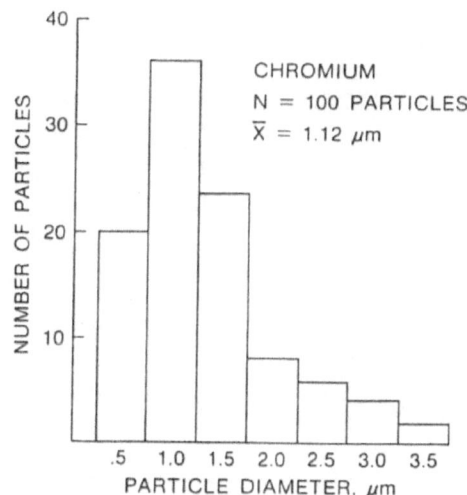

Figure 10   Airborne chromium particle-diameter distribution.

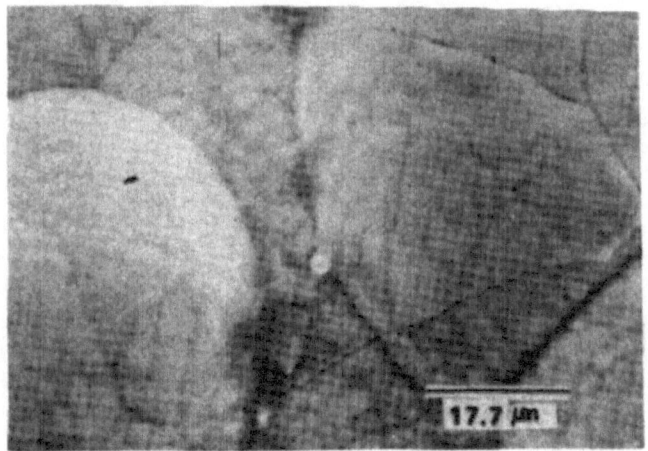

Figure 11.   Cracks in chromium prior to testing.

When chromium is plated, it has a tendency to form microscopic cracks on its surface. An SEM micrograph of a cam shaft taken at 1130X magnification clearly shows the residual surface cracks in the chromium prior to testing (see Figure 11). This same cam shaft shows the presence of chromium nodules that create an unacceptable abrasive surface. This cam shaft was not tested in this condition, but was stripped of the chromium and replated. Surface-fatigue wear is observed during repeated sliding or rolling over the same surface. The repeated loading and unloading cycles to which the materials are exposed may induce the formation of surface cracks or cause existing cracks to propagate, resulting in the fracturing of the chromium. The combination of high surface stresses with the initial surface cracks makes chromium a likely candidate for surface fatigue wear as the load/unload cycles progress.

An SEM micrograph of a chromium particle on a Nucleopore filter is shown at a magnification of 8,000X in Figure 12a. The particle on the filter is clearly an agglomeration of over 100 submicron size spherical

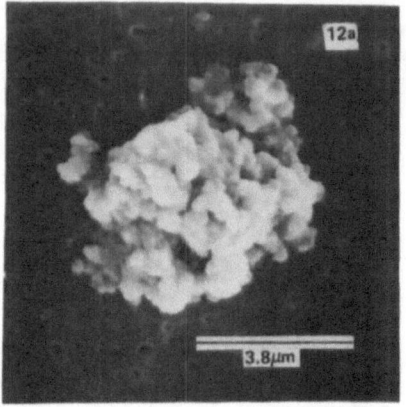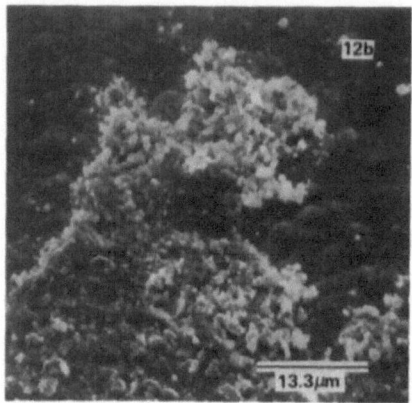

Figure 12.   a) Agglomerates of chromium particles on filter.
              b) Fine abrasive wear particles in the cam shaft wear scar.

chromium particles. This indicates that some of the particles in the powder are compacted by the rolling and rubbing action of the cam against the lever and are later ejected into the airstream.

Figure 12b includes an SEM micrograph of the cam lobe after the wear test. After testing, deep scars are evident and show signs of abrasive wear indicated by the clean straight lines in the wear scar and the presence of submicron debris both in and out of the wear zone.

Nedox-SF2®

The Nedox® process is a synergistic coating which capitalizes on the benefits of the hardness and toughness of electroless nickel and the lubricant properties of fluorocarbons. The two-step Nedox® process reduces the nickel alloy coating on the stainless steel surface. This deposit contains countless micropores which are enlarged in a series of proprietary treatments. Then, the surface is sealed with a controlled infusion of submicron-size particles of polytetrafluoroethylene (PTFE), after which it is heat treated to harden the nickel and create a smooth surface with a low coefficient of friction.

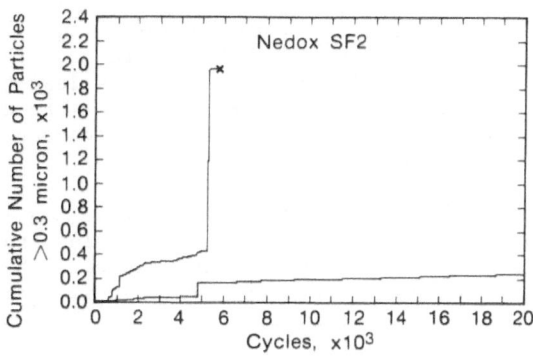

Figure 13   Cumulative particle-counts of Nedox-SF2®

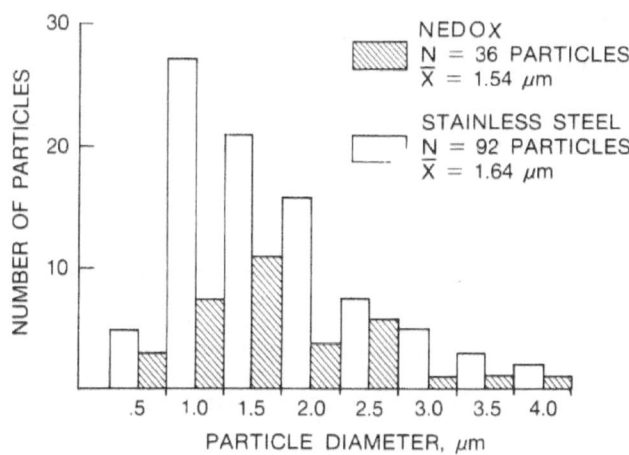

Figure 14   Airborne Nedox-SF2® particle-diameter distribution.

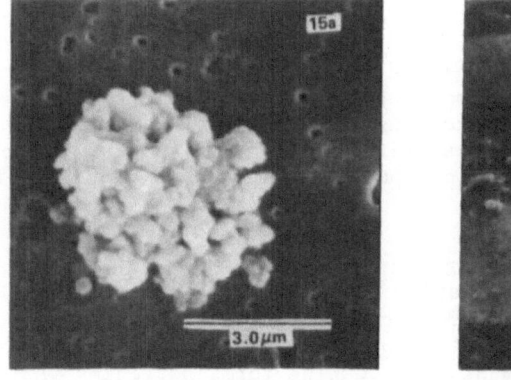
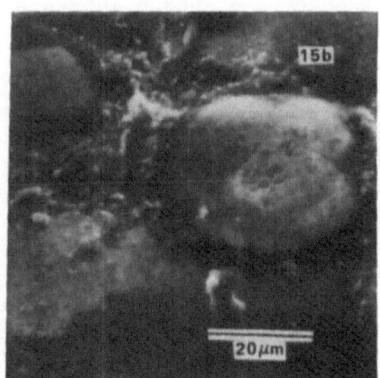

Figure 15.   a) Agglomerates of Nedox-SF® particles on filter.
              b) Fine nickel debris and teflon nodules in the cam shaft
                 wear scar.

Nedox-SF2® plating generates very few particles as shown on the cumulative airborne particle count graph in Figure 13. The average number of particles at 2500 cycles is 190, which is less than 1% of that for stainless steel. Small bursts of particles give the cumulative graph a stair-step effect, because the one-minute particle counts were so low. The failure point is denoted by an X.

The airborne Nedox-SF2® particle-diameter distribution is shown in Figure 14. The elemental composition of the particles indicates two distinct populations because the Nedox eventually wore through, exposing the stainless steel substrate. The mean Nedox particle diameter is 1.54 μm, while the stainless steel particles are greater in number and larger with an mean diameter of 1.64 μm.

An SEM micrograph of a Nedox® particle on a Nucleopore® filter is shown at a magnification of 10,000X in Figure 15a. This single large particle is an agglomeration of 20-30 submicron-size nickel particles.

Figure 15b shows an SEM micrograph of the cam lobe after the wear test. Nedox® has small nodules of PTFE that are exuded from the micropores in the nickel. There is evidence of a small amount of abrasive wear. Also shown is the fine submicron nickel particles that form a powder on the cam lobe at a magnification of 1000X. This phenomenon was also observed with nickel and chromium.

Titanium-Nitride

Titanium-nitride is applied to the stainless steel parts by means of *Physical Vapor Deposition* (PVD). PVD is a vacuum coating process in which the titanium-nitride is vaporized from a source by evaporation or sputtering, then moved through a vacuum by acquired kinetic energy to strike the target parts. The titanium-nitride condenses to form a film on the surface of the stainless steel. The optimal thickness, 2-4 μm, provides wear resistance, while preventing the domination of brittle properties. The main advantage of PVD is that the low application temperature, 150° - 500°C, will neither soften nor distort the substrate.

Titanium-nitride coating generates very few particles as seen on the cumulative airborne particle-count graph in Figure 16. The average number of particles at 2500 cycles is 240 which is less than 1% of that found for uncoated stainless steel. Two of the cam shaft and lever sets did not fail. When the third set did fail at 15,500 cycles, its failure was preceded by a sudden increase in the particle-generation rate starting at about 10,000 cycles. The failure point is denoted by an X.

The airborne titanium-nitride particle-diameter distribution is shown in Figure 17. The elemental composition of the particles indicates two distinct populations because the coating wore through, exposing the stainless steel substrate. The mean titanium-nitride particle diameter is 2.93 μm, while the stainless steel particles are smaller with an mean diameter of 2.52 μm.

Figure 18a shows an SEM micrograph of a titanium-nitride particle on a Nucleopore filter at a magnification of 30,000X. Shown on Figure 18b are the submicron titanium-nitride particles that form a layer of powder on the cam surface at a magnification of 3000 times. As would be expected, the brittle titanium-nitride coating forms particles that are agglomerations of smaller, submicron particles.

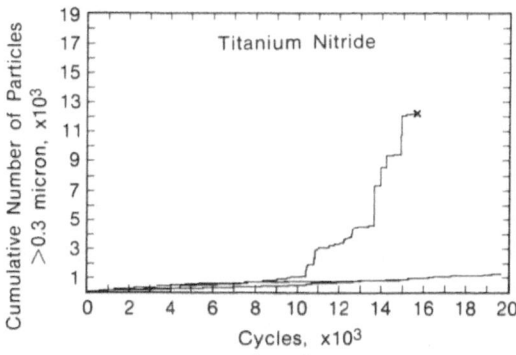

Figure 16.   Cumulative particle-counts of titanium-nitride.

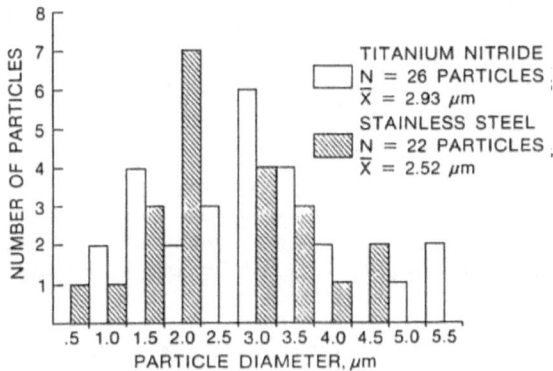

Figure 17.   Airborne titanium-nitride particle-diameter distribution.

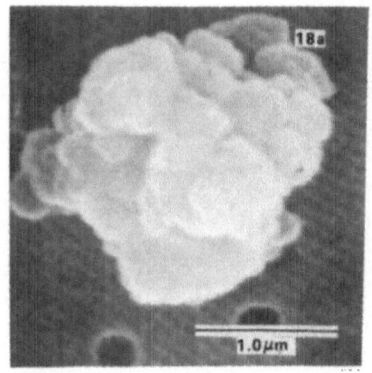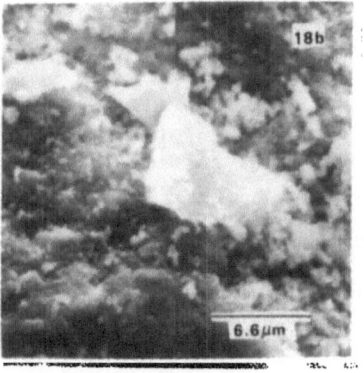

Figure 18.   a) Agglomerate of titanium-nitride partiles on filter.
              b) Fine abrasive wear particles in the cam shaft wear scar.

229

Amorphous Diamond

Amorphous diamond is applied to the stainless steel by means of Chemical Vapor Deposition (CVD). CVD is a vacuum coating process in which methane gas is chemically reacted to remove the hydrogen to form a film of amorphous diamond on the heated surface of the stainless steel. The optimal thickness, 2-3 μm provides wear resistance, while preventing the domination of brittle properties.

Amorphous diamond generates very few particles as shown on the cumulative airborne particle count graph in Figure 19. The average number of particles at 2500 cycles is 101 which is less than 1% of that found for uncoated stainless steel. One of the cam shaft and lever sets began failing catastrophically when the particle generation rate suddenly increased by a factor of five; this occurred at about 9,100 cycles.

The airborne particle-diameter distribution, Figure 20, explains the high particle counts. The amorphous diamond coating wore through exposing the stainless steel substrate to attack by abrasive wear. These fine stainless steel particles have a mean diameter of 0.83 μm. The amorphous diamond particles cannot be detected by EDX.

Figure 21a shows an SEM micrograph of three stainless steel particles. They are the smaller agglomerates of submicron-size particles that were produced during the abrasive wear of the stainless steel by the amorphous diamond. Shown on Figure 21b are the submicron stainless steel particles that form in and around the wear scar.

Iron Boride

Iron boride is created at the stainless steel surface by means of a thermal diffusion process. By submerging the parts in a molten borate-salt bath, the surface of the stainless steel is converted to a very hard iron boride, chromium boride, and nickel boride. The typical thickness is 2-3 μm.

Iron boride generates particles rapidly as seen on the cumulative airborne particle-count graph in Figure 22. The average number of particles at 2500 cycles is 10,900, which is a poor performance compared to soft stainless steel.

The airborne iron boride particle-diameter distribution is shown in Figure 23. The elemental composition is iron, chromium, and nickel with the apparent absence of boron because of EDX detection limitations for elements lighter than sodium. All of the particles collected are iron/chromium/nickel borides for two reasons. First, the wear scars burnish very smooth and show no evidence of breakthrough into the stainless steel substrate. Second, a layer of metal boride crystals project from the surface and are easily broken off. The mean iron boride particle diameter is 0.59 μm, which is half the size of the typical stainless steel particles.

Figure 24a shows an SEM micrograph of a captured iron boride particle at a magnification of 12,000. Also shown are three SEM micrographs 24b, c and d, depicting wear as it progresses from a rough, highly crystalline surface, through partial wear, to a smoothly-burnished wear scar. Fine particles accumulated around the periphery of the wear scar.

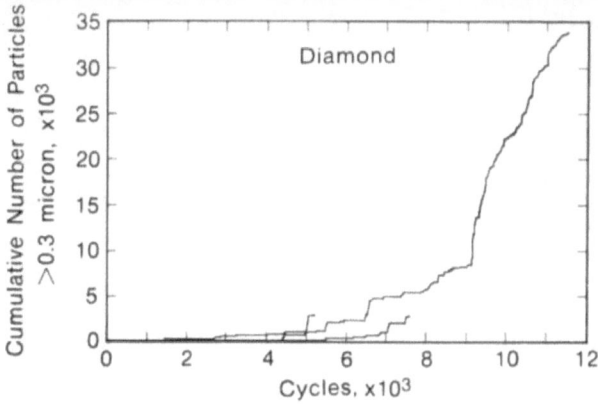

Figure 19.   Cumulative particle-count of amorphous diamond.

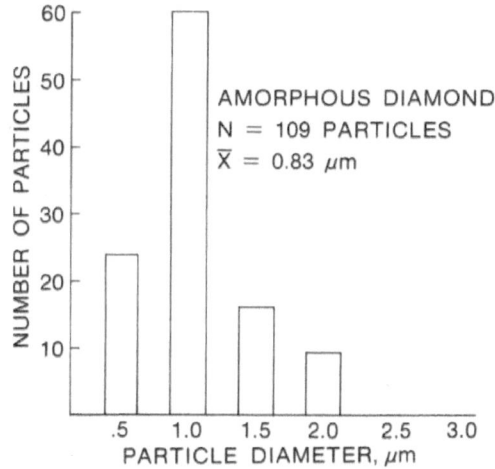

Figure 20.   Airborne amorphous diamond particle-diameter distribution.

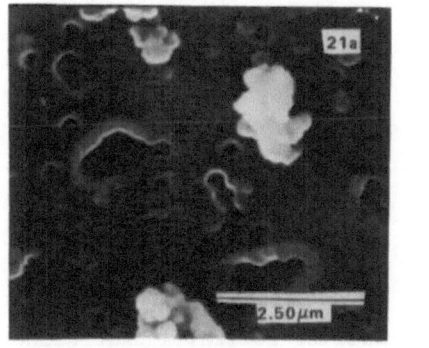

Figure 21.   a) Steel particles from amorphous diamond test on filter.
              b) Fine abrasive wear particles in the cam shaft wear scar.

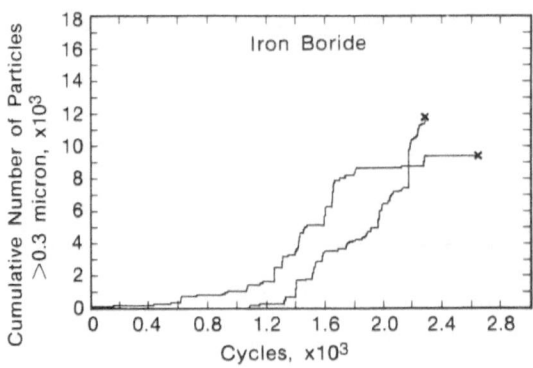

Figure 22.   Cumulative particle-counts of iron boride.

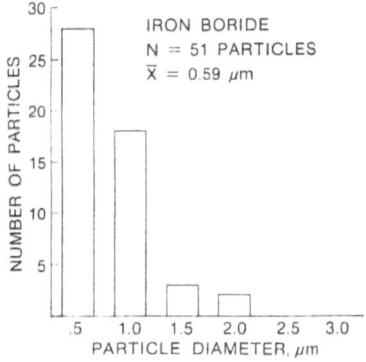

Figure 23.   Airborne iron boride particle-diameter distribution.

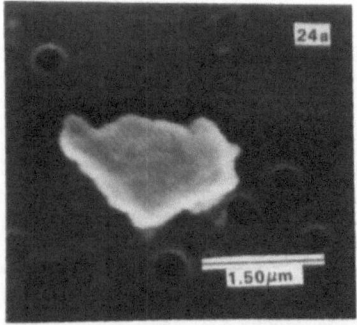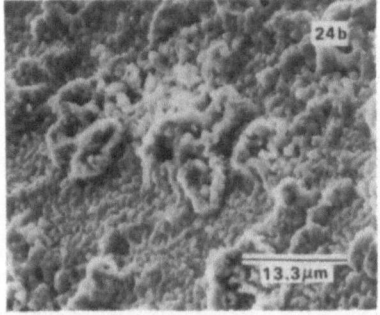

Figure 24.   a) Single iron boride particle on filter.
            b) Cam shaft surface unworn showing crystalline surface.

 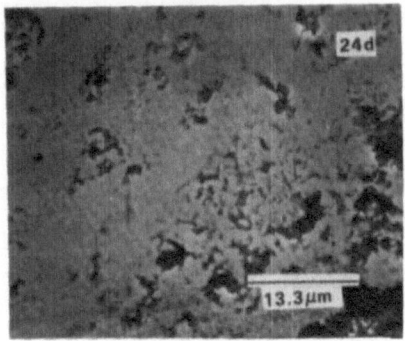

Figure 24. c) Cam shaft wear scar partially worn on high points only.
d) Cam shaft wear scar worn smooth.

CONCLUSIONS

1. The rate of particle generation by metals is much higher than that from ceramics or synergistic coatings.

2. The differences in particle morphology are explained by the wear-mode that generate the particles. Adhesive wear produces single flat particles, while abrasive wear produces spherical submicron-size particles that form agglomerates.

3. The particle-diameter distribution is skewed normal and is dependent upon the type of material. Amorphous diamond and iron boride formed significantly smaller particles, while titanium-nitride formed larger particles.

4. Titanium-nitride is the most suitable material for sliding motion applications in a magnetic disk drive because of its low particle generation rate, and its large particles, see Table II.

Table II. Summary of Results

Cam Shaft Material	Follower Material	Sample Size Tested	Cumulative Particles at 2500 Cycles, x 10³	Mean Particle size, μm
302 Stainless Steel	301 Stainless Steel	5	33.1	1.31
Electroless Nickel	Electroless Nickel	1	204.0	1.20
Chromium	Chromium	3	41.7	1.12
Nedox-SF2®	Nedox-SF2®	2	0.190	1.54
Titanium Nitride	Titanium Nitride	3	0.240	2.93
Amorphous Diamond	Amorphous Diamond	3	0.101	0.83
Iron Boride	Iron Boride	2	10.9	0.59

REFERENCES

1. L. Bailey and G. Rogers, Microcontamination, 3, (11), 80-84 (Nov. 1985).

2. R. Rosler, Microcontamination, 4, (4), 14-18 (April 1986).

3. E. Rabinowicz, "Friction and Wear of Materials", pp. 151-158, John Wiley and Sons, New York, 1986.

4. N. Soda, Y. Kimura, and A. Tanaka, Wear, 35, 331-343 (1975).

5. E. Finkin, ASLE Trans.,7, 377-382 (1964).

6. D. Scott, W. Seifert, and V. Westcott, Scientific American, 88-97 (May 1974).

7. C. Wick, Manufacturing Engineering, 26-31 (December 1986).

8. C. Rain, High Technology, 59-64. (March 1983).

9. R. Edwards, T. Lebiedzik and G. Stone, Scanning, 8, 221-231. (1986).

10. T. Vander Wood and J. Rebstock, Microcontamination, 6, (2), 24-27 (February 1988).

11. D. Bakale and C. Bryson, Microcontamination, 1, (3), 32-35, 63 (Oct/Nov 1983).

PARTICLE REDUCTION ON SILICON WAFERS AS A RESULT OF ISOPROPYL ALCOHOL VAPOR DISPLACEMENT DRYING AFTER WET PROCESSING

Joan W. Koppenbrink, Christopher F. McConnell, and
Alan E. Walter
CFM-Rodel, Inc.
9495 E. San Salvador Drive
Scottsdale, AZ 85258

CFM Technologies, Inc.
501 Gordon Drive
Lionville, PA 19353

The presence of particles on the surfaces of semi-conductor wafers is known to negatively affect device yields. As linewidths continue to shrink, the inherent limitations in many of the commonly used methods for drying wafers after wet processing have created the need for alternatives which offer better particle control.

In this paper the use of IPA vapor displacement drying was examined as a means to dry wafers. Wafers were immersed in water in an enclosed chamber, and IPA vapor was then released into the top of the chamber while the water was pumped out through the bottom. Physical observation of drop formation on the wafers indicated both temperature and drain rate are important factors in effective drying with this method. When a commercial apparatus applying these principles was used to dry wafers in a Class 10 clean room, fewer than two particles greater than 0.5µm in diameter were routinely added to 100mm wafers, confirming the practical applicability of the findings.

It is postulated that the particle reduction associated with this drying system is related to both the surface tension characteristics and the polar/non-polar characteristics of isopropyl alcohol. In addition, the volatility of IPA plays an important role in its drying performance.

INTRODUCTION

During the fabrication stages of silicon wafers, several process steps require contacting the wafers with fluids. Steps based on wet chemistry include etching, photoresist stripping, and prediffusion cleaning. The equipment conventionally used for wet chemical processes generally consists of a series of tanks or sinks into which racks of semiconductor wafers are dipped. A widely used alternative method sprays the chemicals directly onto the wafers in an enclosed chamber[1].

Particulate control is, of course, a critical concern in any wet processing step. Baths in particular are susceptible to particle transfer from the surface of the liquid to the wafer as the wafers are drawn out of the tank through the interface[1,2]. Additional challenges in contamination control arise during the drying step which follows the final rinse in most processes. Evaporation is undesirable since it often leads to spotting, streaking and particle deposits. Even the evaporation of ultra high purity water can lead to problems because such water is very aggressive to the wafer surface and will dissolve traces of silicon and silicon dioxide during even short periods of water contact. Subsequent evaporation will leave residues of the solute material on the wafer surface[3].

Conventionally, semiconductor wafers are dried through centrifugal force in a spin-rinser-dryer. Because these devices rely on centrifugal force to throw water off the wafer surfaces, evaporation of traces of water--with the potential for the problems mentioned above--is difficult to avoid. Also, since the wafers travel at a high velocity through dry nitrogen, static electric charges can develop on the wafer surfaces. Since oppositely charged airborne particles are quickly drawn to the wafer surfaces when the dryer is opened, particle contamination can result[3].

Recently, methods and apparatus have been developed for steam or chemical drying of wafers. A chemical drying process currently used in Japan consists of immersing the wafer-carrying vessel in tanks of deionized water and then suspending the wafers above a tank of boiling isopropanol. The wafer-carrying vessel is then slowly withdrawn from the isopropanol vapor to pull the water droplets off the wafer surfaces[3]. A second type of chemical drying process, the one which is the subject of this paper, is comprised of two steps: first, the rinsing fluid is driven off the wafers and replaced by a nonaqueous drying fluid; second, the nonaqueous drying fluid is evaporated using a pre-dried gas, preferably an inert gas such as nitrogen, at a low flow velocity.

To learn more about the physical parameters that are associated with this type of drying, a series of experiments were done. First a group of drying runs were performed in a see-through vessel which allowed visualization of the process. These findings, coupled with particle data generated in a clean room environment, were used to develop hypotheses to explain the observations.

EXPERIMENTAL RESULTS

The drying technique under investigation is one in which wafers remain completely immersed in water after completion of the final rinse. This water is subsequently displaced by a vapor, in this case isopropanol, which is then followed by a nitrogen purge to complete the process.

Laboratory Testing

The apparatus used to view this process consisted of a quartz bell jar, with an enclosed boat of wafers, incorporated into a pipeline. Water was piped into the chamber from the bottom and allowed to overflow at the top until temperature equilibration was reached. The flow was then stopped, and IPA vapor was released into the top of the vessel while the water was metered out through the bottom. A flash valve was used to keep the IPA vapor pressure as constant as possible. After the rinse water had been totally displaced with IPA, nitrogen gas was used to remove the IPA from the system. A schematic of this system is shown in Figure 1.

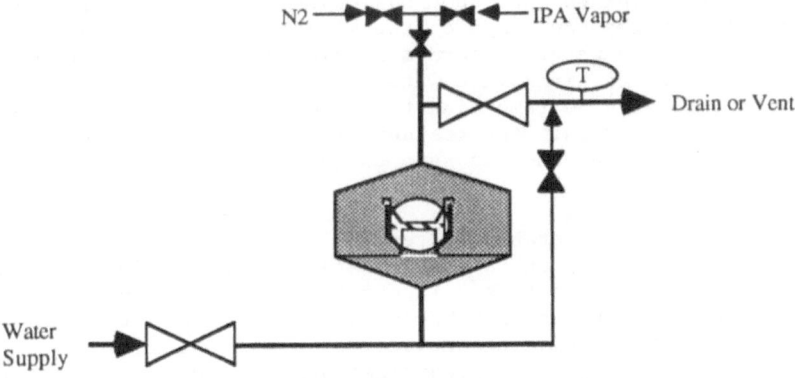

Figure 1A. Hot water flows through vessel until constant temperature is reached.

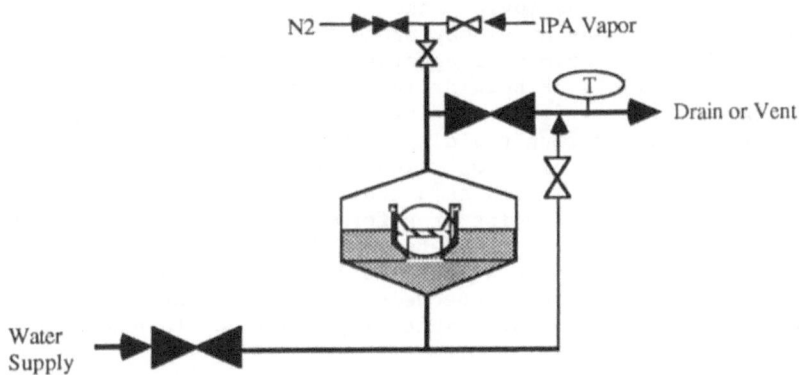

Figure 1B. IPA vapor displaces water at a controlled rate.

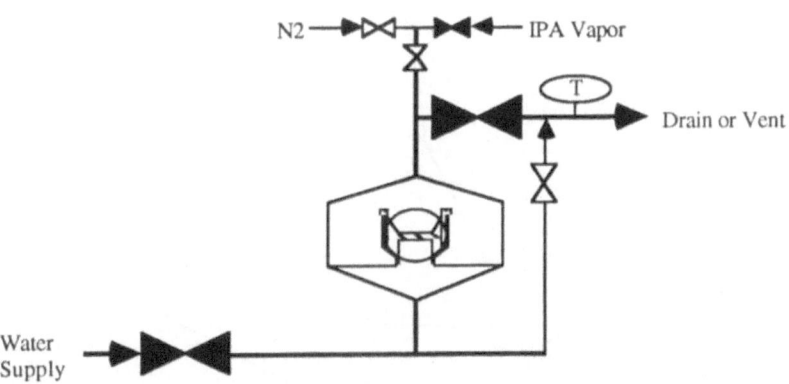

Figure 1C. N2 gas purges IPA from the vessel.

Figure 1. Experimental apparatus. Valves are open (▷◁) or closed (▶◀) as shown.

One of the findings from numerous operations of this apparatus was that IPA actually condensed on the sides of the vessel and on the surface of the rinse water to form a liquid layer which traveled down on the surface of the exiting rinse water. This layer could then act as a drying fluid, dissolving remaining traces of water left on the wafer and carrying it out to drain. Any residual IPA left on the wafers was a highly volatile species and was seen to quickly evaporate in the vapor phase.

This mechanism is notably different from other types of IPA dryers in that the wafers pass through a liquid water/liquid IPA interface and a liquid IPA/vapor IPA interface. In conventional IPA dryers, the wafers are exposed only to IPA vapor. Leenaars has shown both theoretically and experimentally that sub-micron particles can be removed from a silicon substrate by the passage of a liquid/gas phase boundary and that this removal is in theory independent of particle size[4]. The phase boundaries which occurred in the experimental system created the potential for particle removal in a way in which conventional drying techniques cannot.

Leenaars also noted that the kinetics strongly affected the particle removal efficiency of the moving phase boundary, with greater efficiency when the boundary passes at slower speeds[4]. A series of experiments were performed with the experimental apparatus described in this paper to investigate the effects of both drain rate and temperature. In the experimental apparatus, drops of either water or water and IPA sometimes formed on the wafers during the displacement process. As an indicator of performance, the number of drops which formed on the surface of the front wafer were counted and plotted as a function of temperature and drain rate. Profiles were developed showing drop formation as a function of drain rate at temperatures ranging from 70°C to 90°C (Figure 2). The number of drops which formed was found to be a function of both variables: as temperature increased, the number of drops decreased; as drain rate increased, the formation of drops increased. Since drops could contain dissolved

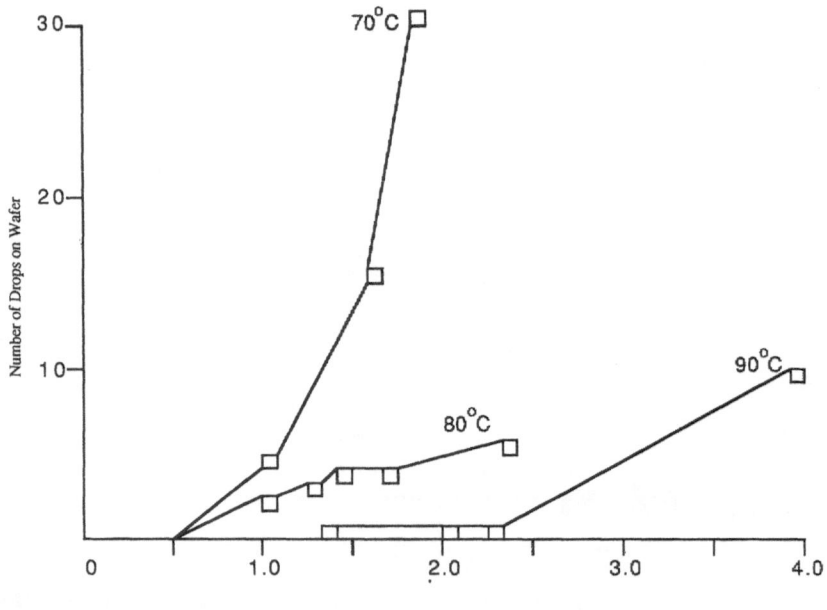

Figure 2. Effects of temperature and drain rate on the number of drops formed on the wafer surface.

contaminants or suspended particles, it could be postulated that the "non-drop" forming portion of the curve would be most advantageous for contamination-free wafers.

Clean Room Testing

To test this hypothesis, additional experiments were performed in a Class 10 clean room with a commercial wafer cleaning unit designed to dry with the IPA displacement technique. To determine the variables which impacted particle deposition or removal, a matrix was designed which included drain rate and temperature at two levels each along with other variables (flash valve opened versus closed, time of IPA exposure, and aqueous versus nonaqueous IPA source). Particle counts across batches of 25 wafers were taken before and after the runs to determine particle change for each cell. Through statistical analysis, the two variables under investigation were found to be significant ones: particle counts were correlated with both drain rate and rinse water temperature. As noted by Leenaars and as could be postulated from the drop-forming experiments, lower particle counts were obtained with slower passage of the phase boundary. Not predicted from the drop-forming study, however, was the finding that an optimum temperature range existed between 60°C and 70°C. At higher or lower temperatures, particle counts increased.

When temperature and drain rate were held in the ranges that were found to yield lower particle counts, results like those shown in Figure 3 have been routinely observed. These data reflect an average particle count change across a batch of 25 100mm wafers (starting counts less than 10) in the range from +3 to −2 when measured at a lower sensitivity level of 0.5μm

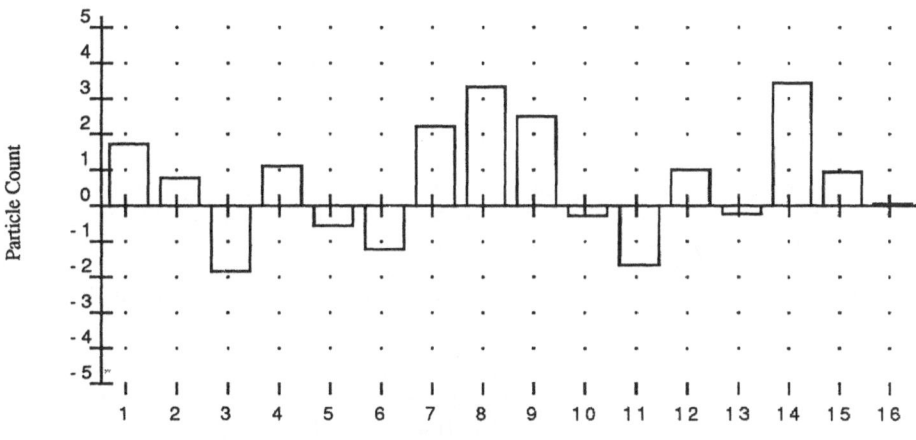

Figure 3. Change in particle counts after IPA drying. Particle data were taken from an Aeronca WIS 150 before and after displacement drying with IPA. Values shown represent the mean of 25 wafers (100mm) for each experiment. Particles > 0.5μm were measured.

Within any given run, particle counts were consistent across the batch of 25 wafers.

All of the experiments described thus far had focused on IPA drying; however, it could be speculated that low particle counts might be achieved without the use of the alcohol. Because the system is completely enclosed, the wafers are drying in a very low particle environment. If drop formation could be prevented by control of the temperature and flow rate of the exiting water, it seemed possible that displacing the water with a hot inert gas might also produce low particle wafers. To determine if IPA was an important factor in particulate control with this system, batches of wafers were dried using IPA vapor to displace the rinse water versus nitrogen gas as the displacing vapor. Again, low particle count starting wafers were used, and particle data were obtained from an Aeronca WIS 150 before and after displacement drying with IPA or nitrogen. The results are depicted in Table I as the mean of 25 wafers (100mm). In these tests, IPA actually caused a net reduction in the number of particles on the wafers, particularly in the very small particle range; N_2, on the other hand, caused a net increase. It should be noted that the nitrogen-dried wafers were still relatively low in particle counts compared to what is often seen with conventional drying techniques; nonetheless, IPA was shown to improve the results even further.

Table I. Particle count change after drying.

	Change in number of particles $< 0.5\mu$m	Change in number of particles 0.5-2.0μm	Change in number of particles $> 2.0\mu$m
Wafers Dried with N_2	+6	+4	+2
Wafers Dried with IPA	−102	−14	−0.5

DISCUSSION

It can be postulated that the efficacy of IPA in this system is related to both surface tension effects and polarity effects.

Surface Tension Effects

Surface tension plays a role in the ability of IPA to displace water and act as a drying agent in the system. Bolster and his colleagues at the Naval Research Laboratory have studied the use of displacement fluids for water removal[5]. The rate of displacement was shown to be proportional to the spreading pressure of the fluids, which, for mutually soluble liquids, is the difference between their surface tensions. The relatively high surface tension of water thus facilitates its displacement by liquids with lower surface tensions.

IPA was not considered to be an effective alcohol for water displacement in the work done at the Naval Research Laboratory because the volatility and solubility of the low molecular weight alcohols limited their persistence. In the apparatus described in this paper, however, IPA vapor was fed into the system; and the receding liquid surface provided a condensation site for continuous replenishing of the liquid IPA layer. Also, as opposed to being a detriment in this system, the volatility of IPA is advantageous in that it allowed evaporation to remove the final traces of the solvent from the wafer.

The particle removal capability of this system may also be related to surface tension and its effect on drop formation. As noted, the tendency to form drops on the surface of the wafer increased as the liquid receded more rapidly. Any drops which remained could evaporate from the surface of the wafer, potentially leaving particles and other contamination behind. This drop formation and subsequent evaporation may be related to Leenaars's observation that particle removal by a passing interface was related to the kinetics of the system. However, in contrast to the findings of Leenaars, whose calculations showed that particle removal was directly proportional to the surface tension of the gas/liquid interface, the experimental results showed that the system with IPA was more effective than that without IPA, despite the lower surface tension of the former system.

The fact that the tendency for drop formation decreases in a system with lower surface tension may help explain the lower particle counts with the IPA system. Figure 4 shows the surface tension of the two liquids in the system, water and IPA. For both liquids, surface tension decreases as temperature increases, a phenomenon which is consistent with the observations in the drop formation experiments. In addition, the fact that the surface tension of IPA is significantly less than that of water may help explain why IPA appears to be an important element in the system. Surface tension is very low at the IPA/vapor interface, and the tendency to leave

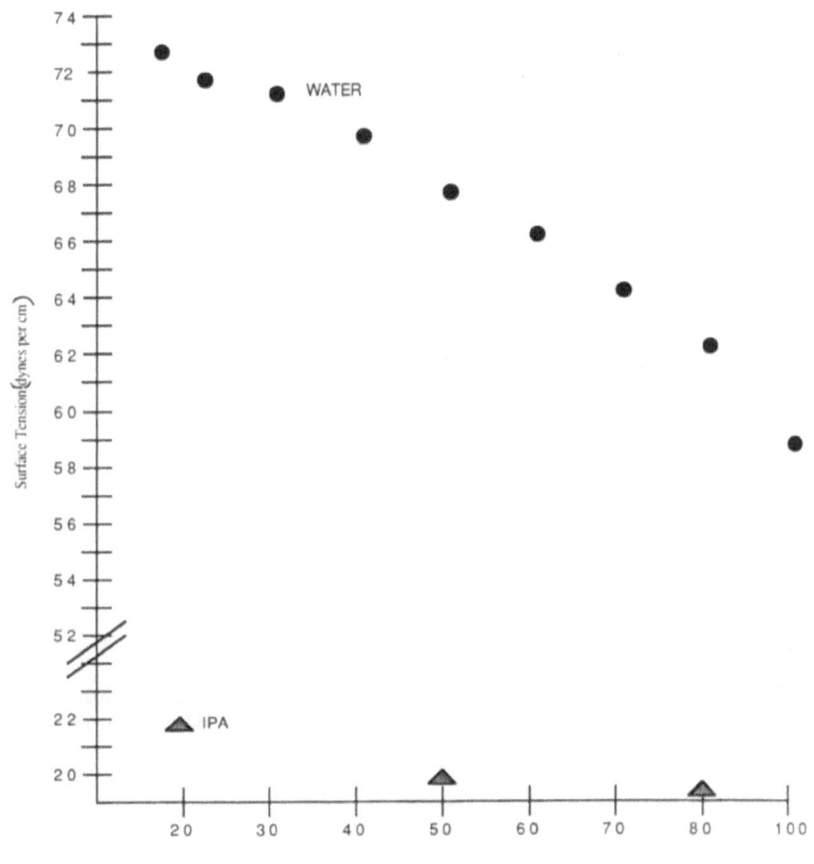

Figure 4. Effect of temperature on the surface tension of water and IPA[6],[7].

Isopropyl Alcohol   Water

Figure 5A: The structures of isopropyl alcohol and water showing the
amphipathic characteristic of the alcohol

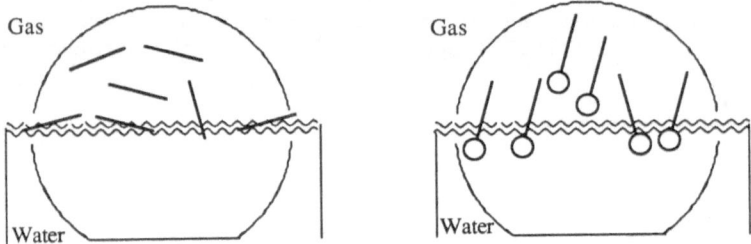

Figure 5B: In a water system, hydrophobic contaminants locate at the gas/liquid
interface and can be left behind on the wafer surface as the liquid recedes.

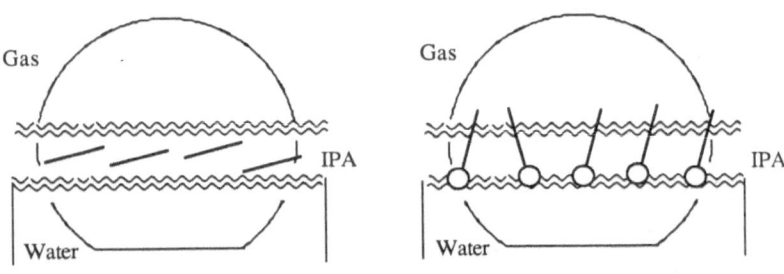

Figure 5. Model for orientation of polar and non-polar substances in water
and isopropanol. ▬ Hydrophobic contaminants. ▬○ Polar/non polar contaminants.

242

droplets behind as the liquid recedes is limited.

Polarity effects

IPA may also play an important role due to its amphipathic, or polar/ non-polar, qualities. In water alone, non-polar contaminants can concentrate at the water surface. As the water recedes (or the wafer is pulled out of the water as in a wet sink), particles can be attracted to the relatively non-polar surface of the silicon wafer. In the IPA system, however, the non-polar contaminants may be distributed in the IPA layer and carried out with the liquid. Contaminants with both polar and non-polar moieties (such as many of the hydrocarbons) may be preferentially oriented at the water/IPA interface with the hydrophilic portions oriented toward the water and the hydrophobic segments toward the IPA. In either case, many of the non-polar species which tend to be left behind as water recedes can be distributed in liquid IPA and carried out with the drain fluids. This idea is shown pictorially in Figure 5.

CONCLUSIONS

These experiments indicate that using isopropyl alcohol to displace rinse water is an effective drying method to decrease particle contaminat on semiconductor wafers. In the apparatus described here, the physical parameters of drain rate and temperature have been shown to be important toward generating the lowest particle results.

The theoretical explanation for the role of IPA in the system could lead to other chemicals which might also be effective drying agents for semiconductor wafers. Important criteria include the following:

1. Volatility to yield clean, dry wafers

2. Miscibility with water to prevent water droplet formation

3. Polar and nonpolar-characteristics for solvency

4. Lower surface tension than water for effective displacement

Of course for commercial applications safety, economy, and availability also play an important part. At this point, isopropyl alcohol appears to be a good choice for a wafer drying agent.

REFERENCES

1. K. Skidmore, Semiconductor International, 9, 80-85 (1987).

2. D. Burkman, Semiconductor International, 4, 103-114 (1981).

3. C.F. McConnell and A.E. Walter, patent application (1988).

4. A.F.M. Leenaars in "Particles on Surfaces 1: Detection, Adhesion and Removal," K.L. Mittal, Editor, pp. 361-372, Plenum Press, 1988.

5. R.N. Bolster in "Surface Contamination: Genesis, Detection and Control," K.L. Mittal, Editor, Vol. 1, pp. 359-368, Plenum Press, 1979.

6. Enjay Laboratories technical publication "Isopropyl Alcohol," p.94, Enjay Chemical Company, New York.

7. R.C. Weast, editor, "Handbook of Chemistry and Physics," 50th ed., p. F-30, The Chemical Rubber Co., Cleveland, 1969.

IMPLICATIONS OF PARTICLE CONTAMINATION FOR THIN FILM GROWTH

Arye Shapiro and Charles M. Falco

Optical Sciences Center and Department of Physics
University of Arizona
Tucson, Arizona 85721

Particles can serve as nucleation sites for the
growth of defects in metal and semiconductor thin films
deposited by evaporation, sputtering, and Molecular Beam
Epitaxy (MBE). For each of these deposition technologies,
examples of particle-induced defects are given. Future work
is outlined with an emphasis on MBE.

INTRODUCTION

Particulate contaminants can serve as nucleation centers,
affecting the structure of metal and semiconductor thin films grown by
various vapor deposition techniques. For example, various groups have
reported that particles cause "oval defects" in GaAs films grown by
Molecular Beam Epitaxy (MBE). A recent review is given by Weng.[1]
Although much less is currently known about MBE-grown Si films, similar,
if not more severe, effects are expected.[2-6] Since in semiconductor
structures these defects can be electronically active, the performance of
devices fabricated on the wafers can be seriously degraded.[7-9] In this
paper we review recent work on the effect of particulate contaminants on
thin films grown by evaporation, sputtering, and MBE techniques.
Finally, we outline future work in this area, with an emphasis on MBE.

BACKGROUND AND RESULTS

Evaporation and Sputtering

Thermal evaporation is the simplest physical vapor deposition
technique, consisting of Joule heating of the coating material in either
a boat or electron-beam source in a high vacuum chamber.[10,11] Some of the
evaporated atoms or molecules are adsorbed on the substrate.

Particle-induced defects occurring in refractory optical
coatings deposited by thermal evaporation have been studied by
Guenther.[12] Scanning Electron Microscopy (SEM) cross-sections of
PbTe/ZnSe multilayers revealed that cone-shaped "nodular defects"
originated at substrate surface asperities and coating material spatter.
Guenther also performed computer modeling of the deposition process,

which showed nodular growth for surface topographic distortions as small as a few atoms.

Sputtering involves ion bombardment of the coating material in a high vacuum system.[10,11] The atoms or molecules which are ejected find their way through the plasma, and impinge on the substrate. Spalvins and Brainard[13] studied defects in thick sputter-deposited metal films using SEM cross-sectioning techniques. They deposited S-Monel, silver, and 304 stainless steel on mica and brass substrates with various surface finishes. They observed conical, cauliflower-shaped, and worm-like nodules in the deposited overlayers.

Molecular Beam Epitaxy (MBE)

MBE is a process used to grow single-crystal thin film structures in ultra-high vacuum using multiple beams of atoms or molecules.[14] Some characteristics of MBE which distinguish it from the above conventional technologies are growth rates of less than 1 Å per second (for precise control of structure and properties), chamber pressures of less than 10^{-10} torr during growth (to avoid chemical contamination of samples during growth), temperature-regulated monocrystalline substrates ($+2^0$C at 1000^0C), in situ sample characterization using a variety of electron and photon probes, and a typical system cost of $1.5 million or more.

Table I gives a partial list of electronic devices grown by MBE. Because of the very high quality films and devices which can be produced, MBE is rapidly gaining acceptance as a thin film deposition technology, as indicated by Figure 1.

GaAs "Oval Defects"

For several years, researchers have been aware of the presence of "oval defects" on the surface of GaAs epilayers grown by MBE, and have attributed these defects to various causes which occur during film growth,[1] such as spitting from the Ga cell, Ga oxides in the Ga melt, surface contamination, and Ga droplets on the epilayer surface due to Ga agglomeration. Recently, Weng et al.[8] used optical microscopy to show that particulates on the substrate surface are an unambiguous cause of oval defects. They obtained photomicrographs of particles on the GaAs substrate prior to introduction into their GaAs MBE system. Photomicrographs taken after GaAs deposition exhibited oval defects at particle locations. Fujiwara et al.[15] have classified oval defects occurring on single GaAs and multiple GaAs/AlGaAs epilayers. They found at least seven representative types, the majority caused by particulates on the substrate. Many researchers have shown concern that the density of oval defects may be seriously detrimental to the success of GaAs integrated circuits.[8,9,16,17]

Table I.   Partial List of Electronic Devices Grown by MBE.

Opto-Electronic Devices:	Microwave Devices:	Other Devices:
Laser	Low-Noise FET	Diodes
Waveguide	Power FET	MIS Capacitor
Bi-Stable Switch	IMPATT	Tunnel Triode
Integrated Optics	Mixer	Solar Cell
Directional Coupler	Varactor	
Light Emitting Diode (LED)		
Photodetector		
Modulation Doped MODFETs		

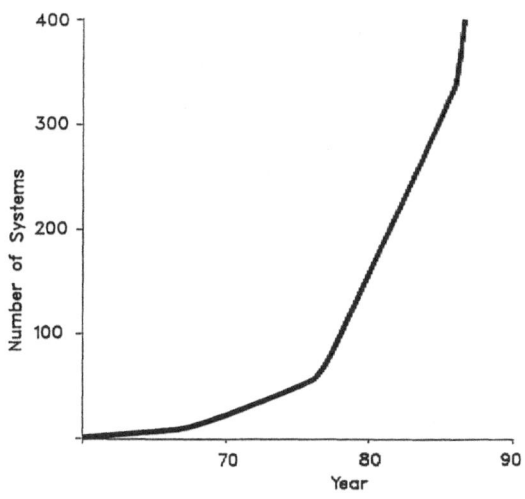

Figure 1. MBE systems installed worldwide.

Si Growth Defects

Particle-induced defects have been observed in MBE-grown Si as well as in GaAs.[2,4,5,6] These workers used optical microscopy to observe particulates of the order of microns in the Si overlayers, most of which appear to be crystalline Si. These particles are generally believed to impinge on the wafer during Si MBE. The Si particulate density has been reduced by shielding the electron beam evaporators. In addition, Chrenko and co-workers[5] have seen interfacial, crystalline particles as small as 50 Å using high-resolution transmission electron microscopy. In a number of cases,[2,5] TEM cross-sections revealed that the particles did not disrupt Si epitaxial growth. In a review of the status of Si MBE, Marsh[3] has observed that "the particulate density [in current Si MBE processing] is still too high for integrated circuit manufacturing".

Silicon MBE System

Figure 2 shows a schematic of the Perkin-Elmer Model 433-S "Silicon/Metals" MBE system at the University of Arizona. The introduction chamber is located under a class 100 HEPA filter in a class 1000 clean room. The growth chamber contains two electron beam evaporators, one dedicated to Si, the other to refractory metals.

This system is equipped with a number of surface science probes which can be utilized to investigate in situ the effect of particle contamination on the nucleation and growth of thin films.

Surface crystal structure can be studied in the growth chamber using Low-Energy Electron Diffraction (LEED) and Reflection High Energy Electron Diffraction (RHEED) instruments. Figure 3 shows a typical LEED apparatus. Note that the electron beam is incident on the sample at close to normal incidence. Atomic-sized defects on a crystalline surface will affect the LEED spot intensity profiles so that one can distinguish between surface structures such as terraces and lattice strain.[18] Figure 4 shows the effect of various surface topographies on the RHEED pattern. The grazing incidence geometry of RHEED makes it useful for studying the growth mode of the thin film during deposition.

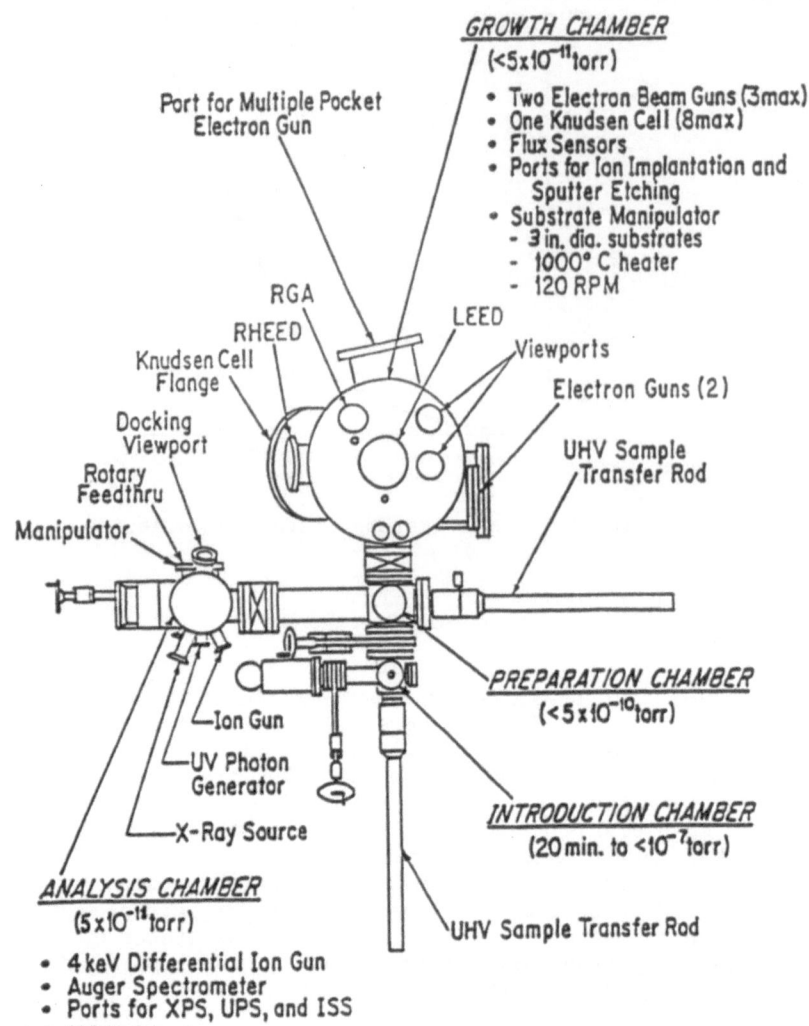

GROWTH CHAMBER
(<5x10⁻¹¹torr)

- Two Electron Beam Guns (3max)
- One Knudsen Cell (8max)
- Flux Sensors
- Ports for Ion Implantation and Sputter Etching
- Substrate Manipulator
  - 3 in. dia. substrates
  - 1000° C heater
  - 120 RPM

Port for Multiple Pocket Electron Gun

RGA
RHEED
Knudsen Cell Flange
Docking Viewport
Rotary Feedthru
Manipulator

LEED
Viewports
Electron Guns (2)
UHV Sample Transfer Rod

Ion Gun
UV Photon Generator
X-Ray Source

PREPARATION CHAMBER
(<5x10⁻¹⁰torr)

INTRODUCTION CHAMBER
(20 min. to <10⁻⁷torr)

ANALYSIS CHAMBER
(5x10⁻¹¹torr)

- 4 keV Differential Ion Gun
- Auger Spectrometer
- Ports for XPS, UPS, and ISS
- 1000° C heater

UHV Sample Transfer Rod

Figure 2. Schematic of the Perkin-Elmer 433-S "Silicon/Metals" MBE system at the University of Arizona.

Surface chemical analysis is performed with Auger Electron Spectroscopy (AES), X-ray Photoelectron Spectroscopy (XPS, or ESCA), and Ion Scattering Spectroscopy (ISS), which are located in the analysis chamber. These complementary techniques can reveal the presence of surface contaminants with concentrations as low as a fraction of a monolayer. Weng and co-workers[19] have used AES to identify non-GaAs particles on GaAs substrates.

SUMMARY

Particles nucleate structural defects in semiconductor and metal films grown by sputtering, evaporation, and Molecular Beam Epitaxy (MBE) techniques. These defects have a variety of deleterious effects on device performance and on the process yield of integrated circuits. For

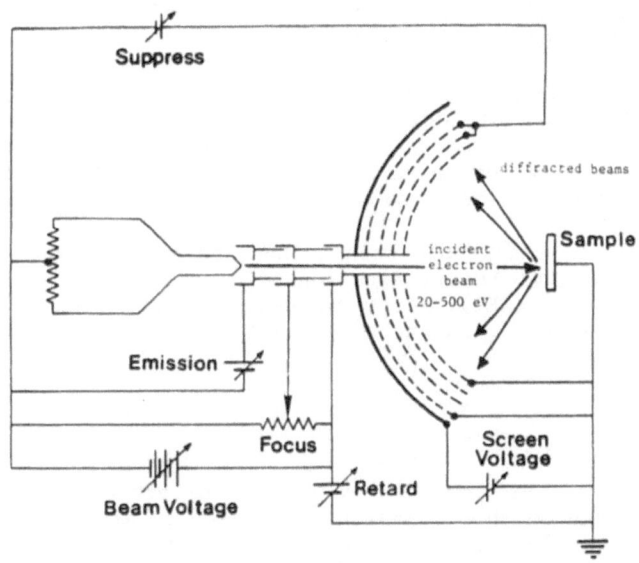

Figure 3. Princeton Research Instruments Reverse View LEED apparatus.

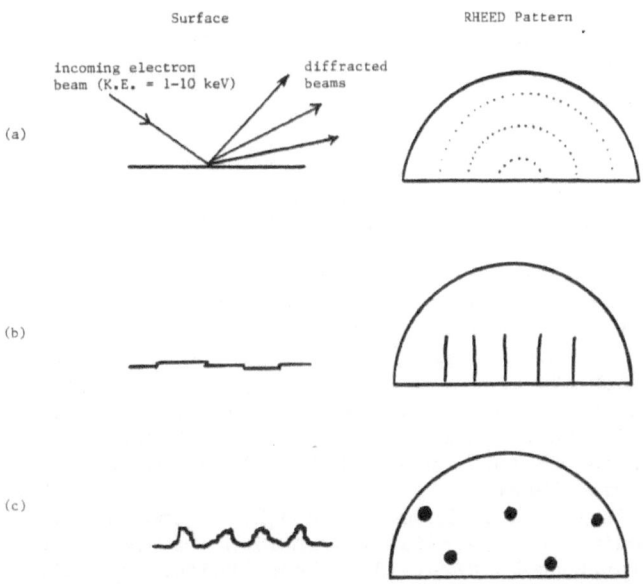

Figure 4. Calculated RHEED patterns for surface which (a) is atomically smooth, (b) consists of steps of atomic height, or (c) contains "large" irregularities (several Å high).

GaAs, the major cause of defects was found to be particulates on the wafer surface.[1] In Si MBE, a variety of particle-induced defects has already been found in Si epilayers, but much of the work is of a preliminary nature. Thus, considerable additional work is needed to understand and control this problem.

Many of these particles are assumed to originate in the vacuum equipment,[2] and efforts are being made to reduce the defect density through better vacuum system design.[20] Unfortunately, the vacuum systems themselves are very expensive (e.g. over $1 million for an MBE system), so these have not been widely available for such types of studies.

Surface analysis tools such as Low Energy Electron Diffraction (LEED), Reflection High Energy Electron Diffraction (RHEED), Auger Electron Spectroscopy (AES), X-ray Photoelectron Spectroscopy (XPS, or ESCA), and Ion Scattering Spectroscopy (ISS) are useful for understanding these problems.

FUTURE WORK

Research is needed to understand the origin of particulates in MBE systems. Based on this research, re-design of mechanical transfer mechanisms, wafer holders, shutter mechanisms, etc. can be undertaken. Considerable work is required to understand the origin, and to reduce the density, of particulates in Si MBE. Basic studies of the nucleation of defects by particles are needed, as is the development of models to understand nucleation and growth of thin films.

ACKNOWLEDGEMENTS

This work was supported by the Center for Microcontamination Control at the University of Arizona.

REFERENCES

1.   S.-L. Weng, J. Vac. Sci. Tech. B 5, 725 (1987).
2.   D. Bellavance and J. Liu, J. Vac Sci. Tech. B 5, 751 (1987).
3.   O.J. Marsh, in "Proceedings of the Second International Symposium on Silicon Molecular Beam Epitaxy," J.C. Bean and L.J. Schowalter, Editors, p. 333, Electrochemical Society, Pennington, New Jersey, 1988.
4.   R.F. Houghton, G. Patel, W.Y. Leong, T.E. Whall, E.H.C. Parker, R.A.A. Kubiak, and R. Nayler, J. Cryst. Growth 81, 326 (1987).
5.   R.M. Chrenko, L.J. Schowalter, E.L. Hall, and N. Lewis, in "Layered Structures and Epitaxy," J.M. Gibson, G.C. Osbourn, and R.M. Tromp, Editors, p. 27, Materials Research Society, Pittsburgh, 1986.
6.   R.A.A. Kubiak, W.Y. Leong, R. Houghton, and E.H.C. Parker, in "Proceedings of the First International Symposium on Silicon Molecular Beam Epitaxy," J.C. Bean, Editor, p. 124, Electrochemical Society, Pennington, New Jersey, 1985.
7.   J.R. Monkowski, in "Treatise on Clean Surface Technology," K.L. Mittal, Editor, pp. 123-148, Plenum, New York, 1987.
8.   S.-L. Weng, C. Webb, Y.G. Chai, and S.G. Bandy, Appl. Phys. Lett. 47, 391 (1985).
9.   D.L. Miller, in "IEEE GaAs Integrated Circuit Symposium Technical Digest 1984," p. 37, IEEE, New York, 1984.
10.   L. Eckertova, "Physics of Thin Films," Plenum, New York, 1986.

11. L.I. Maissel and R. Glang, Editors, "Handbook of Thin Film Technology," McGraw-Hill, New York, 1970.
12. K.H. Guenther, Appl. Opt. 23, 3806 (1984).
13. T. Spalvins and W.A. Brainard, J. Vac. Sci. Tech. 11, 1186 (1974).
14. E.H.C. Parker, Editor, "The Technology and Physics of Molecular Beam Epitaxy," Plenum, New York, 1985.
15. K. Fujiwara, K. Kanamoto, Y.N. Ohta, Y. Tokuda, and T. Nakayama, J. Cryst. Growth 80, 104 (1987).
16. M. Illegems in ref. 14, chap. 5.
17. S. Hiyamizu, K. Kondo, T. Fujii, J. Saito, T. Ishikawa, S. Sasa, and S. Tatsuta, in D.C. Look, J.S. Blakemore, Editors, "Semi-Insulating III-V Materials," p. 364, Shiva, Nantwich (England), 1984.
18. D.G. Welkie, Ph.D. dissertation, p. 28, University of Wisconsin-Madison, 1981.
19. S.-L. Weng, C. Webb, Y.G. Chai, and S.G. Bandy, J. Electronic Mater. 15, 267 (1986).
20. J.F. O'Hanlon, J. Vac. Sci. Tech. A 5, 2067 (1987).

IMPLICATIONS OF PARTICULATE CONTAMINATION IN E-BEAM LITHOGRAPHY

Robert L. Dean

Perkin-Elmer Corporation
Electron Beam Technology
26460 Corporate Avenue
Hayward, California 94545

Experiments were made to compare chromium spot and pinhole defects on finished photomasks written with Manufacturing Electron Beam Lithography Systems (MEBES®). The experiments were performed while adding particulate contamination naturally during blank manufacture, packaging, storage, and by the writing process. In addition, particulate contamination that was added purposely before or after exposure was compared to chromium spot and pinhole defects found on finished experimental plates. The results are as follows:

1. The defect density on process monitor plates exceeds the number of detected particles by greater than a factor of two.

2. Particulate contamination that is on the surface of the plate before exposure and lies in the path of the electron beam is likely to produce chromium spot defects.

However, the study showed that particles of latex smaller than 2.0 μm added to the surface of the mask before exposure did not cause chromium spot defects. Similarly, particles of AC fine dust did not produce defects in a 1:1 correlation between the number of particles in the path of the e-beam and the number of defects detected. As with latex particles, smaller AC fine particles probably do not cause defects.

3. Particulate contamination added after e-beam exposure has a low probability of producing defects. Particles of AC fine or latex deposited on plates just after exposure were shown to have a low probability of producing either chromium spot or pinhole defects.

4. Particulate contaminants of certain chemical compositions degrade the e-beam resist material. This produces chromium spot or pinhole defects regardless of when in the mask-making process the contaminants were added.

INTRODUCTION

This paper discusses the implications of particulate contamination and its effect on defect density in e-beam lithography. A study was performed to examine the problem. The objectives of the study were to show the effect of particulate contamination on the defect density during various stages of the mask-making process. Particulate characteristics including size and composition were also investigated.

CAUSES OF DEFECTS DURING THE MASK-MAKING PROCESS

Mask-making is a single-level process that begins with exposure of polymer material of high molecular weight, which acts as a resist material to a high-energy beam of electrons. After exposure, the process comprises:

- development
- post-baking
- descum
- etching of the chromium
- resist removal
- cleaning
- automatic inspection of geometries and linewidths
- detection of defects
- repair of defects

A defect is either a spot of chromium where there should be none or a clear spot where there should be chromium. Defects can result from several causes, one of which is particulate contamination during the lithography process. When making masks, defects as small as 0.25 μm in diameter are found by an automatic inspection system and are repaired, if possible. Several types of defects are shown in Figure 1. Considerable time and expense are spent in repairing defects during mask manufacture. Some defects are unrepairable and require rejection of the mask.

A primary cause of defects on masks is particulate contamination that is deposited on the plate during the various processes of mask-making. Milner[1] has shown a high correlation between the number of particulates on masks before exposure and the defect density. However, there are several other causes of defects. These include resist problems, substrate problems, and problems with the e-beam machine. Dean[2] has shown that the resist can undergo oxidative degradation, which increases the number of defects. The age of the resist film is also a factor in the observed defect density.[3]

Chemical interactions of some contaminants, including particles, and nonparticulate matter contribute to the number of defects, as well. Thompson, et al[4] have found that sodium contamination leaching from soda lime glass into the resist solution or from soda lime glass masks produces an increase in pinhole defects. Frank, et al[5] have found that resist film thermal history can affect the solubility of the resist, which may also affect the defect density.

Substrate problems are known to cause some defects. These include poorly applied chromium; effects of contaminants during chromium application; mechanical properties of the chromium film, which affect its etching properties; and the polishing of glass or quartz substrates.[6-7] Machine errors account for some defects,[8] and human contamination such as "spit spots," skin flakes, and skin oils also produce defects. Figure 1 includes micrographs of some defects due to human contamination.

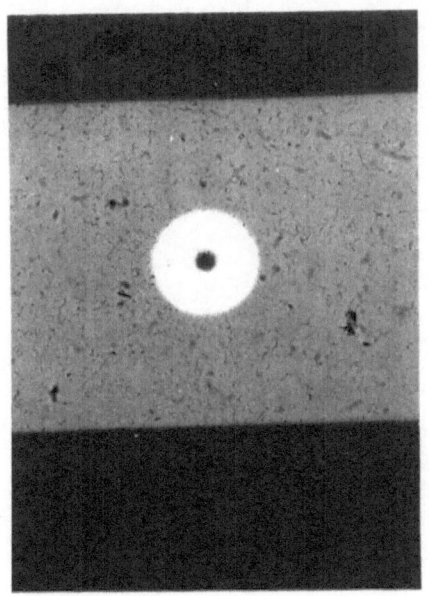

a. Pinhole defect
   (clear spot in chromium).

b. Clear protrusion "mouse nip."

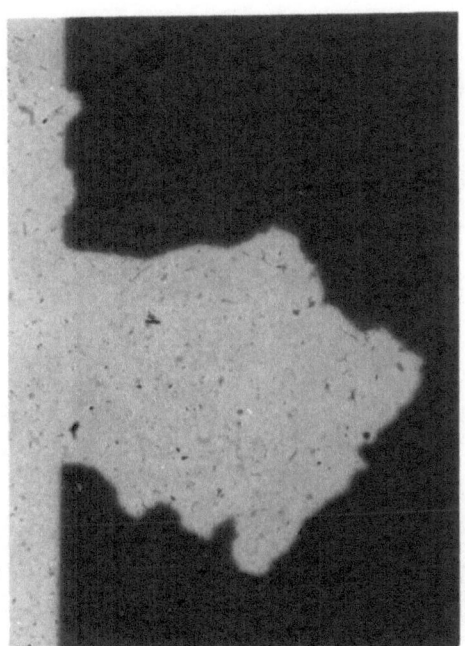

c. Opaque protrusion.

d. Chromium spot due to
   saliva spot.

Figure 1A. Mask defects.

e. Skin oils etch the chromium.

f. Chromium spot defects due to chemical cause.

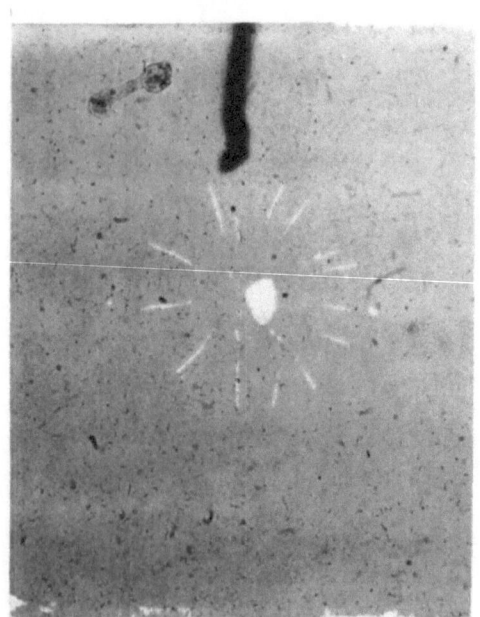

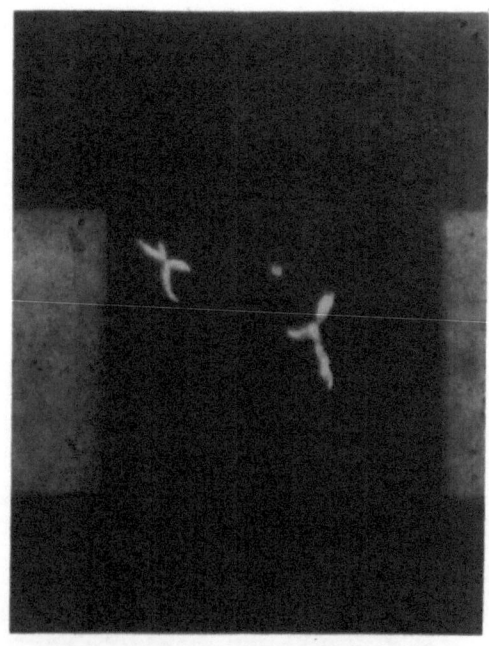

g. Chromium spot defects due to chemical cause.

h. "Birds" chromium spot defects due to chemical causes.

This study has shown the following:

1. There are more defects on finished masks than can be accounted for by the particulate count detected on the mask surface before and after exposure to the electron beam.

2. Surface particles in the path of the electron beam cause chromium spot defects. However, for latex particles added to the mask surface before exposure, there is a correlation between particle size and e-beam spot size. Particles 2.0 μm or smaller do not produce defects with a 1.0 μm e-beam spot size. Latex particles 1.09 μm or smaller do not produce defects when a 0.5 μm e-beam spot size is used, but 2.0 μm particles do. Dust particles (primarily quartz) deposited produce less than a 65 percent correlation between the number of particles and the number of defects. It is assumed that this result is due to the size of the dust; that is, the smaller dust particles do not cause defects.

3. Latex particles or dust deposited after e-beam exposure have a low correlation with the number of defects detected on the mask. Defects amount to less than one percent of the number of particles added.

4. Particles created from elements of the first and second rows of the Periodic Table, including sodium chloride, potassium chloride, calcium chloride, magnesium nitrate, and magnesium oxide, have a high probability of causing defects regardless of their size or when they are deposited on the resist film.

Tests Performed

The following tests were performed to determine how the type and size of particulate contaminants affected the defect density on completed masks.

Test 1. Eighty-six blanks of the best grade were exposed to a process monitor pattern in a MEBES e-beam machine. The blanks were 50 percent clear and 50 percent opaque. The number of particles with a size of 1.0 μm or larger were measured with a Tencor 300 surface particle counter both before and after exposure. The exposed blanks were processed normally and inspected for defects with a KLA 101 or 208 defect inspection system. The number of particles were compared to the number of defects.

Test 2. Particles of AC fine (Arizona road dust, by General Motors) were deposited on the surface of three blanks. The number of deposited particles 0.5 μm or larger were measured with a Tencor 364 surface particle counter with an edge exclusion of 1 cm.

The contaminated plates were exposed to an experimental pattern that produced 169 clear areas of 0.25 cm^2. The number of defects was determined optically. Defects were counted on 25 of the squares. We assumed those squares to be representative and from that calculated the total number of defects for the plate.

Test 3. Particles of AC fine were added to three 5-inch plates before exposure. The plates were exposed at 80 nA with a 0.5 μm e-beam spot size. The plates had an 11x11 array of the 50 percent clear/50 percent opaque process monitor pattern. The exposed plates were processed normally and inspected for defects using a KLA 101 or 208. The original number of particles were compared with the number of defects.

Test 4. Sized latex microspheres (Duke Scientific) were deposited on plates. The plates were exposed to a 13x13 array of 0.25 cm^2 squares.

Defects in the squares were determined optically. Two plates had 1.09 μm particles, one written at 1.0 μm and one written at 0.5 μm e-beam spot size. Two had 2.02 μm spheres written at a 1.0 μm spot; one had 2.02 μm microspheres and was written at 0.5 μm spot. Finally, one had 4.1 μm spheres and was written at a 1.0 μm spot size. A final plate with 1.09 μm spheres was written with an 11x11 array of the process monitor pattern at a 0.5 μm spot and was inspected with a KLA 208 for defects.

The particles were deposited onto the surface of the blanks by suspending them in isopropyl alcohol (IPA) and depositing the solution on the plate as it was spinning on a resist spinner. About 0.05g of AC fine (the amount on the end of a damp micro spatula) was added to 500 ml of IPA. Fifty ml of this solution was diluted with 450 ml of IPA. Another 50 ml of the diluted solution was further diluted with 450 ml of IPA. This solution was shaken and 4-5 ml was deposited on the blank as it was spinning at 2000 rpm. This technique deposits a uniform distribution of particles across the surface of the blank. The particles range in size from about 0.5 μm to 4 μm in diameter. (Larger particles are thrown off the blank during the spinning.)

The latex comes already suspended in an aqueous solution. In this case, one to ten drops of the micron spheres are added to 10 ml of IPA and 4-5 ml of the solution is added to the spinning blank. The number of particles added can be adjusted by changing the number of drops of latex solution or the dilution. The latex may swell in IPA, so the solution must be applied as soon as it is prepared.

Test 5. Particles of AC fine were deposited on two plates that had already been exposed to the process monitor pattern. These were processed normally and inspected. The number of particles added and the number of defects were compared. Two plates were developed, post-baked, and descummed. AC fine was added to the plates before they were chromium etched. The number of particles and the number of defects were compared.

Test 6. Particles of sodium chloride, calcium chloride, magnesium chloride, magnesium nitrate, magnesium oxide, and potassium chloride were deposited on separate plates that had been exposed. Each sample had been crushed in a mortar and pestle, suspended in IPA, and added to the blanks as before. The plates were inspected for defects caused by these particles.

RESULTS AND DISCUSSION

Test 1. Comparison of Defect Density to Detected Particles Deposited During Blank Manufacture and in the MEBES Machine

The results of this experiment showed more defects than could have been caused by the number of detected particles. Table I gives average values. Figure 2 shows a plot of the number of particles detected after exposure and the number of defects. In 68 of the 74 points shown, there are more defects than detected particles. Overall there was a 2.81 ratio of defects to detected particles and a 1.90 ratio of chromium spot defects to detected particles.

The numbers in Table I and Figure 2 were corrected for clear-to-opaque ratio by multiplying total defects by two. The results were also corrected for the total area scanned by the Tencor surface particle detection system relative to the total area inspected by the KLA defect inspection system.

The conclusions that can be drawn from these results are the following: First, there may be more particles than were detected by the Tencor, and that accounts for the excess of defects. Second, there may be other causes of many of these defects.

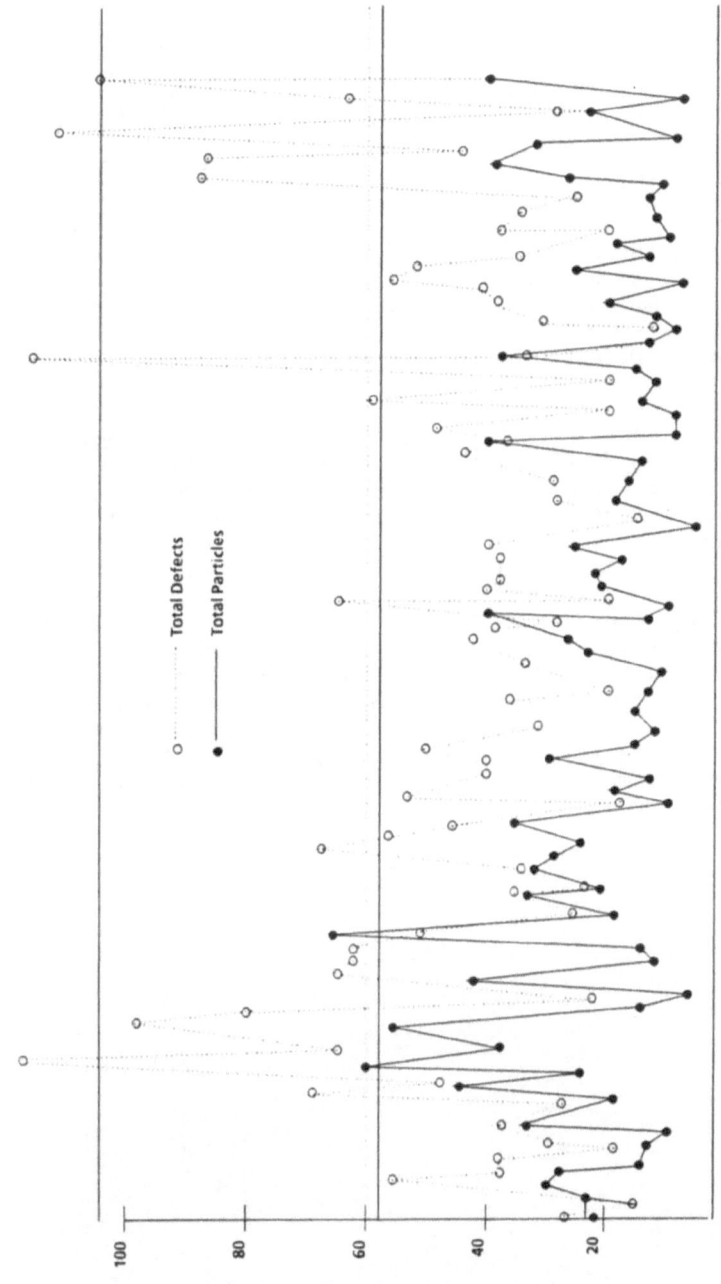

Figure 2. Total defects compared with total particles.

Table I. Comparison of the Number of Particles Added During Manufacture and the Number of Defects.

Average number of particles 1μm or greater ± 3 sigma	Average number of defects ± 3 sigma (corrected for clear-to-opaque ratio)	Defect-to-particle ratio	Average number of chromium spots ± 3 sigma (corrected for clear-to-opaque ratio)	Chromium spot-to-particle ratio
21.2 ± 33.6	59.6 ± 89.4	2.81	40.2 ± 34.8	1.90

The rest of the experiments were designed to determine the contribution of particulates to the defect density. This was done by adding large numbers of particles to plates at various stages of the process.

Tests 2 and 3. Defect Density as a Result of Adding Particles Before E-Beam Exposure

Particles of AC fine were added to plates and the number of defects were detected either optically or by KLA inspection. The results of these experiments show clearly that particles do cause defects, as is seen in Figure 3 where a chromium spot defect and the particle that caused it are shown together. However, these experiments showed fewer defects than added particles. The data are shown in Tables II and III.

For Test 2, we used an array of 169 clear areas of 0.25 cm^2. Defects were detected optically by counting chromium spot defects in 25 squares. The results in Table II show fewer defects than detected particles. However, it is easy to introduce large counting errors when using an optical microscope and counting only five percent of the total area. Therefore, in Test 3, three more plates were contaminated with AC fine, the array exposed with the process monitor pattern, processed, and defects counted with KLA defect inspection systems. The results are shown in Table III.

Figure 3. Chromium spot defect and the particle that caused it.

Table II.   The Effect of AC Fine Particles Added Before Exposure
(Defects Detected Optically).

Plate #	Number of particles added (0.5 μm or greater)	Number of defects in 25 squares (determined optically)	Number of defects corrected for counting area to mask area	Ratio of defects to particles
D	19695	473	9454	0.48
D	19183	508	10167	0.53
D	20704	559	11180	0.54

Table III.   The Effect of AC Fine Particles Added Before Exposure
(Defects Detected by KLA 101 or 208).

Plate #	Number of particles added	Number of defects detected	# of opaque defects	# of clear defects	Corrected number of defects	Ratio of defects to particles
N1	5945	865	863	2	3761	0.63
N2	5946	926	918	8	4026	0.68
N3	6384	884	875	9	3843	0.60

The corrected number of defects column in Table III involves two
corrections. First, there is a correction for the clear-to-opaque ratio in
the pattern. It was assumed that the number of chromium spot defects was
twice the number counted because the pattern is 50 percent opaque and
particles on the part of the mask surface that is opaque do not cause
chromium spot defects. The same is true for pinhole defects, which would
have been observed in clear areas. Second, the ratio of the area scanned by
the KLA to that scanned by the Tencor was 0.46. The numbers were corrected
to account for this.

The average defect-to-particle ratio in Table II is 0.51. In Table III,
it is 0.64. The difference is probably due to undercounting the optically
detected defects. The KLA values should be closer to the true value.

The results of these experiments show the following. First, pinhole
defects are not due to AC fine particulate contamination. The number of
pinhole defects is within the range normally observed on plates that are not
contaminated. This number averages about eight pinhole defects per plate
without deliberately adding contamination. Second, there are fewer defects
than particles. This implies that not all particles are causing defects.
It also reinforces the postulate that many of the defects observed on
manufactured plates are due to causes other than particulate contamination.

There are two explanations for this observation. The first is the size
distribution of the deposited particles. It has been reported that e-beams
at 10 kV can penetrate as deeply as 5 μm.[9] It is possible for the beam to
penetrate smaller or thinner particles, which achieves enough exposure in
the resist below the particle to prevent the formation of an unexposed area.
An unexposed area would result in a defect. The next set of experiments was
designed to test this hypothesis.

The second explanation is that the Tencor is detecting particles that produce defects too small to be detected by the KLA, or the Tencor is overcounting. The sensitivity of the Tencor was set to detect a minimum intensity of scattered light corresponding to 0.5 μm^2 or 0.7 μm in diameter.

Measurement errors and sensitivity differences between the two instruments could explain the results. However, the results of the next set of experiments confirms the particle size hypothesis.

Test 4. Defect Density Resulting from Deposition of Sized Latex Microspheres Before E-Beam Exposure

To confirm the hypothesis that smaller particles on masks before exposure do not cause defects, sized latex particles were deposited on mask surfaces. The masks were then exposed with a 13x13 array of 0.25 cm^2 squares with an e-beam spot size of 1.0 μm and current of 320 nA. The number of defects was determined after processing optically as before. A second set of masks with sized microspheres were exposed to the 13x13 array of squares written with a 0.5 μm spot at 80 nA. A final mask was written with the array of the process monitor pattern at 0.5 μm spot size and 80 nA. It was inspected with a KLA 208. Table IV lists the results.

The results show a particle size dependence and an e-beam spot size dependence. For a 1 μm spot size, no chromium spot defects were detected optically for particles up to 2.02 μm in diameter. For 4.10 μm particles, there is a 1:1 correlation between particles and defects. When a 0.5 μm e-beam spot size was used, 2.02 μm latex microspheres showed a 1:1 correlation between numbers of particles and defects. However 1.09 μm spheres did not print defects. This was confirmed by the results of the process monitor array. Although there were many defects on this plate, none were found that were due to the spheres. Those defects are easily identifiable, because they are images of the round sphere. Figure 4 illustrates a chromium spot defect image of a 2.02 μm microsphere. These results confirm the hypothesis that the particle-defect ratio is affected by the particle size. Other factors include e-beam spot size and may include the density and chemical composition of the particle.

Table IV.   The Effect of Sized Latex Particles on Defect Density.

Particle size	Number of deposited particles	Number of defects counted	Number of defects corrected for area	Ratio of defects to deposited particles	E-beam spot size (μm)
1.09	14235	0	0	0	1.0
1.09	13321	0	0	0	0.5
2.02	14371	0	0	0	1.0
2.02	14040	0	0	0	1.0
2.02	14148	712	14200	1	0.5
4.10	5738	290	5800	1	1.0
Process	monitor	pattern	inspected	by	KLA 208.
1.09	14423	320	695	0	0.5

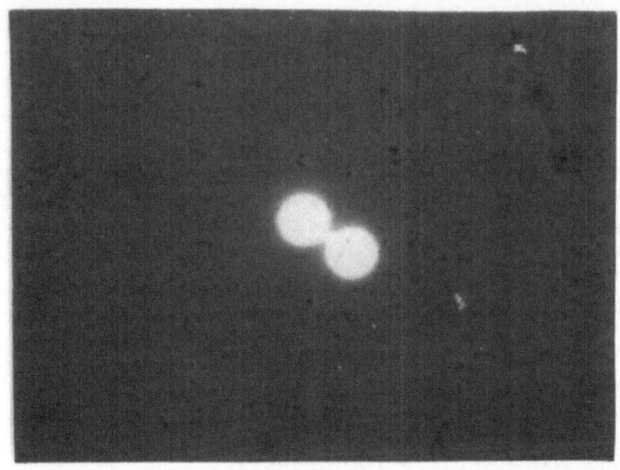

Figure 4. Chromium spot defect image of a 2.02 μm microsphere.

Test 5. Defect Density Resulting from Depositing AC Fine Particles on the Mask Surface After Exposure

Particles were deposited on the surface of masks after they had been exposed with the 11×11 array of the process monitor pattern written at a 0.5 μm spot and 80 nA. A second set of masks was exposed and developed through descum, followed by particle deposition. Each set was then fully processed and inspected with one of the KLA systems for defects. The results are reported in Table V.

These results show that AC fine has only a small effect on the defect density if added after exposure. The corrected number of defects was calculated as in previous tests. The value in the corrected number of defects column was calculated by subtracting the average number of defects found on the mask written for the first experiment. The number of particles added to the set of plates after descum was not counted, as the Tencor is not capable of counting particles on a patterned plate. It is assumed that the counts were similar to previous plates.

Table V. The Effect of AC Fine Particles Added After E-Beam Exposure.

Plate number	Number of particles added	Number of defects	KLA model number	Corrected number of defects	Ratio of particles to defects
Pro 1	19702	77	101	167	0.008
Pro 2	22103	209	101	454	0.021
CPBS1 (after descum)	not counted	74	101	160	---
CPBS 2	not counted	67	208	146	---

Test 6. The Effect of Chemically Corrosive Particles on Defect Generation

In separate experiments, we analyzed particulates deposited on masks that had been cycled through a MEBES machine. The results showed that 18 percent of the particles contained large concentrations of magnesium oxides and chlorides. This was probably from the load and work chambers of the machine, which are cast of magnesium. Other particles were found that contained sodium, calcium, and potassium in large concentrations. Most were chloride salts. The majority of particles from the machine consisted of aluminum wear particles.

Because Thompson[5] had shown that sodium contamination had an effect on the resist, we conducted an experiment that deposited sodium chloride, potassium chloride, magnesium nitrate, calcium choride, and magnesium oxide on the surfaces of exposed plates. No particle counts were taken, due to difficulties in counting these types of particles. The plates were processed and characterized for defects.

The results are shown in Figure 5. Pinholes due to magnesium nitrate are visible, as well as chromium spot defects due to calcium chloride. All of the salts and the oxide tested produced defects in large numbers. We conclude from this that these types of contaminants have a high probability of causing defects regardless of when they are added to the plate. The defects are due to chemical interactions of the resist with the contaminants. We have begun work to eliminate the sources of this type of contamination in the MEBES machine.

CONCLUSIONS

From this study, we can draw the following conclusions regarding defect densities observed in e-beam generated masks.

• There are more defects on average than can be accounted for by the particle count.

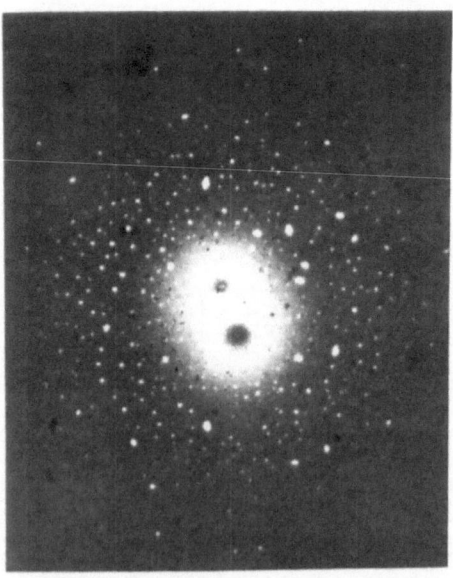 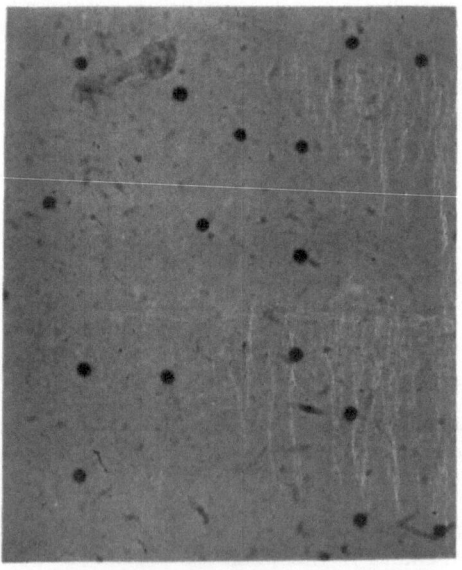

a. Chromium spots due to particles   b. Pinholes due to particles of
    of calcium salts.   magnesium salts.

Figure 5. Defects due to particulate contamination.

- Particles deposited on the surface of the resist before exposure do cause defects, but not in a 1:1 correlation. It has been shown that the size of the particle is important and that smaller size particles do not cause defects, even if they lie in the path of the beam. The size of the particles that produces defects depends on the e-beam spot size.

- Particles deposited after exposure have little or no effect on defect densities if they are inert toward the resist.

- Particles that are chemically corrosive or catalytic toward the resist degradation cause defects if deposited after exposure.

- Defects are produced by a complex set of problems, including particulates. Further investigation of all causes of defects is warranted in the mask-making and chip-making industry.

REFERENCES

1. K. Milner, Paper presented at MEBES Users Group meeting, Monterey, CA, January 1985.
2. R. Dean, SPIE Proc. 773-15, March 1986.
3. R. Dean, Paper presented at MEBES Users Group meeting, Monterey, CA, September 1987.
4. L.L. Kosbar, W.J. Kuan, C.W. Frank, R.F.W. Pease, Photophysical studies of spin cast polymers films, Paper presented at SPIE, March 1988.
5. Discussion of unpublished work with L. Thompson of AT&T Bell Labs, Murray Hill, NJ, 1985.
6. Discussion of unpublished work with Henry Kamberian of Electronic Materials Corporation, Azusa, CA, 1986.
7. Discussion of unpublished work with John Skinner of AT&T Bell Labs, Murray Hill, NJ, 1986.
8. Discussion of unpublished work with LeRoy McCutcheon of Perkin-Elmer EBT, Hayward, CA, 1986.
9. J.W.S. Hearle, B. Lomas, J.T. Sparrow, J. of Microscopy 92(3), 205, (1970).
10. R. Dean, Paper presented at MEBES Users Group Meeting, Monterey, CA, November 1985.

PART IV.   PARTICLE REMOVAL

ENHANCED REMOVAL OF SUB-MICRON PARTICLES FROM SURFACES

BY HIGH MOLECULAR WEIGHT FLUOROCARBON SURFACTANT SOLUTIONS

Robert Kaiser

Entropic Systems, Inc.
P.O. Box 397
Winchester, MA 01890-0597

A novel patented method of removing unwanted sub-micron particles from solid surfaces is described. A key component of this approach is the use of solutions of high molecular weight fluorinated surfactants (M.W.> 2,000 Daltons) in an inert perfluorinated carrier liquid to effect particle removal. The unique characteristics of the high molecular weight surfactants used in these solutions are that they:

a) can result in the formation of a relatively thick solvated shell around the particles, which promotes the detachment of solid particles from a solid substrate, and

b) are reversibly adsorbed on the solid surfaces being treated, and that adsorption is controlled by controlling their concentration in the liquid phase, and which can be made as low as desired.

In this process, a contaminated surface is first contacted under conditions of shear, as in an ultrasonic bath, with a cleaning solution consisting of a dilute solution of a high molecular weight highly fluorinated surfactant in a highly fluorinated carrier liquid to effect the removal of particles from the surface. The surface is then rinsed with a surfactant-free volatile perfluorinated liquid to effect the removal of residual surfactant from the surface, and finally dried to evaporate the rinsing liquid.

Preliminary results are presented which indicate that sub-micron particles are more effectively removed from a solid surface by washing the surface with a 1% solution of a high molecular weight highly fluorinated surfactant in a highly fluorinated carrier liquid than by washing it in the carrier liquid without surfactant.

BACKGROUND

Particle Adhesion Mechanisms

Sub-micron particles are generally observed to adhere tenaciously and non-specifically to other solid surfaces, and can not be removed by simple mechanical means. There is a large body of physical evidence that suggests that their adhesion is due to relatively weak secondary valence forces, or van der Waals forces, that act between adjacent solid bodies, and whose magnitude becomes much larger than that of other forces that can be applied to a solid as its characteristic size decreases to the order of 1 um (10^{-4} cm) or less. Adhesion of fine particles to solid surfaces, in general, is believed to be due mainly to secondary valence interaction between the particles and these surfaces, rather than by primary valence bonding, electrostatic charges, or surface tension forces due to capillary condensation of a liquid.

General explanations for the nature of Van der Waals forces were first proposed by London [1] in 1930, and later by Lifschitz [2]. These explanations are based on quantum theory which considers that all atoms, even in their ground state, possess rapidly fluctuating dipole moments which lead to an attraction between them. The force between two solid objects, which are viewed as assemblies of many atoms, is obtained from the summation of the individual interactions of all the atoms in the two bodies with the force between any two atoms being considered independent of the presence of other atoms.

For a sphere of diameter, d, at a distance of closest approach, h, from a semi-infinite solid of the same material, for small separations ($h \ll 10^{-5}$ cm), the attractive force can be expressed as follows [3]:

$$F_a = -Ad/3h^2 \qquad (1)$$

In the above equation, A is the Hamaker constant of the material, which ranges from about 10^{-11} to 10^{-14} ergs, depending on the substance. For a silica sphere interacting with a planar silica surface in a vacuum, A has been found to be approximately 10^{-12} erg [3].

If the atoms are an appreciable distance apart, the time taken for the electrostatic field from one instantaneous dipole to interact with a neighbor may be comparable with the "lifetime" of the dipole. In that case, the interactions are out of phase which leads to a smaller force of attraction. To take into account the retardation effect, the interaction between a sphere and a plate for large distances ($h > 10^{-5}$ cm) of separation is more accurately expressed by the following alternate equation [3]:

$$F_a = -Bd/3h^3 \qquad (2)$$

where B is a material constant, whose value is [4,5] of the order of 10^{-19} erg/cm.

Even though the force between a particle and an adjacent surface decreases with particle size, it becomes more difficult to remove a solid particle from a solid surface because the value of the ratio, F_a/W, i.e., of the force of attraction to the weight of a particle, increases rapidly as the diameter of a particle decreases, as indicated in the table below for a spherical silica particle based on an assumed value of $B = 10^{-19}$ erg/cm:

Table 1

Surface Separation, h/d	10^{-3}	10^{-2}	10^{-1}	1
Particle Diameter, cm	Normalized Attraction, F_A/W			
0.001	7.6×10^{1}	7.6×10^{-2}	7.6×10^{-5}	7.6×10^{-8}
0.0001	7.6×10^{6}	7.6×10^{3}	7.6	7.6×10^{-3}
0.00001	$7.6. \times 10^{11}$	7.6×10^{8}	7.6×10^{5}	7.6×10^{2}

The effective surface-to-surface separation between a particle deposited from the atmosphere and a flat surface, such as a semiconductor wafer, is mainly determined by the surface roughness of the particle and of the surface, and by the thickness of a gas film adsorbed on these surfaces. For particles of the size range of interest, h/d is estimated to range from approximately 0.001 to 0.1. Based on the values given in the above table, one would expect that a 10 μm particle to be readily removed by inertial forces with an acceleration of less than 1 g. For particles of the order of 1 μm in diameter, secondary valence forces are one to four orders of magnitude greater than the weight of a particle. In general, such particles would be expected to adhere to a substrate, but, under favorable circumstances, such particles might be removed by the application of inertial forces with an acceleration of many g's. For particles of the order of 0.1 μm in diameter, secondary valence forces are more than five orders of magnitude greater than the weight of a particle. Such particles can not be removed from a substrate by inertial forces that can be practically applied.

Whereas it is extremely difficult to modify the interaction between the surface of a fine particle and that of an adjacent solid in the gas phase, this can be readily accomplished if the solids are immersed in a liquid phase with certain characteristics. As in the gas phase, solid particles immersed in a liquid phase are also subject to secondary valence forces of attraction, which can result in their adherence to other solid surfaces. However, in the liquid phase, it is possible to induce the generation of repulsive forces between the adjacent surfaces that are larger than the secondary valence forces responsible for particle adhesion, as evidenced by the existence of stable colloidal dispersions of particles that are significantly smaller than 1 μm in diameter.

The stability of colloidal dispersions in a liquid phase is due to the development of repulsive forces between particles. The stability of hydrosols (aqueous colloids) is usually attributed to the formation of an electrostatic double layer around a particle resulting from the adsorption of charged species on or around the particle surface [6] Stabilization occurs because of the electrostatic repulsion between two similarly charged particles.

A second, more general, mechanism of stabilization that applies to both aqueous and non-aqueous colloidal dispersions results from the entropic repulsion of large, solvated molecules that are adsorbed on the particle surfaces. These adsorbed molecules form a thick film of essentially bound liquid. When two particles collide, they are separated by a minimum distance equal to twice the film thickness. If the film thickness is sufficiently large, the secondary valence interaction between particles becomes negligible in comparison to their thermal energy, kT, and which, under these circumstances, becomes dominant factor. It has been estimated [7] that the condition necessary for the formation of a stable dispersion is that the ratio of film thickness to particle diameter be >= 0.15. This mechanism of entropic repulsion predominates in the formation of colloidal suspensions in non-aqueous media, in which sub-micron particles are stably suspended in a liquid due to the presence of polar molecules that can both adsorb on the surface of the solid particles and be solvated by the liquid medium.

Current Surface Cleaning Technology

Removing sub-micron sized particles from a solid substrate in the electronics industry is not a new problem, and has been widely recognized for several decades [8,9]. Wet cleaning of wafers is not a new concept. A variety of devices are commercially available which claim an ability to remove particles from solid substrates. Most existing cleaners use mechanical force as the means of obtaining particle detachment. Devices used include ultrasonic baths, megasonic (high intensity ultrasonic) baths, or mechanical scrubbers, such as high energy sprays or rotating brushes [10,11].

Very few cleaning fluids are used in conjunction with these devices, with the major ones being deionized water, trichlorotrifluro- ethane (Freon[R] TF), and Freon[R] TF/alcohol mixtures [10 - 13]. Solid carbon dioxide has recently been proposed as a cleaning agent as well [14]. These are all low molecular weight fluids that are not capable, by themselves, of forming thick solvated films on solid surfaces that may be present.

Aqueous surfactant solutions are commonly used to remove or suspend sub-micron particles from solid surfaces in many washing and cleaning applications. For reasons of solubility and detergency, surfactants used in commercial washing applications typically contain from twelve to eighteen carbon atoms. These have not been successfully used to clean in-process wafer surfaces because they introduce another contamination problem, namely, the presence of residual surfactant molecules adsorbed on the wafer surface.

Mechanical force is the principal means of particle detachment with these systems. As discussed above, this is not an effective means of detaching micron sized particles from solid substrates. In those cases where particles may be momentarily detached, nothing prevents particles from being reattached upon collision with another section of the surface being treated.

Some attempts have been made to remove particles from wafer surfaces in the dry state by using high intensity acoustic fields [15] or electrostatic fields [16] without significant success.

An interesting approach was recently proposed by Leenaars [17], which is based on displacement of a particle from a surface by a moving liquid/gas interface. A major limitation of this technique is that, in order to effect particle removal, the rate of displacement of the interface has to be small, less than a particle diameter per second, making this approach impractical.

In the author's opinion, none of the approaches to wafer cleaning now used in the semiconductor IC industry effectively meet the industry's requirements with regards to sub-micron particle removal.

272

ENTROPIC PROCESS APPROACH

General Concept

The fact that fine particles can be stably dispersed in a surfactant solution (as compared to their adherent properties in the gas phase) forms the basis of a novel patented method[18] of removing undesirable foreign particles from a surface, such as that of a wafer during the fabrication of semiconductor integrated circuits. A key component of this approach is the use of solutions of high molecular weight surfactants in an inert carrier liquid to effect particle removal. The unique characteristics of the high molecular weight surfactants used in these solutions are that:

a) they can result in a relatively thick solvated shell around the particles, which promotes the detachment of solid particles from a solid substrate, and

b) they are reversibly adsorbed on the solid surfaces being treated, and that adsorption is controlled by controlling their concentration in the liquid phase, and which can be made as small as desired.

The proposed method of wafer cleaning is described below and is outlined in Figure 1.

A dirty wafer is immersed in a washing solution which neither dissolves nor reacts irreversibly with any of the materials present on the surface of the wafer. The washing solution contains a surfactant which adsorbs on the surfaces of the wafer and of any foreign particles present on the surface of the wafer, and forms solvated films which reduce the interaction between the particles and the wafer. Because of thermal energy, the solvated particles are detached from the underlying surface. Mechanical agitation, as provided, for example, by an ultrasonic bath, or flow of the liquid past the wafer surface promotes the dispersion of these detached particles into the bulk fluid. Once suspended, these solvated particles would not reattach to the wafer surface, as could occur if the solvated films were not present. The washing liquid which contains the foreign particles in suspension is then physically separated from the wafer.

The wafer is then rinsed with a liquid that has an affinity for the surfactant used in the previous step to effect the quantitative removal of any surfactant that is adsorbed on the surface of the wafer. This rinsing liquid should be inert with regards to the wafer surface. Preferably, this rinsing liquid should have the same composition as the carrier liquid used to prepare the surfactant solution used in the first step. The rinsing process is continued, by the addition of fresh rinse liquid, until there are no detectable traces of surfactant on the surface of the wafer.

Any residual rinsing liquid still adhering to the wafer surface can be removed mechanically (i.e. by centrifugation) and/or by evaporation, either at room temperature or in a drying oven at a temperature that is not harmful to the wafer.

The product of the process is a clean wafer from which foreign particles have been removed, but is otherwise unchanged.

As also indicated in Figure 1, all the fluid streams that come into contact with the wafer are filtered to remove any foreign particles suspended in the process liquids, and which would otherwise recontaminate the wafer surface. These process filters should be capable of removing all particles of a size large enough to be of concern to the manufacturing process.

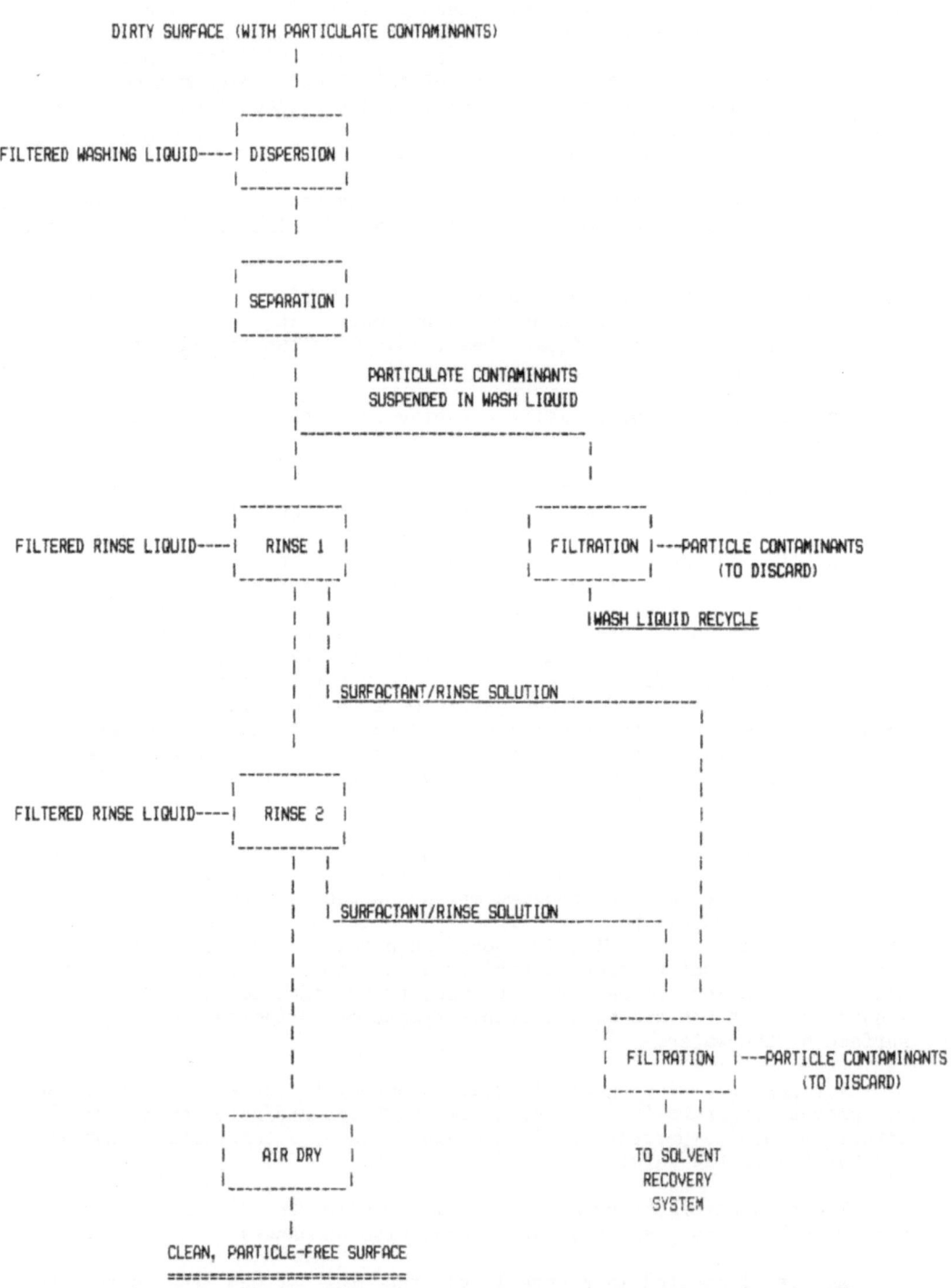

Figure 1. Flow Diagram, Entropic Cleaning Process

While the various liquids used in the above process could be
discarded after one pass, for economic reasons, it will be desirable to
be able to recover and recycle them. The used washing liquid is
recovered by filtration through a membrane filter to remove any
suspended particles. Make-up surfactant and carrier liquid are added to
maintain a washing liquid of reasonably constant composition. The
filter, and the captured suspended solids, is discarded periodically and
replaced by a fresh filter.

Used rinsing liquid, which contains traces of surfactant, is most
easily recovered by distillation, since the rinsing liquid is volatile,
and the surfactant is not. The overhead product is pure rinse liquid
that would be returned to the process. The still bottoms are a solution
of surfactant in rinse liquid that could be recycled as makeup washing
solution.

Preferred Cleaning Agents

The choice of cleaning agents that may be used is constrained by
the very strict requirements that any wafer cleaning method must meet if
it is to be successfully implemented. These include:

a) The removal of sub-micron particles without leaving behind any
   residues, and in a manner that does not alter the physical and
   chemical properties of the wafer being processed.

b) Compatibility with existing established operating practices of the
   semiconductor fabrication industry.

c) No introduction of any new risk or hazard to the work place in
   terms of flammability, toxicity or environmental compatibility.

All the above requirements can be satisfied by the use of a select
group of perfluorinated liquids that would be used as the wash solution
carrier liquid and as the rinse solution, in conjunction with high
molecular weight perfluorinated surfactants. These perfluorinated
liquids are obtained by replacing carbon bound hydrogen atoms with
fluorine atoms in a variety of organic compounds. The perfluorinated
liquids of interest do not contain chlorine. These perfluorinated
liquids show, as a class, a remarkable degree of inertness and
resistance to chemical attack. They are non-polar and have essentially
no solvent action. They principally have an affinity for other highly
fluorinated or chlorinated fluids, and certain oxygenated solvents.
They are non-flammable and have a low order of toxicity. Because of
these combinations of properties, perfluorinated liquids have been used
as test media for electronic components and devices, and have been
specified as the test media of choice in a number of military standards,
such as MIL-STD 833, MIL-STD 750, and MIL-STD 202.

Preferred perfluorinated liquids for the proposed process have a
boiling point at ambient pressure in the range of from $35^{\circ}C$ to $150^{\circ}C$ to
facilitate removal of carrier liquid at temperatures that do not harm
the object being cleaned. Examples of these liquids include a
perfluoroalkane, predominantly perfluorooctane (C_8F_{18}) with an average
molecular weight of 435 Daltons , marketed by the 3M Corporation as
FC-104 Fluorocarbon Electronic Liquid; a perfluorinated cyclic ether
containing eight carbon atoms with an average molecular weight of 420
Daltons, marketed by the 3M Corporation as FC-75 Fluorocarbon Electronic
Liquid, and FC-77 Fluorocarbon Electronic Liquid, an azeotropic mixture
of FC-75 and FC-104, also marketed by the 3M Corporation.

By themselves, these perfluorinated liquids are not good dispersing media. However, as outlined in U.S. Patent 3,784,471 [19], stable dispersions of sub-micron particles of widely varying composition can be formed in a mixture of a perfluorinated liquid and a high molecular weight perfluorinated surfactant which contains preferably 20 to 100 carbon atoms and one or more polar groups capable of interacting with a solid surface. These polar groups include species with an active hydrogen atom, such as carboxylic acids and alcohols. The surfactant preferably has a much higher boiling point than the carrier liquid with which it is used to facilitate the recovery of used process solutions. It is also important that the surfactant can be easily removed from the surface to be cleaned by rinsing with surfactant-free fluorocarbon liquid. Otherwise the cleaning process will merely result in the substitution of one contaminant for another. Surfactants with a high molecular non-polar moiety are more readily desorbed than otherwise comparable lower molecular weight surfactants, and thus are preferred. These high molecular weight surfactants also have an extremely low vapor pressure.

Surfactants which are suitable for the proposed process include polyhexafluoropropylene oxide (HFPO) carboxylic acids which can be represented by the following formula:

$$\text{F--[CF--CF}_2\text{--O]}_n\text{--C--COOH}$$

where the structure shows CF_3 groups attached.

where n is an integer from 3 to 50. Preferred surfactants are HFPO carboxylic acids with average molecular weights of from 2,000 to 8,000 Daltons, marketed under the Krytox 157FS trade designation by E.I. du Pont de Nemours & Co., Inc.

Experimental Results

The ability to form stable colloidal dispersions of a wide variety of solid particles that are smaller than one micron in diameter in dilute solutions (< 1% by weight) of HFPO carboxylic acids in perfluorinated liquids by ball milling or ultrasonic agitation at ambient conditions, has been demonstrated and documented in the literature [19-21]. It should be noted that stable colloidal dispersions of magnetite, where one has to contend with interparticle magnetic dipole forces as secondary valence forces, have been in prepared in perfluorinated liquids.

Removal of Sub-micron Particles from Solid Substrates: A key finding is that dilute (1%) solutions of HFPO carboxylic acids in perfluorinated liquids, such as FC-77 or FC-104, more effectively remove a variety of sub-micron particles from a variety surfaces than either Freon TF, a standard cleaning agent now used for precision cleaning, or a fluorinated liquid that does not contain any dispersing agent.

In these experiments, a cylinder, about 1 cm in diameter, was immersed in a surfactant-free suspension of a fine (e.g. sub-micron) powder in Freon TF. The suspension was agitated ultrasonically, and the cylinder removed. After the Freon TF had evaporated, the tube was covered with fine particles. The tube was then immersed in a 25 mm diameter vial that contained a control liquid, usually Freon TF, of known turbidity. The control liquid was filtered through a 0.2 μm filter before being placed in the vial to eliminate foreign particles. The vial was placed in an ultrasonic bath for 3 min. After the cylinder was removed from the control liquid, its turbidity was then measured. The change in the turbidity of the control liquid was a direct indication that particles had been removed from the test cylinder. The test cylinder was then immersed in a second sample of control liquid, and treated in the same manner as with the first sample. The procedure was repeated until no noticeable change in turbidity was observed with at least two consecutive liquid samples, which was taken as an indication that significant numbers of particles were no longer being removed.

The test cylinder was then immersed in a solution of about 1 wt-% HFPO carboxylic acid in a perfluorinated liquid (i.e. cleaning solution), but otherwise repeating the same test procedure. It should be noted that the turbidity of the filtered surfactant solution was not significantly different than the turbidity of the filtered surfactant-free carrier liquid. A significant increase in turbidity of the first two aliquots of cleaning solution was usually observed. The turbidity levels of the exposed samples were often higher than the turbidity of the control liquid after initial contact with the dirty surface. These results indicate that:

a) a significant number of particles were still present after ultrasonic cleaning with Freon[R] TF, or other surfactant-free liquid medium, and that

b) a significant fraction of these residual particles were removed by ultrasonic cleaning with a solution of HFPO carboxylic acid in a perfluorinated liquid.

Tests were run with a variety of sub-micron particles and test surfaces. Particles examined included fumed lead oxide, silica, ferric oxide, and carbon black. Substrates that were examined included glass (pyrex), quartz, teflon, polypropylene, steel and brass.

Representative results are summarized in Table I for the removal of fumed lead oxide (with a median particle diameter of 0.55 μm) from a pyrex cylinder. The cylinder was sequentially exposed to aliquots of Freon[R]-TF, of 1% HFPO carboxylic acid with a molecular weight of approximately 5,000 Daltons, further designated in this paper as surfactant K(M), in Freon[R] TF, and 1% surfactant K(M) in FC-104. As emphasized in Figure 2, 1% surfactant K(M) in Freon[R] TF is a more effective particle removal agent than pure Freon[R] TF, and, in turn, 1% surfactant K(M) in FC-104 is a more effective particle removal agent than 1% surfactant K(M) in Freon[R] TF.

A measure of the size of the particles removed in each sequential step was obtained from the change in the measured turbidity with time which provided an indication of the rate of sedimentation of the particles in suspension. The equivalent Stoke's particle diameter corresponding to various settling times in FC 75 or FC 104, and in Freon[R] TF, are presented in Figure 3. These measurements were facilitated by the large difference between the density of lead oxide (9.53 g/cm^3) and that of the fluorocarbon liquids (1.58 g/cm^3 for Freon[R] TF and 1.75 g/cm^3 for FC 75, FC 77, and FC 104).

It should be noted that the turbidity of the liquid samples was still significantly higher than that of the initial solution even after 1,000 minutes of elapsed time after sonolation, indicating that a

Table I. Summary of Results - Effect of Liquid Composition on the Removal of Fumed Lead Oxide From a Pyrex Glass Test Tube by Ultrasonic Agitation at Ambient (18°C) as Measured by the Turbidity of the Washing Liquid

Sample Sequence No.	1	2	3	4	5	6	7	8	9	10	11	12	13	14
Liquid Sample No.	TF-1	TF-2	TF-3	TF-4	TF-5	TF-6	TF-7	TF-1M	TF-2M	TF-3M	TF-4M	TF-5M	104-M9	104-M10
Carrier Liquid	Fr-TF	Fr-TF	Fr-TF	Fr-TF	Fr-TF	Fr-TF	Fr-TF	Fr-TF	Fr-TF	Fr-TF	Fr-TF	Fr-TF	FC-104	FC-104
Surfactant	None	None	None	None	None	None	None	K(M)	K(M)	K(M)	K(M)	K(M)	K(M)	K(M)
Surfactant Conc., Wt-%	0	0	0	0	0	0	0	1.048	1.037	1.044	1.022	1.053	0.977	0.977
Solution Turbidity, T_b, NTU	0.19	0.19	0.19	0.20	0.20	0.21	0.20	0.24	0.23	0.21	0.21	0.28	0.24	0.24
Sonolation Time, min	1	1	1	1	5	6	5	1	1	5	6	1	5	1

Liquid Sample Turbidity After Removing Substrate, T, As a Function of Settling Time

Liquid Sample Turbidity, T, NTU

Settling Time, min.														
1	0.62	0.9	0.58	0.32	0.54	0.32	0.33	1.5	0.68	0.92	0.33	0.29	0.66	0.29
100	0.36	0.56	0.27	0.23	0.36	0.23	0.24	0.87	0.49	0.6	0.3	0.28	0.56	0.26
1,000	0.21	0.26	0.19	0.2	0.22	0.21	0.2	0.32	0.26	0.3	0.24	no data	0.44	0.25

Relative Sample Turbidity After Removing Substrate, T/T_b, as a Function of Settling Time

Relative Sample Turbidity, T_b/T

Settling Time, min.														
1	3.26	4.74	3.05	1.60	2.70	1.52	1.65	6.25	2.96	4.38	1.57	1.04	2.75	1.21
100	1.89	2.95	1.42	1.15	1.80	1.10	1.20	3.63	2.13	2.86	1.43	1.00	2.33	1.08
1,000	1.11	1.37	1.00	1.00	1.10	1.00	1.00	1.33	1.13	1.43	1.14	no data	1.83	1.04

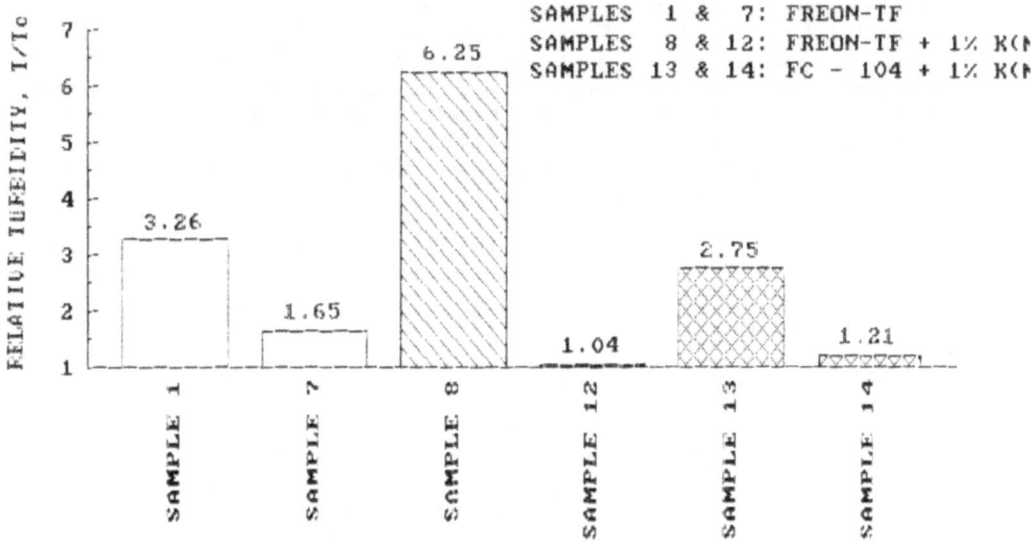

Figure 2. Effect of fluorocarbon liquid composition on the removal of fumed lead oxide particles from pyrex glass.

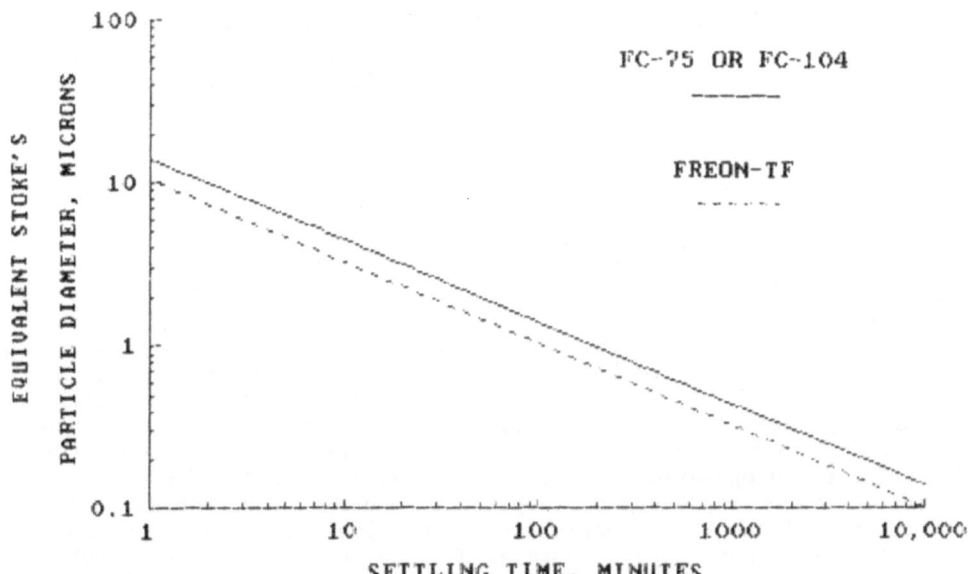

Figure 3. Sedimentation of lead oxide particles in selected fluorinated liquids.

significant fraction of the particles in suspension were smaller than 0.4 um. In particular, it was observed that settling occurred very slowly in Sample 104-M9 (Sequence No. 13), which contained 1% surfactant K(M) in FC-104. The turbidity of the sample after 1,000 minutes was approximately 2/3 of the initial turbidity. After 5,600 minutes, when observations were ceased, the turbidity of the sample was 0.30 NTU, still 25% higher than the turbidity of the initial solution. This indicated the presence of a significant number of particles in suspension that were smaller than 0.2 um, and that could only have come from the surface being cleaned.

Surfactant Partition: The equilibrium partition of HFPO carboxylic acids containing from 10 to 40 mer units between a silica substrate and a perfluorinated liquid was measured by liquid chromatography. These results indicate that:

a) Complete monomolecular coverage of the surface by the surfactant occurs when the liquid phase
concentration of the surfactant is equal to or greater than approximately 0.3% by weight, and

b) Surface adsorption of the HFPO carboxylic acid molecules is reversible, and that these can be completely eluted from the substr by rinsing with surfactant-free solution.

Silicon Wafer Compatibility: Solutions of 1% HFPO carboxylic acids in a perfluorinated liquid are compatible with silicon wafers. To determine whether their was any reaction between these solutions and the surface of a silicon wafer, a weighed coupon (2 cm by 8 cm) cut from a virgin wafer was half immersed in 50 ml of a 1% solution of surfactant K(M) in FC-77. Duplicate test samples were placed in a constant temperature bath maintained at 60 °C for two weeks. The coupons were then removed from the test solution and rinsed with surfactant-free FC-77. The samples were then examined for any changes in weight or visual appearance. No weight changes were observed. The portion of the wafer that had been immersed was visually indistinguishable from the upper portion that had not been immersed.

FURTHER DATA REQUIRED

The ability to remove particles from solid surfaces under conditions of interest to the semiconductor IC industry, or other precision industries, is still to be demonstrated. The ability of this method to effectively remove particles from silicon wafers and other semiconductor wafer materials, such as gallium arsenide, under clean room conditions, is yet to be performed. For example, the adsorption characteristics of the active agents on semiconductor material surfaces need to be established. The ability to remove particles from in-process wafers at the low contamination levels characteristic of semiconductor IC manufacturing operations is still to be demonstrated. The effects of various IC processing steps on the effectiveness of the proposed process, as well as the compatibility of the proposed cleaning method with these wafer manufacturing steps, need to be explored. Last but not least, the impact of this cleaning method on product yield is still to be demonstrated.

CONCLUSIONS

Contaminating sub-micron particles are more effectively removed from the surface of a solid object by washing in a dilute solution (about 0.3 to 1.0 wt-percent) of a high molecular weight (e.g. M.W.> 2,000 Daltons) fluorinated surfactant in a fluorinated liquid than if the object were washed in surfactant-free fluorinated liquid. The sub-micron particle removal capability of a solution of fluorinated surfactant in a perfluorinated carrier liquid is greater than that of a solution of fluorinated surfactant in a partially fluorinated carrier liquid, such as Freon^R TF. Enhanced particle removal is ascribed to the formation of relatively thick solvated layers of adsorbed surfactant on the surfaces of the solid object and of the contaminating particles.

Adsorption of these high molecular weight surfactants on silica surfaces is reversible, indicating that adsorbed surfactant can be removed from the surface of a solid object being washed by subsequently rinsing with surfactant-free fluorinated liquid.

The above establish the basis for a novel industrial process for the removal of sub-micron particles from solid surfaces.

REFERENCES

1. F. London,, Z. Physik, $\underline{63}$, 245 (1930)
2. E.M. Lifschitz, J. Exp. Theor. Phys. USSR, $\underline{29}$, 94 (1954)
3. D. Tabor, J. Colloid & Interface Sci., $\underline{31}(3)$, 364 (1969)
4. J.G.V. De Jongh, Doctoral Dissertation, Utrecht (Netherlands), 1958.
5. H.B.G.Casimir, and D. Polder, Phys. Rev., $\underline{73}$, 360 (1948)
6. E.J.W. Verwey and J. Th. G. Overbeek, "Theory of the Stability of Lyophobic Colloids", Elsevier Publishing Co., Amsterdam, 1948.
7. E.L. Mackor, J. Colloid Sci., $\underline{6}$, 492 (1951)
8. "Cleaning of Electronic Device Components and Materials", STP No. 246, ASTM, Philadelphia, 1959.
9. "Cleaning and Material Related Processing for Electronics and Space Apparatus", STP No. 342, ASTM, Philadelphia, 1963.
10. A.H. Szkudlapski, in "Environmental Control in Electronic Manufacturing", P.W. Morrison, Editor, Chap. 7, Van Nostrand Reinhold Company, New York, 1973.
11. A. Khilnani, in "Particles on Surfaces 1: Detection, Adhesion, and Removal", K.L. Mittal, Editor, pp. 17-35, Plenum Press, New York, 1988.
12. O. Hamberg and E.M. Shon, "Particle Size Distribution on Surfaces in Clean Rooms", Proc. 30th Annual Technical Meeting, Institute of Environmental Sciences, pp. 14-19, May 1984.
13. R.P. Musselman, and T.W. Yarbrough, J. Environmental Sci., $\underline{30}(1)$, 51 (1987)
14. S. Hoenig, in "Particles on Surfaces 1: Detection, Adhesion, and Removal", K.L. Mittal, Editor, pp. 3-16, Plenum Press, New York, 1988.
15. K.W. Montz, J.K. Beddow, P.B. Butler, and A.F. Vetter, Paper Presented at the 17th Annual Meeting of the Fine Particle Society, San Francisco, July 28-August 2, 1986
16. D.W. Cooper, H.L. Wolfe, and R.J. Miller, in "Particles on Surfaces 1: Detection, Adhesion, and Removal", K.L. Mittal, Editor, pp. 339-349, Plenum Press, New York, 1988.
17. A.F.M. Leenaars, in "Particles on Surfaces 1: Detection, Adhesion, Removal", K.L. Mittal, Editor, pp. 361-372, Plenum Press, New York, 1988.
18. R. Kaiser, U.S. Patent 4,711,256, "Method and Apparatus for Removal of Small Particles from a Surface", December 8, 1987.

19. R. Kaiser, U.S. Patent 3,784,471, "Solid Additives Dispersed in Perfluorinated Liquids with Perfluoroalkyl Ether Dispersants", January 8, 1974.
20. R. Kaiser, "Preparation of Stable Colloidal Dispersions in Fluorinated Liquids", NASA Tech Brief, B-72-10529, August 1972.
21. R. Kaiser, and R.E. Rosensweig, "Study of Ferromagnetic Liquid", NASA Contractor Report, NASA CR-1407, August 1969.

COMPARISON OF FREON WITH WATER CLEANING PROCESSES

FOR DISK-DRIVE PARTS

Ming Ko

IBM Corporation
General Products Division
San Jose, California 95193

This paper documents the evaluation of three parts-cleaning
processes for disk-drive parts: Freon TF/ultrasonics,
water/ultrasonics, and water spray systems. Particulate,
organic, and ionic contaminants, along with moisture residue,
were measured and compared. The data showed that a water
cleaning system was a feasible replacement for a Freon
cleaning system for disk-drive parts.

INTRODUCTION

In March of 1988, the NASA Ozone Trends Panel announced that a new
analysis of measurements confirmed a two percent decrease in protective
stratospheric ozone during the past 17 years.[1] Fully halogenated fluorocarbons (CFCs) may have contributed to the observed decrease and probably
are the major contributors to a larger decrease over Antarctica. In
response to this news, the DuPont Company called for an orderly transition to the total phaseout of fully halogenated CFCs for long-term protection of global ozone.[2] Also in March, the United States Senate
ratified the Montreal Protocol, which calls for a freeze at 1986 CFC
production levels, followed by reduction to 80 percent of those levels
beginning in mid-1993 and to 50 percent beginning in mid-1998.

Freon TF[3] is a key solvent used for cleaning most metal parts in the
disk-drive industry. In an effort to reduce the amount of Freon used,
and possibly to provide enhanced cleaning processes, a study was undertaken to investigate aqueous cleaning processes that combined ultrasonics, spray, and hot-air drying to clean parts at high throughput.
Previous reports had compared particulate removal by using optical surfaces,[4] metallic coupons,[5] semiconductor wafers,[6] or printed wiring
assemblies.[7] This study evaluated single-solvent systems such as
Freon/ultrasonics, water/ultrasonics, and water spray for their ability
to remove particulate, organic, and ionic contaminants. In addition,
drying rate and moisture residue were also measured using typical disk-
drive parts.

EXPERIMENTAL

Equipment and Processes

The following cleaning systems were used.

1. Freon/Ultrasonic Cleaning System

   The four-stage Freon cleaner used was comprised of two Branson ultrasonic degreasers, Model WSD-2024-W, 40 kHz. Each degreaser had a
   boiling sump and an ultrasonic tank containing Freon TF. The
   degreasers were equipped with 0.5-micron filtration and solvent distillation accessories. Parts were immersed in the ultrasonic tank of
   the first degreaser unit for four minutes. Following a 15-second
   drip-dry period, the parts were transferred to the second degreaser
   unit and immersed in the ultrasonic tank for two minutes. In the
   final step, the parts moved to the vapor zone of the degreaser unit,
   where they remained for two minutes. The parts went through a
   15-second drip-dry period before exiting the second degreaser unit.
   The parts were stopped before entering the cleanroom and removed and
   sealed in cleanroom packaging material.

2. Water/Ultrasonic Cleaner

   The water/ultrasonic cleaner used was Branson Ultrasonic Cleaning
   Systems' Model BC 1824, 40 kHz, with 0.02% Triton X-100[8] surfactant
   at 50°C to wash for four minutes and deionized water to rinse for two
   minutes. The cleaned parts were blown dry with ambient air at 50 psi
   and packaged in a clean bag.

3. Water Spray Cleaner

   The water spray cleaning process (Atcor SRD 3410 Cleaning System) had
   two stages. The part to be cleaned entered the first chamber and was
   sprayed with an aqueous Tergitol 1X[9] solution (0.02% by volume), then
   sprayed with deionized water at approximately 15-20 psi. The part
   then passed through a curtain of clean, dry air. This air curtain
   (50 psi) dissipated any large water droplets left on the part.
   Then, low-pressure (10-20 psi) clean, dry air, heated to about 66°C,
   was used to evaporate any remaining moisture. The part was then
   packaged in a clean bag.

Test Methods

     The detailed procedures of the four test methods used are described
in the following paragraphs (Figure 1). The reported values were
obtained after the background readings had been subtracted.

Microscopic Count (M) and Liquid Particle Counting (LPC) Tests. Both
microscopic and liquid particle counting tests employed ultrasonics to
extract particles for 30 seconds without using surfactant to reduce the
artifact errors caused by air bubbles. A Hiac/Royco Series 4300 particle
counter was used to measure particulate population. The microscopic particle counting portion of the test was primarily based on the modified
method of ASTM F24.[4] Duplicate measurements were performed.

Turbidity (T) and Enhanced Turbidity (ET) Tests. Particulate contaminants were extracted with diluted (0.02%) Triton X-100 in water by using
an ultrasonic cleaner for one minute. The ultrasonic cleaner was filled
to a depth of 2.5 cm with deionized water. Turbidity readings (in For-
mazin Turbidity Units, FTUs) were recorded directly from the turbidim-
eter. The enhanced turbidity test uses filtration and filter-dissolving

WATER EXTRACTION 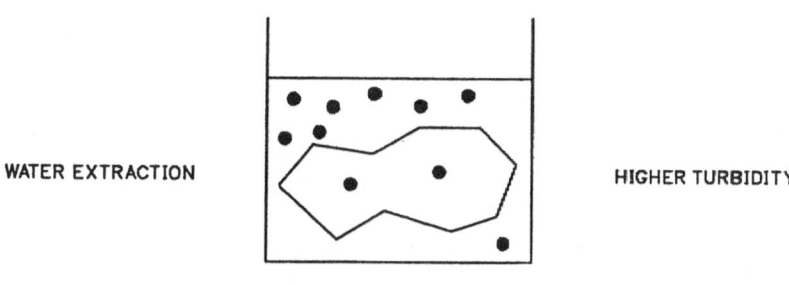 HIGHER TURBIDITY

FREON EXTRACTION 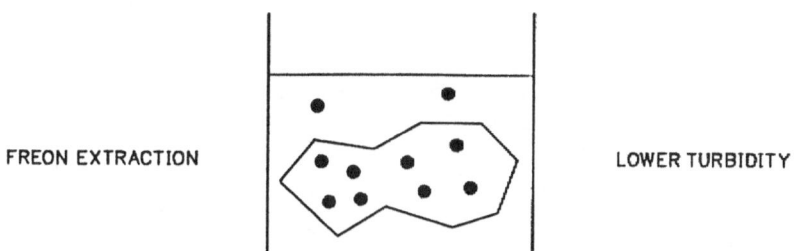 LOWER TURBIDITY

Figure 1. Particulate contamination test methods

techniques to increase the sensitivity of the turbidity test. The results are presented in Enhanced Turbidity Units (ETUs). The enhanced turbidity technique provides a more concentrated particulate extract for turbidimetry measurement. The sample size of parts tested was ten.

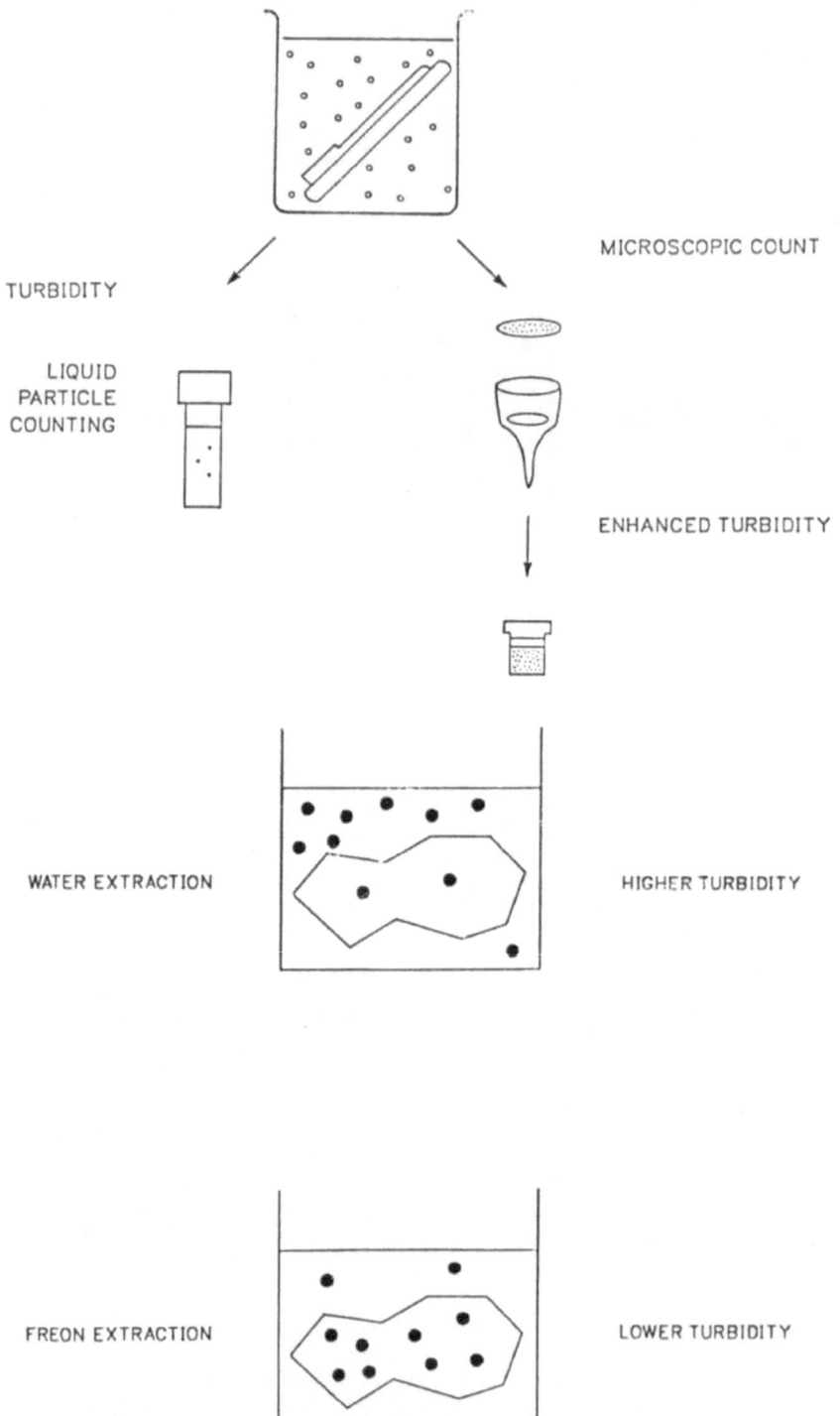

Figure 2. Multiple extraction: Water/ultrasonics versus Freon/ultrasonics

Multiple Extraction Technique (Figure 2). This technique is useful in determining the relative cleaning efficiency of various solvents used in the same ultrasonic cleaner. It provides a direct comparison of particulate removal by water/ultrasonics and Freon/ultrasonics while holding all the other variables constant. The equipment used was a Branson ultrasonic cleaner, Model DHA 1000. The parts were placed in a beaker. The ultrasonic cleaner was filled with the appropriate test solvent (either Freon TF or 0.02% Triton X-100/water). Sequential extractions were applied to the same part, followed by the enhanced turbidity test. Each extract was filtered and measured.

Organic Contaminants. Part surfaces were precleaned using Freon TF and methanol. Measured amounts of typical organic contaminants found in the disk-drive environment were applied with a micro-syringe onto the parts being tested, in order to set a standard by which to assess the results. Gas chromatography/mass spectrometry (GC/MS) analysis was used to measure trace quantities of organic residues in chloroform and hexane extract of the parts.

Ionic Contaminants. All parts were sprayed with a spray rinser at approximately 10 psi with 100 ml de-ionized water per test. Ionic contaminants were monitored by a conductance meter (Omega Instruments, Model CDB-70).

Dryness Evaluation. Both visual inspection and gravimetric methods were used to examine parts for residual water left after cleaning. The gravimetric method used predried desiccants in a sealed polyethylene bag. Water residue was absorbed by the desiccants and showed up in the form of a weight increase. Comparison of the before and after weights of the sealed desiccant bags yielded a measure of residual moisture left on the surfaces of the cleaned parts.

RESULTS AND DISCUSSION

The results discussed here apply only to the particular type of contaminants encountered in our manufacturing processes. Representative samples of head-disk assembly part types were selected for evaluation as shown in Table I.

Particulate Contaminant

Direct microscopic inspection was not feasible due to the great variation and roughness of the surfaces of the disk-drive parts. Instead, measurements were made on particles extracted from the parts. All four test methods -- turbidity, enhanced turbidity, microscopic count, and liquid particle counting -- were applied to evaluate the three cleaning processes investigated in this study.

Both microscopic and liquid particle measurements gave the same general trend in descending order of effectiveness: water spray, water/ultrasonics, and Freon/ultrasonics (see Table II).

A further study of particulate removal was performed using turbidity or enhanced turbidity tests. Enhanced turbidity tests were not performed on Parts D and F because the turbidity test was sufficiently sensitive for these parts. Table III shows the test results for the parts as they were received and after cleaning. A high standard deviation for particles was found on the parts before cleaning, even though the level of cleanliness was relatively good. After using Freon or water cleaning methods, no significant differences in levels of particles could be demonstrated. Particulate counts for Part G were low because this was a plastic part. On average, all the cleaned parts had particulate levels suitable for use in disk drives.

Table I. Summary of Parts Selection for Cleaner
Evaluation.

Part	Material/Finishes
A	E-coated, aluminum die casting
B	Stainless steel
C	E-coated, aluminum die casting
D	Zinc-plated chromate conversion coated steel
E	E-coated, magnesium die casting
F	Stainless steel
G	Polycarbonate
H	E-coated, aluminum die casting, stainless steel

Table II. Microscopic and Liquid Particle Count Results.

Part	Microscopic Count				Liquid Particle Count			
	As Rec'd	Water Ultra-sonics	Freon Ultra-sonics	Water Spray	As Rec'd	Water Ultra-sonics	Freon Ultra-sonics	Water Spray
A	1588	619	660	280	1167	406	654	147
B	839	130	138	149	331	63	69	62
C	492	1855	1507	90	408	1433	1278	85
D	362	66	559	41	177	36	370	25
E	828	289	479	465	645	176	233	355
F	245	34	615	18	60	19	152	9
G	29	55	195	3	27	42	162	3
H	740	---	---	3	814	---	23	2

Measurement unit: Thousand particles per part, size range 5-100 μm, average of duplicate samples.

Table III. Enhanced Turbidity and Turbidity Test Results for Various Cleaning Processes.[a]

Part	Enhanced Turbidity (ETU)				Turbidity (FTU)			
	As Rec'd	Water Ultra-sonics	Freon Ultra-sonics	Water Spray	As Rec'd	Water Ultra-sonics	Freon Ultra-sonics	Water Spray
A	80.3 (25.0)	83.9 (13.3)	81.6 (9.1)	106.3 (15.0)	2.3 (0.4)	1.9 (0.5)	1.7 (0.2)	2.2 (0.4)
B	39.9 (11.8)	50.7 (8.3)	67.7 (24.7)	51.5 (23.0)	4.9 (1.9)	6.1 (1.2)	7.1 (2.8)	4.8 (2.5)
C	84.9 (31.5)	68.1 (18.9)	86.6 (5.6)	113.9 (20.0)	1.4 (0.3)	1.1 (0.2)	1.3 (0.2)	1.2 (0.2)
D	Not Measured				11.6 (2.3)	4.9 (2.0)	11.7 (2.9)	5.4 (1.7)
E	68.8 (22.6)	63.5 (25.3)	71.3 (13.7)	84.1 (11.2)	0.7 (0.2)	0.6 (0.1)	0.5 (0.1)	0.6 (0.1)
F	Not Measured				29.4 (5.2)	10.2 (1.5)	23.6 (3.0)	17.7 (0.9)
G	4.0 (2.1)	3.2 (1.4)	6.9 (4.8)	1.7 (1.4)	Not Measured			
H[b]	93.2 (8.0)	115.1 (7.3)	92.5 (8.9)	108.0[c] (18)	2.1 (0.2)	2.1 (0.2)	2.6 (1.3)	2.1[c] (0.2)

Notes:

a. Sample size was 10 and standard deviations are in parentheses.
b. One liter extract volume was used for Part H.
c. Two subassemblies were connected, cleaned, and tested together. Normalized ETU should be approximately half of the reported values.

Negative cleaning efficiencies, such as those for Part C in Table II and Parts B and E in Table III, were not uncommon. This may be due to either or both of the following:

1. The incoming parts showed wide variation in contaminants. Some parts had lower particulate contaminants when received than the other parts after cleaning.

2. All the particulate testing used ultrasonic extraction for sample preparation. Ultrasonic extraction may dislodge or erode some of the tightly-bonded particles (such as oxides on metallic surfaces) to produce higher particle count readings after cleaning.

Multiple-extraction results showed that Triton X-100/water/ultrasonics treatment liberates approximately 10 times more particles than Freon TF cleaning under the same conditions. The results for Part A are shown as an example (Table IV).

Cross extraction reconfirmed the above observations. After the fourth water/ultrasonics extraction, the fifth extraction was made with Freon and showed few particles (4.0 ETU). Similarly, after the fourth Freon/ultrasonics extraction, the fifth extraction with water still showed a great many particles (105 ETU).

Organic Contaminants

Table V summarizes the results of the analyses for removal of typical organic contaminants from disk-drive parts. The values in the table are micrograms of Dotriacontane and Stearamide recovered from the surface of each part before and after the cleaning operation. The values are also equivalent to the percentage of each compound recovered, since 100 micrograms of each compound were used. Chloroform and hexane were used to extract Dotriacontane and Stearamide. The recoverable amounts before cleaning were approximately 82 percent (or micrograms) for both compounds.

The average amounts of the two compounds remaining on the parts after the various cleaning procedures show that Freon was more effective than water with ultrasonics, which, in turn, was more effective than water spray without ultrasonics. Significantly, however, the data do show that water with ultrasonics was effective in removing hydrophobic organic materials used for a disk-drive manufacturing process. The water spray process was also effective for the geometry of parts used in this study. Use of surfactant in the first step of both water cleaning processes should contribute to the effective cleaning of organics.

Table IV. Multiple and Cross Extractions: Measurement of Particles Extracted From the Surface of Part A.

	1st	2nd	3rd	4th	5th (Cross Extraction)
Triton X-100/water	104	75	70.0	50.0	4.0
Freon TF	15	8	6.5	5.5	105.0

Each number = average readings of 3 parts in units of Enhanced Turbidity Units (ETUs).

Ionic Contaminants

Ionic contamination is a concern because of its potential for causing corrosion problems. The experiment used a metal part (Part D) with a common water-based cutting fluid as an artificial contaminant. This experiment showed that water spray cleaning was superior to the Freon process for removing ionic contaminants. The results confirmed that ionic contaminants were easily removed by the water spray cleaning process (see Table VI).

Table V. Evaluation of Cleaning Procedures for Organic Contamination.

Part	Before Cleaning		Freon Ultrasonics		Water Ultrasonics		Water Spray	
	DOTR	STEA	DOTR	STEA	DOTR	STEA	DOTR	STEA
A	82.6	93.5	2.3	3.9	12.3	2.4	3.9	4.7
B	85.4	79.2	7.4	<1	11.7	4.4	28.5	5.3
C	90.6	66.2	25.5	1.9	36.6	<1	83.1	26.1
D	100.3	103.3	<1	<1	0.8	1.1	5.4	0.8
E	66.3	78.6	<1	1.6	9.9	3.3	8.8	2.7
F	87.8	89.5	<1	1.8	9.6	0.7	---	---
G	65.1	65.0	<1	<1	6.1	3.8	67.9	25.6

Values are percentages (or micrograms) of 100 micrograms of contaminant applied.
Values which are less than 1.0, but not definitely resolved, are reported as <1.
DOTR = Dotriacontane, a saturated hydrocarbon.
STEA = Stearamide, a typical antistatic agent on packaging.

Dryness

The results of the dryness evaluation indicated that the typical water cleaner needed up to 8 minutes to dry parts, whereas the Freon cleaner required only 2 minutes.

Comparing parts dryness based on weight increase of the desiccant, the water cleaning process provided drier parts than the Freon cleaning process. The Atcor SRD system, using a drying cycle of 7 minutes, dried approximately 1.5 times as well as the Freon cleaner when drying for 2 minutes in the Freon vapor zone (see Table VII).

Table VI. Cleaning Efficiency of Freon/Ultra-
           sonics vs. Water Spray Cleaners for
           Ionic Contaminants Removal.

Part D Artificially Contaminated with a Solution of 10% by Weight Cimcool Oil Five Star 40			
	Before Cleaning	Freon-Cleaned	Water-Cleaned
Lot A	3.60	2.35	0.82
	3.90	2.00	0.79
	2.05	2.21	0.74
	2.09	1.88	0.80
	3.01		0.76
Mean	2.90	2.10	0.78
Lot B	3.20	2.15	0.85
	2.01	1.96	0.85
	2.09	2.02	1.19
	1.60	1.86	0.81
	1.53	1.86	0.91
Mean	2.50	2.0	0.85

Measurement unit = microsiemens.

The vapor zone of the Freon ultrasonic tank was usually at 47°C, which allowed some moisture in the vapor zone. Although a higher temperature of clean, dry air was required to remove moisture from water-cleaned parts, there was less water residue than with the Freon-cleaned parts under the experimental conditions followed.

Two key factors controlling drying rate are temperature and air velocity past the part surface. The drying rate was optimized for this particular case using these two parameters.

Table VII. Comparison of Surface Dryness Using the
Gravimetric Method: Freon-Cleaned vs.
Water-Cleaned Parts.

Part	Freon-Clean/Dry	Water-Clean/Dry	Delta %
A	0.38568	0.11452	237
B	0.50148	0.09617	421
C	0.30379	0.17297	76
D	0.18254	0.08918	105
E	0.26133	0.12872	103
F	0.10926	0.08845	24
G	0.20439	0.13701	49
Average	--	--	144

Average weight increase in grams.

CONCLUSIONS

Among the three cleaning processes, the water cleaning processes were equivalent to the Freon process in removing particulate contaminants from disk-drive parts. The water cleaning processes worked better than the Freon process in removing ionic contaminants, but were less effective than Freon cleaning in removing hydrophobic organic contaminants. The drying rate of Freon cleaned parts was approximately 2-3 times faster than parts cleaned by an off-the-shelf aqueous cleaner. However, residual moisture could be effectively removed after either process.

In conclusion, a water process is a feasible alternative for cleaning and drying parts, if incoming parts cleanliness is carefully controlled and if part geometry is suitable for drying. The data reported here show that a water cleaning system comprising ultrasonics and spray can be considered as an alternative to using Freon to clean parts, and is an environmentally compatible alternative.

ACKNOWLEDGMENTS

The author wishes to thank IBM management for support and encouragement. The author also wishes to thank the members of the project for designing the experiments and for performing the majority of the tasks. Thanks to Mike Carroll, Rose Cruz, Alice Fukushima, Amelia Logue, Gary Nolan, David Shull, Elizabeth Siderewicz, and Mike Zepeda. The author acknowledges Chuck Hignite, who wrote the section on organic contaminants, and June Andersen, who reviewed and edited the paper.

REFERENCES

1. Stratospheric ozone is decreasing, Science, 239, 1489 (March 25, 1988).

2. (a) The changing atmosphere: implications for mankind, a C&EN special issue, Chemical & Engineering News, (November 24, 1986); (b) CFC production: DuPont seeks total phaseout, Chemical & Engineering News, 4-5 (April 4, 1988).

3. Freon is a registered trademark of the DuPont Company. The chemical name for Freon TF is 1,1,2-trichlorotrifluoroethane (commonly known as CFC113); it belongs to the chemical family of chlorofluorocarbons (CFCs).

4. I. F. Stowers, Advances in cleaning metal and glass surfaces to micron-level cleanliness, J. Vac. Sci. Technol., 15, 751 (1978).

5. Q.T. Phillips, G. J. Stone and J. M. Baldwin, Parts Cleaning: An evaluation of ultrasonic systems, Proceedings of 30th Annual Technical Meeting of the Institute of Environmental Sciences, p.1, 1984.

6. V. B. Menon, L. D. Michaels, R. P. Donovan, V. L. Debler and M. B. Ranade, Particle removal from semiconductor wafers using cleaning solvents, in "Particles in Gases and Liquids 1: Detection, Characterization and Control," K. L. Mittal, editor, pp.259-271, Plenum Press, New York, 1989.

7. W. G. Kenyon and J. J. Daly, A comparison of solvent and aqueous processes for cleaning printed wiring assemblies, Proceedings of the Technical Program of International Electronic Packaging and Production Conference, p.50, 1980.

8. Triton is a registered trademark of Rohm and Haas Company.

9. Tergitol 1X is a surfactant supplied by Union Carbide Corporation.

ULTRASONIC AND HYDRODYNAMIC TECHNIQUES FOR PARTICLE REMOVAL FROM SILICON WAFERS

V. B. Menon, L. D. Michaels, R. P. Donovan, and D. S. Ensor

Research Triangle Institute
P.O. Box 12194
Research Triangle Park, North Carolina 27709

Ultrasonic and hydrodynamic cleaning techniques were used to remove particulate contaminants from bare silicon wafers and wafers with surface oxide layers. The variation of cleaning efficiency with particle size, composition and cleaning solvent was quantified. In comparison to submicrometer polystyrene and glass particles, silicon dust was found to be a particularly difficult contaminant to clean. The RCA standard cleaning solution (SC-1), which is a blend of 5 H_2O : 1 NH_4OH: 1 H_2O_2, when used in the ultrasonic or hydrodynamic cleaner, was less effective than DI-water in removing silicon particles, especially at low cleaning energies. Both types of cleaning systems resulted in particle removal efficiencies above 80% for submicrometer particles, the ultrasonic technique being more sensitive to particle and wafer surface composition.

INTRODUCTION

Recent advances in air filtration technology have significantly reduced the contribution of airborne particles in the cleanroom to silicon wafer contamination. Particles in liquids and those generated within process equipment are now the major sources of wafer contamination. While the use of ultrapure chemicals and point-of-use filtration has reduced the level of liquid-borne contaminants, it has not eliminated the need for wafer cleaning. In fact, wafer cleaning is still the most frequently repeated step in the semiconductor fabrication sequence. This paper describes the results of our study evaluating the particle removal efficiencies of two surface cleaning systems. These systems are (i) a focused acoustic, or ultrasonic cleaner, and (ii) a liquid spray, or hydrodynamic cleaner.

DI-water is, by far, the most commonly used liquid in the cleanroom. Most wet and many dry processing sequences are followed by a DI-water rinse. Water is also used in ultrasonic cleaning tanks and centrifugal spray rinse/dryers for particle removal. The most common wet chemical clean is often referred to as the RCA cleaning procedure[1] and involves the following sequence of operations:

1.	removal of gross organic contamination using a 4 H_2SO_4 : 1 H_2O_2 solution,

2.	stripping of the native oxide layer on silicon using 1 HF : 100 H_2O, to remove mobile ions and metals trapped in the oxide,

3.	removal of trace organic contamination and regrowth of native oxide on the silicon surface with 1 NH_4OH : 1 H_2O_2 : 5 H_2O (75 - 85°C), and

4.	reduction of metal ions with a solution of 1 HCl : 1 H_2O_2 : 5 H_2O (75 - 85°C).

There has been considerable debate about the optimum sequence of using the above cleaning steps[2] and most users have developed their own variations. It is clear from the above list that wet chemical cleaning methods by themselves are ineffective for removing particulate contaminants, because the force required to detach particles from the surface is not supplied by the immersion process. In certain instances, the alkaline clean (step 3), which is also referred to as the standard cleaning solution (SC-1) has been used in conjunction with megasonic energy to remove particles.[3]

In spite of the frequent use of DI-water and RCA cleaning solutions, very little is known about their efficacy for removing submicrometer size particles from wafer surfaces. The force of adhesion of particles to surfaces depends strongly on particle and surface composition, particle size, and chemistry of the cleaning medium. Since particulate sizes and compositions vary widely depending on the processing steps preceding the clean, this implies that a wafer cleaning process must be designed keeping in mind the requirements before and after the clean. A knowledge of the effects of particle composition and wafer surface character on cleaning efficiency will be very useful in optimizing the operating parameters of the cleaner. In earlier papers[4,5] we reported on the effect of various organic solvents such as Freon-TF, Freon-TMS, ethanol, and acetone on particle removal efficiency in an ultrasonic cleaner. It was found that DI-water, in combination with ultrasonic agitation, was a far superior cleaning medium to the freon-based solvents. This paper describes the results of a quantitative evaluation of ultrasonic and hydrodynamic cleaners using the SC-1 solution and DI-water. Since adhesion theory predicts a strong dependence of system composition on particle-surface attachment, particles of widely varying compositions were selected for this study.

EXPERIMENTAL APPARATUS AND PROCEDURE

The cleaning energy produced by conventional ultrasonic tanks is not uniform across the cross-section of the tank. This can often lead to wafer surface damage. For the purpose of this study, a focused acoustic cleaner was assembled in our laboratory. This is a single wafer cleaner in which the diameter of the vibrating horn is approximately equal to the wafer diameter, thus providing uniform ultrasonic intensity across the entire wafer surface. Figure 1 depicts a schematic diagram of the ultrasonic apparatus. The unit vibrates at a constant frequency of 20 kHz. Figure 2 is a diagram of the spray cleaning unit. This unit has two nozzles, one located directly above the wafer, and the other at an angle of 45° to it. In both systems, the cleaning liquid is recirculated through 0.2 μm point-of-use filters. Detailed descriptions of these units are provided elsewhere.[4,5]

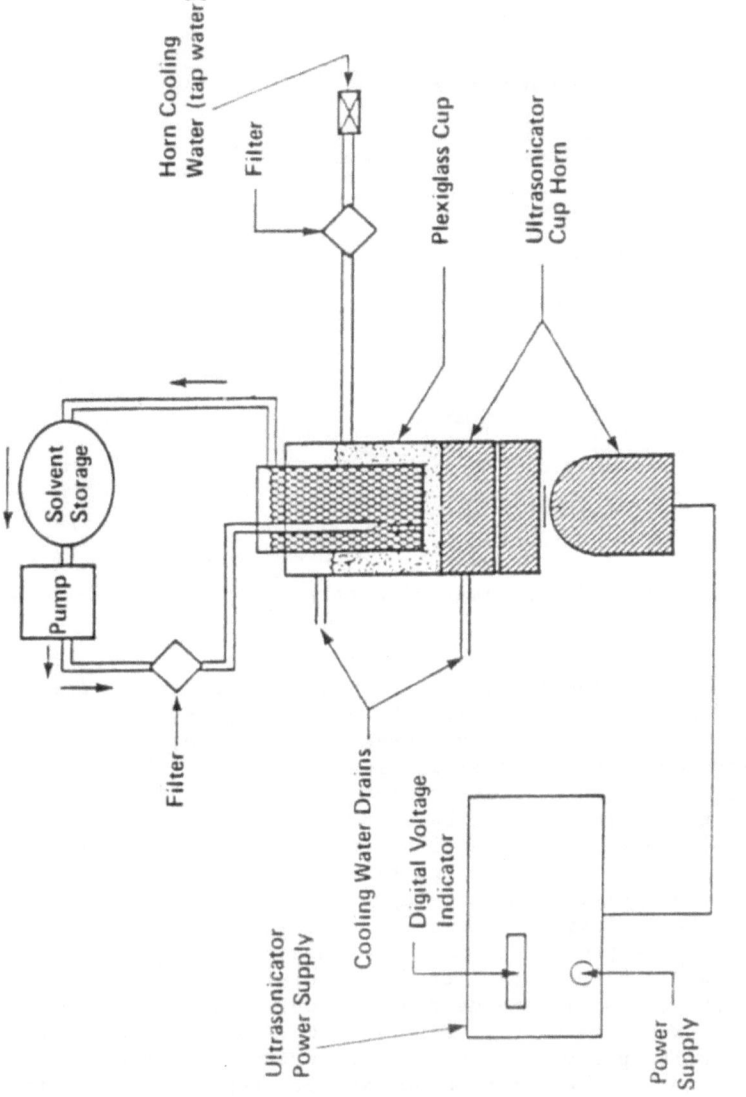

Figure 1. Schematic of experimental apparatus for ultrasonic wafer cleaning.

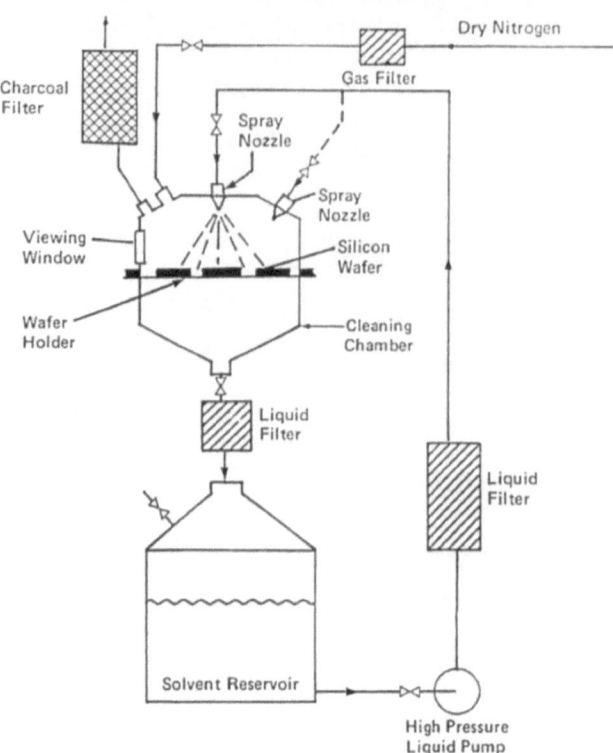

Figure 2. Schematic of experimental apparatus for hydrodynamic wafer
cleaning.

The force of adhesion is, by convention, defined as the force corresponding to 50% cleaning efficiency. Both cleaning systems have features
which allow for the ultrasonic/hydrodynamic force to be gradually
increased. This enables the determination of cleaning efficiency for any
applied force. Figure 3 shows the variables of the study. Polystyrene
latex (PSL) particles were chosen because they are spherical in shape,
have a relatively narrow size distribution, and their composition is not
unlike that of photoresist. Glass particles were selected because of
their sphericity and similarity in composition to SiO_2, which is a common
contaminant in chemical vapor deposition processes. Silicon dust was
chosen because of its prevalence as a contaminant during wafer polishing
and scribing. While PSL spheres were commercially available in submicrometer sizes with very small standard deviation, 1 μm glass spheres had to
be classified from a broader size distribution. Silicon dust was
obtained by grinding pieces of a silicon wafer with a mortar and pestle.
These particles were irregular in shape and are represented in this study
by two size classes, particles of size between 0.5 and 2.0 μm, and
between 2 and 20 μm. These are two of the size classes in which the
laser surface scanner (Estek Standard WIS-600) reports surface particle
counts. PSL and glass particles were deposited on wafer surfaces in a
special aerosol deposition chamber. An aqueous dispersion of the particles in water was nebulized through an aerosol generator into an aerosol
cloud, which was then neutralized, dried and allowed to diffuse onto the
wafer surface in a chamber where HEPA-filtered air was circulated. This
procedure produces a uniform deposit of single particles on the wafer.
The silicon dust particles were deposited on wafers in the same chamber,
but from a cloud of particles, generated by striking a membrane containing the particles.[4,6] The initial number count of particles on a 4" (100
mm) wafer was approximately 600.

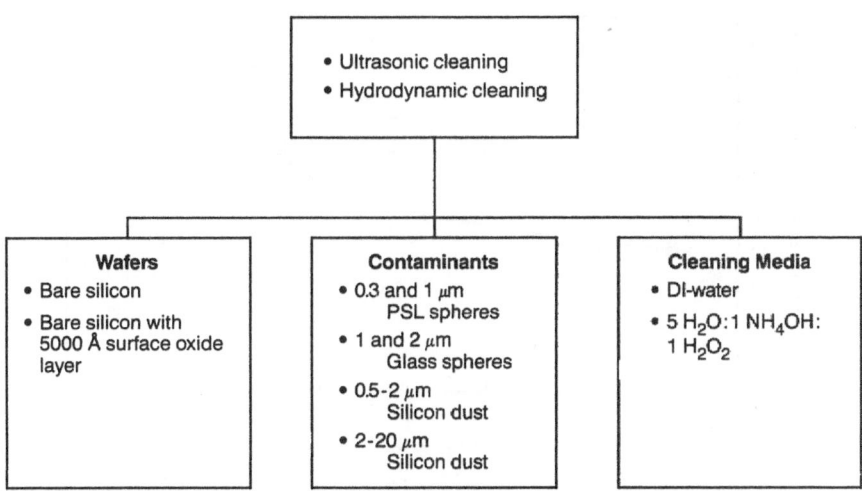

Figure 3. Summary of system variables

All experiments were conducted within the cleanroom at the Micro-electronics Center of North Carolina. The SC-1 chemicals were all of semiconductor grade and low particle count. The house DI-water particle count, as measured using a Particle Measuring Systems Liquid Particle Counter, was typically less than 10/mL particles greater than 0.3 μm. Each wafer was subjected to the cleaning step for two minutes using DI-water or a blend of $5H_2O$: $1H_2O_2$: $1NH_4OH$ (SC-1), rinsed by immersion in a DI-water bath, then spun dry and scanned. The contact angles of water on bare and oxidized wafers were measured using a Kernco goniometer.

RESULTS AND DISCUSSION

The effect of particle composition on ultrasonic cleaning efficiency is depicted in Figure 4. Ultrasonic acceleration is expressed as log ($1000*V_T$) where V_T is the voltage on the ultrasonic transducer at a frequency of 20 kHz. Since the frequency of the vibrating horn is fixed at 20 kHz, any variation in voltage (and therefore, amplitude) is indicative of a change in ultrasonic force. Figure 4 reveals that cleaning efficiency increases with increasing ultrasonic acceleration for all three types of particles (silicon, PSL, and glass), each system having its own characteristic profile. The force of adhesion, as represented by the relative acceleration at 50% cleaning efficiency, was greatest for silicon dust and least for PSL particles. Thus, the ease of particle removal was: PSL > glass > silicon dust.

The theoretical force of adhesion for a particle/substrate/cleaning medium is described by van der Waals equation, which relates the force to (i) the atomic composition of the interacting materials, (ii) the diameter of the particle, and (iii) the separation distance between the particle and the substrate. The term representing the contribution due to atomic composition is called the Hamaker constant, and this constant has a unique value for a specific system. From a search of the literature on adhesion, it was found that silicon dust has a higher value of Hamaker constant (24.8×10^{-13} ergs) than SiO_2 (8.5×10^{-13} ergs) and PSL (6.5×10^{-13} ergs). Glass can have Hamaker constants as low as 2.3×10^{-13} ergs, owing to the presence of components other than SiO_2.[5,7] Thus, the higher value of the force of adhesion for silicon dust may be

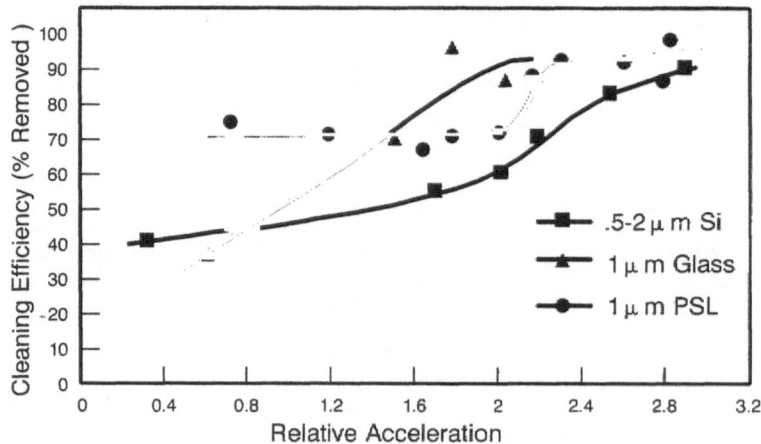

Figure 4. Variation of cleaning efficiency with ultrasonic acceleration in DI-water for silicon, glass and polstyrene latex particles on bare silicon (with native oxide layer) wafers.

attributed to its high value of the Hamaker constant, and to particle nonsphericity, which can enhance adhesion by increasing the area of contact.

Figure 4 also reveals that a significant number of particles was detached from the surface of a wafer by mere immersion in DI-water. For 1 μm PSL particles, over 70% cleaning efficiency was observed with no ultrasonic agitation. For silicon dust (0.5-2 μm) this value was about 40%, and for 1 μm glass particles around 10%. Leenaars[8] attributes this mechanism of particle removal to surface forces acting on the particle as the wafer is immersed or withdrawn from a liquid bath, across the air-liquid interface. Michaels et al.[9,10] have also demonstrated the importance of the air-liquid interface on particle deposition and removal. Figure 5 depicts a schematic diagram of the forces acting on a particle at the interface. The force of adhesion, F_A is opposed by the surface force, $F\gamma_{MAX}^x$, which is dependent on liquid surface tension γ_{lg}, contact angles of the particle (θ) and substrate (α), and particle radius R. If the surface force is greater than adhesion force, particle detachment will be favored during wafer transport across the interface, or, the larger the ratio of these two forces, the greater the tendency for detachment. This ratio was calculated for the systems of interest to this study and are reported in Figure 5. The ratio is highest for PSL, intermediate for silicon, and lowest for glass. PSL particles would, therefore, have the greatest tendency to spontaneously detach from a wafer surface. This trend, predicted by theory, is in agreement with our experimental observations.

Replacement of DI-water by the SC-1 solution for cleaning silicon dust from bare wafers resulted in significant changes in the profiles of particle removal efficiency. Figures 6 and 7 show these profiles for ultrasonic and hydrodynamic cleaning and Si particles of two size classes. At low ultrasonic acceleration, Figure 6 reveals that the cleaning efficiency using SC-1 was considerably lower than that using DI-water. As the acceleration was increased, the cleaning efficiency increased, until finally, there was no distinction between the results

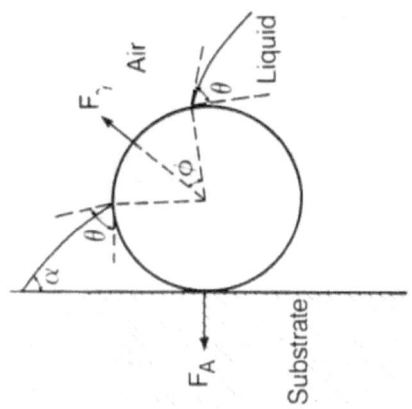

$$F^x_{\gamma,\text{MAX}} = 2\pi R\gamma_{lg}\sin^2\!\left(\frac{\theta}{2}\right)\cos\alpha \qquad (\alpha < 90°)$$

$$F^x_{\gamma,\text{MAX}} = -2\pi R\gamma_{lg}\sin^2\!\left(90°+\frac{\theta}{2}\right)\cos\alpha \qquad (\alpha > 90°)$$

System	θ	α	$F^x_{\gamma,\text{MAX}}$ (dynes)	F_{Ad} (dynes)	$F^x_{\gamma,\text{MAX}}/F_{Ad}$
PSL*–Bare Si–Water	91°	62°	0.0055	0.0001	55
Glass*–Bare Si–Water	8°	62°	0.00005	0.00009	0.56
Si Dust*–Bare Si–Water	62°	62°	0.0028	0.0006	4.7

*Average particle diameter = 1.0 μm
Surface tension of water = 70 dynes/cm

Figure 5. Contribution of the air-liquid interface to particle detachment from a surface[8].

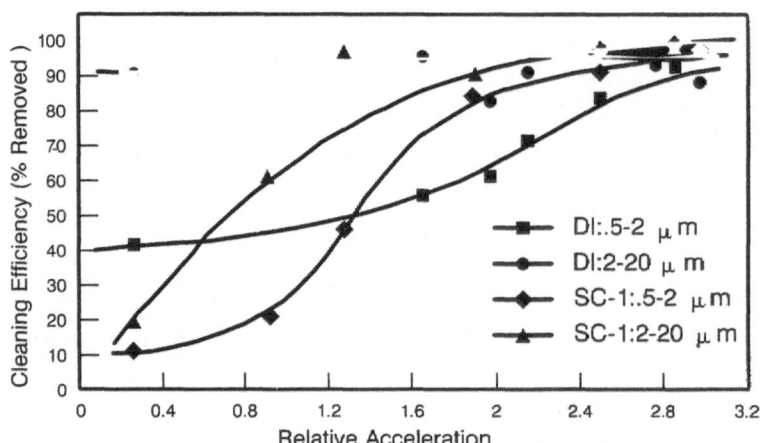

Figure 6.   Comparison of DI-water with SC-1 solution for removing silicon
dust particles from bare silicon wafers in an ultrasonic cleaner.

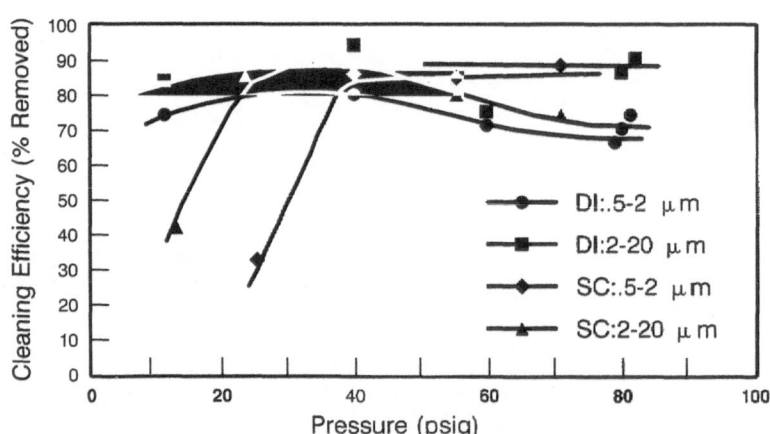

Figure 7.   Comparison of DI-water with SC-1 solution for removing silicon
dust particles from bare silicon wafers in a hydrodynamic wafer
cleaner.

for the two cleaning liquids. Figure 6 also shows that no spontaneous detachment of silicon dust was observed when water was replaced by the SC-1 solution. The differences between DI-water and SC-1, observed during ultrasonic cleaning, were also apparent during hydrodynamic cleaning, as shown in Figure 7. For particles of PSL and glass, very little difference was observed between water and SC-1.

The lower cleaning efficiency for silicon dust particles in the SC-1 solution may be attributed to (i) the formation of a chemical oxide bond between the silicon wafer surface and the particles, and (ii) a reduction in the surface charge of silicon particles in SC-1 owing to its high pH value (11.8), which reduces electrostatic double layer repulsion forces. Depending on the thickness of the initial oxide layer, an oxide bond can form between silicon particles and the wafer surface due to the oxidation potential of hydrogen peroxide in the SC-1 solution.[11] At high values of cleaning energy, this bond may be broken to yield greater cleaning efficiencies. The true mechanism of adhesion enhancement in SC-1 is uncertain and further studies need to be conducted. If chemical bonds do indeed form between particles of materials such as silicon and the wafer surface, they could also enhance adhesion for other metallic and semiconductor particulate contaminants.

An important implication of the results of this study is that mere immersion of silicon dust (and perhaps, other metal particulate) -laden wafers in SC-1, with no additional cleaning energy, will result in no cleaning. Furthermore, the enhancement of particle adhesion in SC-1 may make additional cleaning steps more difficult. In general, our studies have shown that water, in combination with some form of external mechanical force, is an effective cleaning medium for removing most types of particulate contaminants. However, the efficiencies of ultrasonic and hydrodynamic cleaning drop rapidly as particle size decreases below 0.5 μm. Also, this study did not include initial particle concentration on wafers as a variable, the value being maintained around 600 particles. When the initial particle count on a 4" diameter wafer decreases to below 50-70 particles, cleaning effectiveness can deteriorate further.

CONCLUSIONS

Silicon wafer cleaning efficiencies using ultrasonic and hydrodynamic cleaning techniques were found to be strongly dependent on particle composition and size. Silicon particles were more difficult to remove than glass and polystrene latex particles. This trend was in general agreement with the van der Waals adhesion forces for these systems.

Immersion and withdrawal of contaminated wafers across the air-liquid interface resulted in spontaneous detachment of some particles. This detachment was most prevalent for PSL, and least for glass particles. Surface forces that are a function of liquid surface tension and contact angles of particles and substrate are responsible for this mechanism of surface cleaning.

At low ultrasonic or hydrodynamic energies, the SC-1 solution was found to be inferior to DI-water for removing silicon particles from a bare silicon wafer. While the exact mechanism for this behavior is unclear, it emphasizes the importance of system chemistry for surface cleaning. The spontaneous detachment observed with DI-water during wafer transport across the interface was also not seen with SC-1.

ACKNOWLEDGEMENT

This research was sponsored by the Semiconductor Research Corporation through a Manufacturing Science Contract with the Microelectronics
Center of North Carolina.

REFERENCES

1.   W. Kern and D.A. Puotinen, Cleaning solutions based on hydrogen
      peroxide for use in silicon semiconductor technology, RCA Review, 31,
      187 (1970).

2.   D.S. Becker, W.R. Schmidt, C.A. Peterson, and D.C. Burkman, Effects of
      various chemistries on silicon-wafer cleaning, ACS Symp. Series No.
      295, pp. 366-376 (1985).

3.   S. Shwartzman, A. Mayer, and W. Kern, Megasonic particle removal from
      solid-state wafers, RCA Review, 46, 81 (1985).

4.   V.B. Menon, L.D. Michaels, R.P. Donovan, M.B. Ranade, and V.L. Debler,
      Particle removal from semiconductor wafers using cleaning solvents, in
      "Particles in Gases and Liquids 1: Detection, Characterization, and
      Control," K.L. Mittal editor, pp. 259-271, Plenum Press, New York,
      1989.

5.   V.B. Menon, L.D. Michaels, R.P. Donovan, L.A. Hollar, and D.S. Ensor,
      Removal efficiencies of organic and inorganic particles from silicon
      wafers, in proceedings of the Institute of Environmental Sciences,
      Annual Meeting, King of Prussia, Pennsylvania, May 2-6, 1988.

6.   A.D. Zimon, "Adhesion of Dust and Powder," Plenum Press, New York,
      1982.

7.   J. Visser, Adhesion of colloidal particles, in "Surface and Colloid
      Science," E. Matijevic editor, vol. 8, p. 3, John Wiley, New York,
      1976.

8.   A.F.M Leenaars, A new approach to the removal of colloidal particles
      from solid (silicon) substrates, in "Particles on Surfaces 1:
      Detection, Adhesion, and Removal," K.L. Mittal editor, pp. 361-372,
      Plenum Press, New York, 1988.

9.   L.D. Michaels, V.B. Menon, R.P. Donovan, and D.S. Ensor, Mechanisms of
      liquid-based particle deposition onto silicon wafers, in proceedings
      of the Symposium on Particulate Contamination Control in Micro-
      electronics, Annual Meeting of the American Association for Aerosol
      Research, Chapel Hill, North Carolina, October 10-14, 1988.

10.   L.D. Michaels, V.B. Menon, A.C. Clayton, and R.P. Donovan, Particle
      deposition and removal at the air-liquid interface, in proceedings of
      Microcontamination Conference, Santa Clara, California, November
      15-18, 1988.

11.   V.B. Menon, L.D. Michaels, L.A. Hollar Jr., and R.P. Donovan, Inter-
      facial effects associated with the separation of particles from
      semiconductor surfaces, paper presented at the Annual Meeting of the
      American Institute of Chemical Engineers, Washington D.C., November
      27-December 2, 1988.

NEW SONIC CLEANING TECHNOLOGY FOR PARTICLE REMOVAL FROM SEMICONDUCTOR SURFACES

D. M. Berg, T. Grimsley, P. Hammond and C. T. Sorenson

ESTEK / A Kodak Company
3000 Winton Road South, Rochester, NY 14623

Several sonic cleaning technologies are currently used in the semiconductor industry. An overview of these technologies is presented, comparing their cleaning mechanisms, applications and suitability. Using an analogy to photography, we then discuss a "reciprocity principle" in megasonic cleaning and describe an experiment we performed to test this hypothesis. Equivalent cleaning was observed for the same average acoustic power, independent of peak intensity. In addition, cleaning was approximately dependent on total energy exposure. Implications for the design of megasonic cleaning equipment are discussed.

INTRODUCTION

Several cleaning processes using sonic energy are being used today for particle removal in the semiconductor industry [1]. Each process must generate sufficient force to overcome the adhesion forces holding particles to the wafer surface. Once particles are removed, they must be prevented from readhering. Since the mechanisms, advantages and applications of each process are different, we will first present an overview of these technologies to provide a context for our discussion.

We will then discuss a "reciprocity principle" for megasonic cleaning. In photography, the reciprocity principle states that decreasing exposure time and increasing light intensity by the same ratio will result in an identical photograph [2]. For megasonic cleaning, decreasing exposure time and increasing the intensity of the megasonic beam results in the same cleaning performance. This is a surprising result; it would be easy to imagine that more intense megasonic waves would dislodge more particles than less intense waves. We describe an experiment which demonstrates this principle and discuss its implications.

OVERVIEW OF SONIC CLEANING PROCESSES

Ultrasonic Tank Cleaning

Ultrasonic tank cleaning has been used in applications from metal fabrication to dentistry since the 1940's [3]. A traditional tank uses ultrasonic cavitation generated by low power density (0.5 - 0.6 watts/ cm^2 of radiating surface) waves to clean a batch of wafers submerged in a stagnant tank [4] . The frequencies used are around 20 kHz which corresponds to a wavelength of 7.5 cm in water. Ultrasonic cavitation is a process in which microscopic bubbles or voids form in the cleaning solvent during the low pressure half cycle of the acoustic wave. These bubbles collapse violently during the high pressure half cycle, causing very large local pressures which scrub particles from a surface. Cavitation bubbles form regardless of surface topography, so cleaning is effective even on deeply patterned surfaces.

However traditional ultrasonic tanks have generally been discarded by the semiconductor industry. Because the sonic energy distribution in the tank is random and stationary, there can be "hot spots" which cause circuit damage and stacking faults in subsequent layers. In addition, stagnant tanks rely on the wetting properties of the solvent solution to prevent particles from readhereing. These problems, especially the possibility of damage, have lead to the development of improved sonic cleaning technologies.

Focused Acoustic Cleaning

A focused acoustic cleaning system passes single wafers through an area of intense acoustically induced cavitation while they are in a counterflowing solvent stream. The frequency used is the same as in ultrasonic tanks, but the power density is many times higher (7.5 - 15 watts/ cm^2).

Because individual wafers are moved continuously through the cavitation region, exposure is more controlled than in a traditional tank. The entire wafer surface receives the same amount of energy, with no possibility of damaging hot spots. In addition, the counterflowing solvent bath allows solutions with poor wetting properties (e.g. DI water) to clean effectively.

Focused acoustic cleaning overcomes many of the problems experienced with ultrasonic tanks. In the semiconductor industry, it is applicable where deep topographies must be cleaned. If DI water is the only permissible solvent, focused acoustic cleaning is more effective than other cleaning methods. When other chemical solutions can be used to enhance particle wetting, megasonic cleaning may provide an alternative.

Megasonic Cleaning

Megasonic cleaners use millimeter wavelength (1 MHz) acoustic waves directed tangent to the wafers' surfaces to shear or "rock" particles loose. Wafers are cleaned a batch at a time, in either a stagnant or a cascading tank.

The megahertz acoustic frequency is too high for cavitation to occur. Instead of cavitation scrubbing, megasonic cleaners rely on the wetting ability of the solvent to prevent loosened particles from readhering [5,6] . Because the particles are sheared off by waves traveling along the surface, cleaning is ineffective in deep topographies.

Megasonic cleaning is useful during "front end" wafer fabrication, when the use of oxidizing chemicals is possible. For other applications, better wetting agents need to be identified. DI water alone is not effective in this process.

This brief overview of three different sonic cleaning technologies provides a context for discussion of a specific investigation into megasonic cleaning.

"RECIPROCITY PRINCIPLE" FOR MEGASONIC CLEANING

In photography, the "reciprocity principle" states that increasing the light intensity and decreasing the exposure time by the same ratio will leave the photograph unchanged.
The film response is independent of the peak light intensity.

In a megasonic cleaning system, the same average power can be provided by a continuous wave, or by a modulated wave of higher peak intensity. If the "reciprocity principle" holds, the cleaning should be independent of the peak intensity, and depend only on the average intensity. This is analogous to taking a picture with several bright flashes, or with a continuous dim light.

EXPERIMENTAL

To test this hypothesis, we performed an experiment which compared the cleaning performance using continuous and modulated transducer waveforms. The experiment was designed using the ECHIP [TM] experimental design software package [7]. The ECHIP [TM] package provides statistically optimized experimental designs to determine the effects given parameters in terms of selected polynomial models. The package also analyzes the experimental data by fitting the chosen polynomial to the data, displaying contour plots of the result and providing statistical measures of the quality of the fit.

The design is shown in Table I. The Run Number shows the order in which the trials were performed, while the Type Number is an index

Table I. Design for Megasonic cleaning experiment comparing continuous and modulated transducer waveforms.

Run #	Type #	Duty Factor (%)	Peak Power (watts)	Exposure Time (min)	Average Power (watts)
1	12	20	250	10	50
2	13	20	250	20	50
3	3	20	100	10	20
4	1	20	200	10	40
5	4	100	100	10	100
6	5	50	100	15	50
7	6	100	200	10	200
8	7	100	150	20	150
9	2	20	200	20	40
10	1	20	200	10	40
11	8	100	100	20	100
12	9	100	200	20	200
13	10	50	150	10	75
14	2	20	200	20	40
15	11	20	150	15	30

referring to the parameters for each trial. The transducer waveform was a square wave with a frequency of 15 Hz (period = 67 milliseconds) and Duty Factor as shown. A 100% duty factor means the transducer was "on" continuously. Peak Power gives the power supplied to the transducer during its "on" time. The Exposure Time is the duration that the batch of wafers was in the cleaning tank.

For example, the first trial (Run 1) was of Type 12. During each 67 millisecond period, the transducer was "on" for 13 milliseconds at 250 watts. The wafers were in the megasonic tank for 10 minutes.

The experiment was conducted in ESTEK's Class 10 applications laboratory using an ESTEK MCS-6000-1 Megasonic cleaning system modified to allow modulation of the transducer, an SRD-1000 to rinse and dry the wafers after cleaning and WIS-600B laser inspection system to measure particle levels before and after cleaning. SC-1 (1:1:5 Ammonium Hydroxide, Hydrogen Peroxide, DI Water) was the solvent in the megasonic tank [8]. Control wafers were run along with process wafers to prevent

Table II. Summary of wafer conditions, boat loading order and processing procedures for Megasonic cleaning experiment.

Wafers:
         Process: - Bare Si; 100 mm dia; < 50 PPW
                              - Batch Dipped in City Water, Spin Dried

         Control: - Bare Si; 100 mm dia; < 100 PPW

Boat Load Order:
               P, C, P, P, C, P   (P = Process, C = Control)

Procedure:
           1.   Load Process and Control Wafers
           2.   Inspect
           3.   Megasonic Clean (per experiment design)
           4.   DI Rinse 90 sec, 300 RPM; Spin Dry 4 min, 1700 RPM
           5.   Inspect

handling or equipment failures from biasing the data. The test conditions for all trials are shown in Table II. The average initial contamination levels for each trial are shown in Figure 1. The average level was 8078 Particles Per Wafer (PPW) - which corresponds to 103 particles/cm^2 over the 100 mm diameter wafer surface - with standard deviation 1014 PPW.

The results were analyzed using the ECHIP [TM] software. A "factorial" polynomial was fit to the data, and contour plots of this polynomial were generated to present the results graphically. The factorial model was fit to the logarithm of the number of remaining particles to reflect the logarithmic nature of the cleaning process.

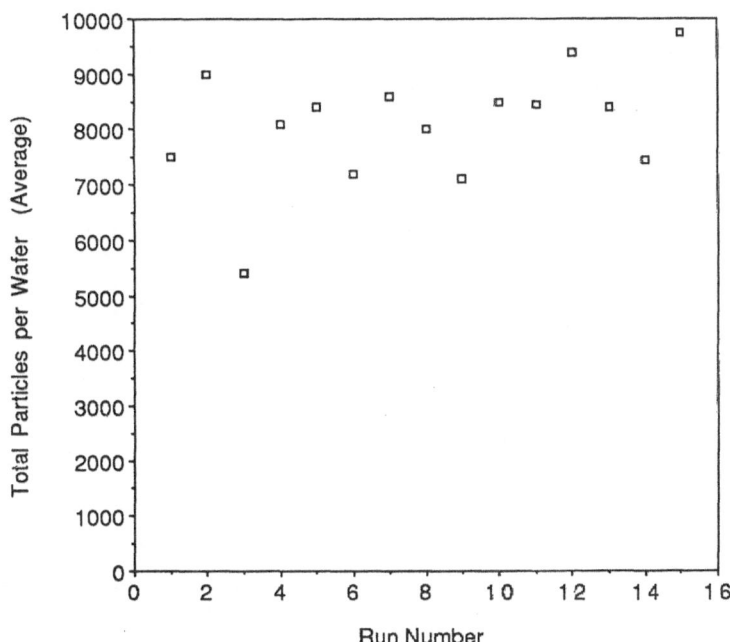

Figure 1.   Average initial wafer contamination level for each trial.

RESULTS

Figures 2 and 3 show contour plots of PPW remaining after cleaning as a function of Peak and Average Power. The contours are logarithmically spaced. In the "Non-physical Regions", the average power would exceed the peak power so the calculated values have no meaning. Figure 2 shows particles remaining after 10 minutes in the megasonic tank. Particle levels decrease steadily for increasing average power, but are effectively independent of peak power. Figure 3 shows particles remaining after 20 minutes exposure. Note that the overall level has decreased from 300 PPW to 125 PPW. There also appears to be decreased cleaning at high peak powers. It is not clear if this is a significant effect.

Figure 4 shows particles remaining after cleaning as a function of average power and exposure time. Increasing either parameter improves cleaning. In fact, we found that cleaning is approximately dependent on acoustic exposure (= average power x exposure time) as illustrated in Figure 5. Thus higher average acoustic power reduces the time required to clean to a given level.

The limitations of these results must be remembered. This experiment was conducted using bare silicon substrates contaminated in city water, and using an SC-1 cleaning solution. Other substrate/contaminant combinations may behave differently. In addition, the replicate standard deviation of the data is 65%. Thus differences of less than a factor of 1.6 are not statistically significant. Finally, this experiment measured a limited range of powers and exposure times. We have previously observed significant cleaning due to submerging contaminated wafers in SC-1 solution without power [9] to the transducers.

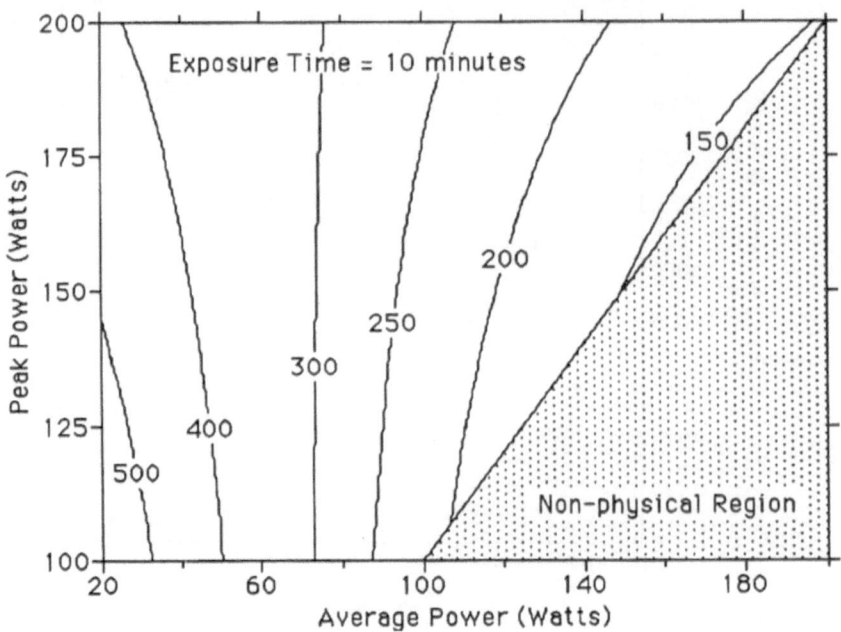

Figure 2.   Particles remaining after cleaning as a function of Average
Power and Peak Power - 10 minutes Exposure Time.

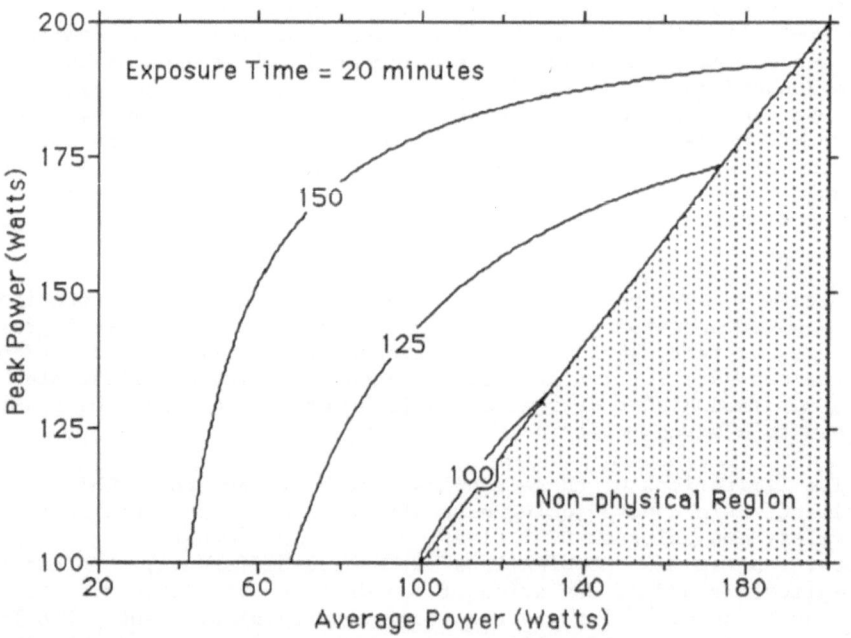

Figure 3.   Particles remaining after cleaning as a function of Average
Power and Peak Power - 20 minutes Exposure Time.

In this case, we have cleaning with zero exposure. This demonstrates that the reciprocity principle does not hold at extremely low average powers.

IMPLICATIONS OF THE "RECIPROCITY PRINCIPLE" FOR MEGASONIC EQUIPMENT DESIGN

The reciprocity principle has some important implications for the improvement of megasonic cleaning equipment. Because a modulated waveform will provide the same cleaning as a continuous wave, transducers may be turned off for some fraction of the time during cleaning without harm. This provides time to test, adjust and actively tune the transducers during cleaning. Of course the average power must be maintained, by applying a higher power to the transducers during the part of the time they are "on".

Because the intensity of the acoustic wave does not itself improve cleaning, a transducer which produces a narrow beam would perform identically to one with a wide beam as long as the acoustic exposures are identical.

However, the improvement of cleaning with exposure indicates that shorter cleaning times can be used with higher average power transducers. The development of transducers capable of higher average powers would enhance productivity.

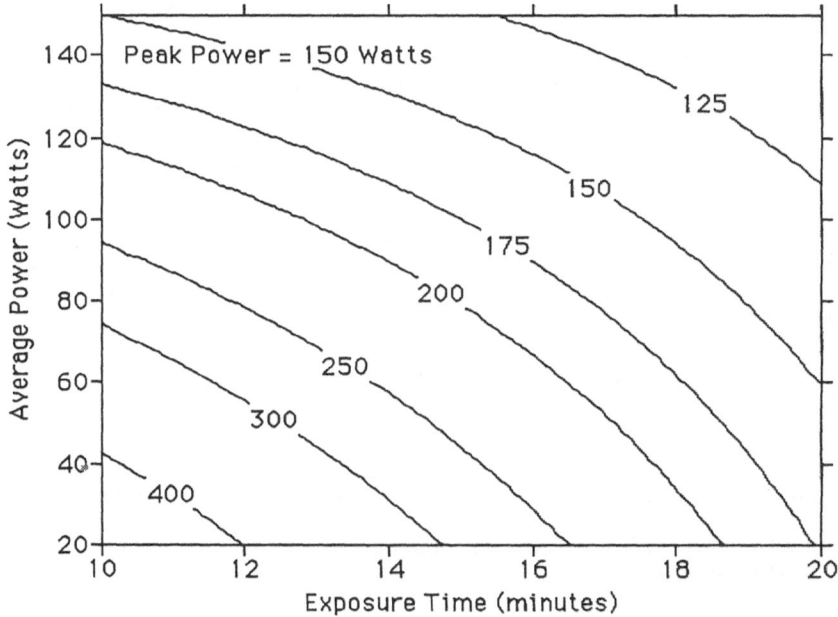

Figure 4.   Particles remaining after cleaning as a function of Exposure Time and Average Power. Results interpolated for 150 watts Peak Power, but are typical for all Peak Powers measured.

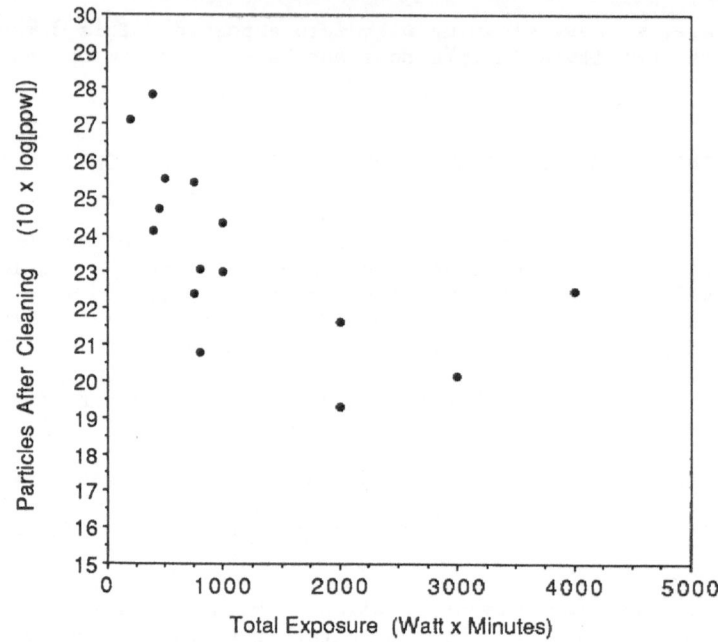

Figure 5. Particles remaining after cleaning as a function of Total Megasonic Exposure. Greater exposures cause better cleaning regardless of peak megasonic power or exposure time.

SUMMARY

We have briefly described three different acoustic cleaning technologies used in the semiconductor industry: ultrasonic cleaning tanks, focused acoustic cleaning and megasonic cleaning.

We then discussed a "reciprocity principle" for megasonic cleaning: that cleaning performance is dependent only on the average acoustic power, not on the peak intensity. This principle was demonstrated with a statistically designed experiment. In addition, we found evidence to support the stronger statement that cleaning depends only on the acoustic exposure (average power x exposure time), rather than on power or time separately.

These results indicate that beam shaping or complicated transducer waveforms will probably have little effect on cleaning performance. Megasonic cleaning equipment can best be improved by providing higher average power transducers.

REFERENCES

1.   K.L. Mittal,Editor, "Surface Contamination: Genesis, Detection and Control," Vols. 1 and 2, Plenum Press, New York, 1979.

2.   H. Baines, "The Science of Photography," p.199, John Wiley & Sons, New York, 1971.

3.   G.W. Willard, Ultrasonically induced cavitation in water: A step-by-step process, J. Acoustical Soc. Amer., $\underline{29(4)}$, 669 (1953).

4.   M. O'Donoghue, The ultrasonic cleaning process, Microcontamination, p.63 (Oct/Nov 1984).

5.   A. Mayer and S. Shwartzman, Megasonic cleaning: A new cleaning and drying system for use in semiconductor processing, J. Electronic Mater., $\underline{8(6)}$, 73 (1979).

6.   S. Shwartzman, A. Mayer and W. Kern, Megasonic particle removal from solid-state wafers, RCA Review, $\underline{46}$, 81 (1985).

7.   ECHIP Software, Expert in a Chip, Inc., 518 Holly Knoll Rd., Hockessin, DE   19707,   (302) 239-5429 .

8.   W. Kern and D.A. Puotinen, Cleaning solutions based on hydrogen peroxide for use in silicon semiconductor technology, RCA Review, $\underline{31}$, 187 (1970).

9.   T. Grimsley, unpublished data (1988).

ABOUT THE CONTRIBUTORS

DAVID M. BERG is presently Staff Scientist at ESTEK (which he joined in 1987) working on the improved methods for characterizing wafer cleaning, and also on development of laser surface inspection equipment. He received his M.S. in Optics from the University of Rochester in 1983 and in 1983 he joined GCA Corporation where he developed lens test, optical metrology and microlithographic equipment.

WILLIAM S. BICKEL has been a professor in the Department of Physics at the University of Arizona since 1965, and a full professor since 1975. He received his Ph. D. from Pennsylvania State University in 1965. He has been a leading researcher in the field of atomic physics and beam-foil spectroscopy. His present research interests include experimental areas of elastic polarized light scattering from particulates, adhesion forces between small particles, and the physics of music. He is the recipient of the 1987 University of Arizona Foundation Creative Teaching Award. He has over 80 refereed publications.

KENNETH D. BOMBEN is a Senior Laboratory Scientist at the Perkin-Elmer Physical Electronics Division in Mountain View, CA, where he is using XPS, AES and SIMS to characterize surfaces. He received his Ph.D. in Chemistry from Oregon State University, Corvallis, and did postdoctoral work in the Chemistry Department of the University of California, Berkeley, and at the Materials and Molecular Research Division of Lawrence Berkeley Laboratory.

J.K. (KIRK) BONNER is currently with Allied-Signal/Baron-Blakeslee in Melrose Park, IL and is Manager of Applications and Process Development Engineering. Before joining Allied Corporations's Buffalo Research Laboratory in 1984, he worked for Digital Equipment Corporation, Martin Marietta Aerospace, and City of Baltimore. He received his Ph.D. from the Johns Hopkins University in Baltimore in 1974. He has published over 15 papers in the areas of printed wiring technology and factory automation.

RAY BOWEN is with Analytical Technologies Division of Eastman Kodak Company in Rochester. He graduated in 1976 from Alfred State College with an AAS in Chemical Technology and has 12 years experience with scanning electron microscopy at Eastman Kodak Company.

ROBERT DEAN has been with Perkin-Elmer Electron Beam Technology in Hayward, CA as an engineer since 1984, where he has conducted research on machine generated particulates and e-beam resists. He attended graduate school at the University of Arizona in organic and analytical chemistry from 1978 to 1981. In 1981 he was employed by the Department of Electrical Engineering at the University of Arizona as a research associate working with a group that become the founders of the Center for Microcontamination Control.

LAWRENCE DeMEJO is Senior Chemical Engineer in the Small Particle Technology Laboratory (Copy Products R&D Division) at Eastman Kodak Company in Rochester where he joined in 1978. He received his Ph.D. degree in Polymer Science and Engineering in 1977 from the University of Massachusetts. He is coauthor of four publications and one patent in the polymer field and six patents on toners for dry electrophotographic developers.

R.P. DONOVAN is a research physicist at Research Triangle Institute in Research Triangle Park, North Carolina. He received an M.S. (1959) in physics from the University of Pennsylvania. He is currently project leader on a program sponsored by the Semiconductor Research Corporation, whose purpose is to evaluate the deleterious effects of particles on silicon device manufacturing.

DAVID S. ENSOR is director of the Center for Aerosol Technology at Research Triangle Institute in Research Triangle Park, North Carolina. He earned his Ph.D. in engineering at the University of Washington in 1972. His present responsibilities include organizing and supervising research in microcontamination, particulate control devices, air pollution control, and chemical defense.

CHARLES M. FALCO is with the University of Arizona which he joined in 1982 as a full professor, with joint appointments in the Physics Department, the Optical Sciences Center, and the Arizona Research Laboratories. Before that he was group leader of the Superconductivity and Novel Materials Group at Argonne National Laboratory, He received his Ph.D. degree in Physics from the University of California at Irvine in 1974. His research in recent years has concentrated on studies of various physical properties of artificial metallic superlattices. In 1986 he established the Laboratory for X-ray Optics to fabricate and study metallic multilayers for use in the soft-x-ray region. He has published over 100 scientific articles. He is a Fellow of the American Physical Society.

MARVIN FEIN is a Staff Engineer at IBM in Endicott, NY where he is working in contamination control of multilayer circuit board processing which includes testing of cleanroom consumables, cleanroom designs and defect analysis. He received his B.S. in Mechanical Engineering from City College of New York in 1963 and an M.S. in Management Science from SUNY at Binghamton in 1972.

ZINOVY FICHTENHOLZ has been with Siliconix in Santa Clara, CA since 1984. He was responsible for the start-up of photolithography (pattern transfer) lines in the Class 16-in wafer fab. As a Staff Engineer he also manages a photolithography engineering group. He received the M.S. degree in Electrical Engineering from the State University of Saransk, USSR in 1975. Since then he has worked in design and process development of bipolar and MOS discrete power devices and IC's.

N. FUJINO is Director of Semiconductor Research Laboratory, Kyushu Electronic Metal Co., Saga, Japan. Graduated from Osaka University and has Sci. Dr. degree.

THOMAS GRIMSLEY is currently heading the Applications and Testing Group at Estek. He has worked at National Semiconductor as a Product Engineer in the Transistor Group. He received his B.S. degree in Microelectronic Engineering from Rochester Institute of Technology in 1986 and is currently finishing M.S. degree in Electrical Engineering from the same institution.

PETER HAMMOND is a Project Engineer at Estek - A Division of Eastman Technologies. His activities include the project management and design of the Estek megasonic cleaning system. Prior to Estek he worked at Xertronix Inc. where he was involved in the formation of the company and the development of the XR-2200 Spin Rinser Dryer. He is currently working towards a degree in physics at the Rochester Institute of Technology.

H. HORIE is Member of Material Research Section of Semiconductor Research Laboratory, Kyushu Electronic Metal Co., Saga, Japan. Graduated from Saga University.

JYH-PING HSU has been an Associate Professor of Chemical Engineering at National Taiwan University since 1984. He earned his Ph.D. degree in Chemical Engineering from Kansas State University in 1984. He has special interest in biotechnology and bioengineering, fuzzy control, waste water treatment and has published over 30 papers in the areas of flocculation, adsorption and adhesion, crystallization, and bioengineering.

WALTER JOHN has been Research Scientist, Air and Industrial Hygiene Laboratory, California Department of Health Services, Berkeley, CA since 1974. Before that he was Professor, Chairman, Physical Sciences Department, California State College, Stanislaus. He received his Ph.D. in Physics in 1955 from University of California, Berkeley. He has been doing research dealing with particulate matter, and is currently studying acidic particles and gases in California air, particle-surface interactions and aerosol sampling techniques. He is a Fellow of the American Physical Society and Director, American Association for Aerosol Research.

WENDY JONES is a Senior Associate Engineer at IBM in San Jose which she joined in 1985. She received her B.S. in Materials and Metallurgical Engineering from the University of Michigan. She is presently in HDA procurement manufacturing engineering.

ROBERT KAISER is President and Founder of Entropic Systems, Inc. in Winchester, MA which was created in 1985 to pursue the development and commercialization of a novel method of removing undersirable particles from solid surfaces. He is also President and Founder of <<ARGOS>> Associates, Inc. which was created in 1977. He also holds an appointment as a Visiting Scientist in the Center for Materials Processing at MIT. He received his Sc.D. degree in Chemical Engineering from MIT. His research has included investigations of fine particle agglomeration and pelletization, as well as the dispersion of fine particles to form stable colloidal suspensions. He received two NASA invention awards, one of which for his development of stable colloidal dispersions in perfluorinated liquids. He is the author/coauthor of numerous technical publications and reports, as well as the recipient of 13 U.S. patents.

J.E. KATON is currently Director of the Molecular Microspectroscopy Laboratory and Professor of Chemistry at Miami University, Oxford, OH where he has been since 1968. He received his Ph.D. in Physical Chemistry from the University of Maryland. Most of his research has been in the area of vibrational spectroscopy and he has authored or coauthored well over one hundred major articles. He has served as President of the Society for Applied Spectroscopy, Member of the Board of Management of the Coblentz Society and Chair of the Federation of Analytical Chemistry and Spectroscopy Societies (FACSS). He is currently a member of the Editorial Board of Microchemical Journal and Book Review Editor of Applied Spectroscopy.

RUDOLPH F. KLIMA is currently with Allied-Signal/Baron-Blakeslee in Melrose Park, IL and is a member of Application and Process Development Engineering Group. He is conducting research on developing new cleaning technology (equipment and solvent) for the metal cleaning and defluxing industry. He received his Ph.D. in Physical Chemistry from Iowa State University. Before joining Allied-Signal, he developed procedures for quality control and quality assurance of hot melt, computer ribbon and other products pertaining to the ink industry.

MING KO is with IBM Corporation in San Jose, CA where he is working on cleaning technology, contamination control and new test method development for new products. He received his Ph.D. degree in Chemistry from Georgia Institute of Technology and MBA in Management from Golden Gate University. He has published a dozen of papers and patents in the areas of polymers, coatings and contamination control.

JOAN W. KOPPENBRINK is currently serving as the technical marketing manager for CFM-Rodel, and she began working in the semiconductor industry with Rodel Products Corporation in 1984. She had held positions as a Research chemist at Sigma Chemical Company, Monsanto Chemical, and Armour-Dial Corp. She is a graduate of the University of Missouri, Washington University and Arizona State University with degrees in chemistry, biochemistry and business administration.

PATRICIA L. LANG is currently an Assistant Professor of Chemistry at Ball State University, Muncie, Indiana. She earned a Ph.D. degree in Physical Chemistry from Miami University, Oxford, OH. While at Miami University, she received the Federation of Analytical Chemistry and Spectroscopy Society Student Award, the Coblentz Society Student Award and The Miami University Dissertation Fellowship. She authored the first dissertation on the subject of molecular microspectroscopy. She has coauthored eight publications dealing with the use of molecular microspectroscopy in the identification of, among others, pigments from medieval manuscripts, and dust particulates.

CHRISTOPHER F. McCONNELL is the president and co-founder of CFM Technologies and has been responsible for the technical design, development and implementation of Full-Flow^{TM} technology. Prior to founding CFM, he worked with Dow Chemical Company. He holds degrees from Dartmouth, Purdue and Harvard in chemical engineering and business administration. He holds a number of patents in the fields of semiconductor processing and chemical handling, and has presented several papers on wet processing techniques.

JOHN McDOWELL is Senior Designer at IBM in San Jose which he joined in 1970. He received his A.A. from Pennsylvania State University. He is presently working in the Advanced Component Development Department.

VENUGOPAL B. MENON is currently at Sematech in Austin, Texas. Before his current position he was a research engineer in the Center for Aerosol Technology, Research Triangle Institute (RTI), Research Triangle Park, NC. He received his Ph.D. (1986) degree in chemical engineering from Illinois Institute of Technology, Chicago. At RTI he was the project leader on a program, sponsored by the Semiconductor Research Corporation, concerned with the cleaning of silicon wafers and other semiconductor surfaces. He was also involved in projects related to advanced materials processing and hazardous waste management.

LINDA D. MICHAELS is an enviromental scientist in the Center for Aerosol Technology at Research Triangle Institute. She received an M.S. in enviromental science from the University of North Carolina in 1984 and is currently involved in contamination control studies involving the measurement and removal of particles from semiconductor processing liquids.

KASHMIRI LAL MITTAL * is presently employed at the IBM US Technical Education in Thornwood, N.Y. He received his M.Sc. (First Class First) in 1966 from Indian Institute of Technology, New Delhi, and Ph.D. in Colloid Chemistry in 1970 from the University of Southern California. In the last 16 years, he has organized and chaired a number of very successful international symposia and in addition to this volume, he has edited 26 more books as follows: Adsorption at Interfaces, and Colloidal Dispersions and Micellar Behavior (1975); Micellization. Solubilization. and Microemulsions, Volumes 1 & 2 (1977); Adhesion Measurement of Thin Films. Thick Films and Bulk Coatings (1978); Surface Contamination: Genesis. Detection. and Control, Volumes 1 & 2(1979); Solution Chemistry of Surfactants, Volumes 1 & 2 (1979); Solution Behavior of Surfactants: Theoretical and Applied Aspects, Volumes 1 & 2 (1982); Adhesion Aspects of Polymeric Coatings, (1983); Physicochemical Aspects of Polymer Surfaces, Volumes 1 & 2 (1983); Surfactants in Solution, Volumes 1, 2 & 3 (1984), Adhesive Joints: Formation, Characteristics, and Testing (1984), Polyimides: Synthesis, Characterization and Applications, Volumes 1 & 2 (1984); Surfactants in Solution, Volumes 4, 5 & 6 (1986); Surface and Colloid Science in Computer Technology (1987); Particles on Surfaces 1: Detection, Adhesion and Removal, (1988); and Particles in Gases and Liquids 1: Detection, Characterization and Control (1989). Also he is Editor of the Series, Treatise on Clean Surface Technology, the premier volume appeared in 1987. In addition to these books he has published about 60 papers in the areas of surface and colloid chemistry, adhesion, polymers, etc. He has given many invited talks on the multifarious facets of surface science, particularly adhesion, on the invitation of various societies and organizations in many countries all over the world, and is always a sought-after speaker. He is a Fellow of the American Institute of Chemists and Indian Chemical Society, is listed in American Men and Women of Science, Who's Who in the East, Men of Achievement and many other reference works. He is or has been a member of the Editorial Boards of a number of scientific and technical journals, and is the Editor of the Journal of Adhesion Science and Technology, which made its debut in 1987.

S. MIYAZAKI is a Member of Material Research Section of Semiconductor Research Laboratory, Kyushu Electronic Metal Co., Saga, Japan. Graduated from Saga University.

R. NAGARAJAN is a Staff Engineer at IBM Corporation in San Jose which he joined in August 1988. Before joining IBM, he was Research Assistant Professor at West Virginia University. He received his Ph.D. in Chemical Engineering from Yale University in 1986. Among his research interests is the study of deposition of particles.

LEON L. PESOTCHINSKY is presently with San Jose State University and Vitaphore Consulting Group in Palo Alto, CA. He received his Ph.D. degree in Statistics from the Leningrad State University. Since 1969 he has worked as an industrial statistician, taught at the University of California and acted as a consultant to various companies on applied statistics and quality control.

---

* As the editor of this volume.

WALTER PRATER is currently an Advisory Engineer at IBM in San Jose which he joined in 1978. he received his M.S. in Mechanical Engineering from San Jose State University. He is presently in the Advanced Actuator Development Department.

LINDA A. PSOTA-KELTY is a Member of Technical Staff in the Materials Reliability and Electrochemistry Research Department of AT&T Bell Laboratories. She is responsible for the analysis of contamination associated with electronic components and assemblies; in particular, water soluble contamination analyzed by ion chromatography. She received a masters degree in physical chemistry in 1985 from New York University.

JAMES REED is a Research Officer in the Research Division of Berkeley Nuclear Laboratories, Central Electricity Generating Board, Berkeley, U.K. Most of his time has been spent studying the fundamental nature of interactions of particles with surfaces regarding deposition and resuspension in order to be able to predict the amount of radioactive particulate released from a nuclear reactor following an accident. He received his Ph.D. in the field of liquid structure using neutron scattering techniques from University of Kent at Canterbury. Recently he became a Committee Member of the UK Aerosol Society which has been in its third year of existence.

DONALD S. RIMAI is currently a Research Associate in Copy Products R&D at Eastman Kodak Company in Rochester. He received his Ph.D. in Physics from the University of Chicago in 1977. Prior to joining Eastman Kodak, he was at Purdue University (1977-1979). He holds 3 patents and has authored about 30 publications in condensed matter physics.

DALE A. SCHEER is Senior Engineer, Material Laboratories, McDonnell Douglas Corp., in St. Louis which he joined in June 1980 as a Test Engineer in the Analytical Chemistry Laboratory. He received his B.S. in Chemistry in 1980 from the University of Missouri. His activities have included development of techniques for trace analysis of organic compounds in fuel, oil and water, and procedure development for analyzing outgassing components of various resin systems.

ARYE SHAPIRO is a graduate research assistant completing the Ph.D. degree in Physics at the University of Arizona. His current research interests include thin film growth and surface analysis of metals and semiconductors by molecular beam epitaxy.

HELEN C. SHIELDS is a Senior Staff Technologist at Bellcore where she has been working on indoor air research since March 1985. The aim of the indoor air studies is to identify the major chemicals (condensed and/or vapor phase) found within the environment of a telephone office. She has worked in several private laboratories and hospitals, and earned a B.S. in chemistry from Monmouth College in December 1984.

T. SHIRAIWA is Executive Technical Consultant of Osaka Titanium Co., Osaka, Japan. Has Sci. Dr. degree and graduated from Osaka University.

J.D. SINCLAIR is Head of the Materials Reliability and Electrochemistry Research Department at AT&T Bell Laboratories in Murray Hill, New Jersey. He received his Ph.D. in Inorganic Chemistry from the University of Wisconsin in 1972. Since coming to Bell Laboratories in 1972 he has studied the corrosion, contamination, and reliability of electronic equipment, devices and materials and has published extensively in this area.

ANDRE J. SOMMER has served for the past two years as the Assistant Director of the Molecular Microspectroscopy Laboratory at Miami University in Oxford, OH. He received his Ph.D. in Analytical Chemistry from Lehigh University. After leaving Lehigh in 1985 he joined the Systems Technology Division of IBM as a postdoctoral research scientist. His research interests involve the application of infrared and Raman microspectroscopies for materials characterization and has authored or coauthored 12 journal articles.

CONRAD T. SORENSON is presently working with ESTEK as a product marketing manager and consulting applications engineer. He has worked at Motorola in Austin and NCR in Fort Collins, CO. While at NCR he was a member of a team of process engineers that won an award for improving, among others, product quality. He received a B.S. degree in Physics from the University of Texas at Austin in 1979.

PETER STEWART is Head of Department of Pharmacy, University of Queensland, Australia where he received his Ph.D. in 1976. During the last ten years, he has researched in the area of solid mixing processes and particle interactions and has published some 50 papers and commissioned research reports. He has worked in the College of Pharmacy University of Iowa and was a visiting professor at Pharmacy Research at the Squibb Institute for Medical Research in New Jersey.

WILLIAM F. STICKLE is a Senior Laboratory Scientist in the Analytical Laboratory of the Perkin-Elmer Physical Electronics Division in Mountain View, CA. where he is using XPS, AES and SIMS to investigate and characterize material surfaces. He received his Ph.D. in chemistry from the University of Notre Dame, Notre Dame, IN.

GARVIN STONE is Associate Engineer at IBM in San Jose which he joined in 1982. He received his B.A. in aquatic biology from U.C. Santa Barbara. He is presently working on particle analysis in the Technology Assurance Laboratory.

MARK B. STUTMAN has been with the technical staff of W.L. Gore and Associates since 1984, providing research and marketing support to the Clean Room Products Group. He holds an M.S. in Climatology from the University of Delaware. He recently presented a paper on "Measuring Performance of Clean Room Garments."

S. SUMITA is Manager of Material Research Section of Semiconductor Research Laboratory, Kyushu Electronic Metal Co., Saga, Japan. Has Eng. Dr. degree and graduated from Kyushu University.

M. TAKESHITA is Member of Materials Research Section of Semiconductor Research Laboratory, Kyushu Electronic Metal Co., Saga, Japan. Graduated from Saga University.

DAVID R. TALLANT has been with Sandia National Laboratories in Albuquerque, NM since 1976. Since 1980 he has become involved in the areas of Raman spectroscopy, microRaman spectroscopy, waveguide Raman spectroscopy and pulsed laser fluorescence spectroscopy as applied to materials identification and structural characterization. He also has extensive experience in failure analyses for optical, mechanical and electrical components due to contamination and material degradation mechanisms. He received a Ph.D. in Analytical Chemistry in 1976 from the University of Wisconsin at Madison.

STEPHEN M. WALL has been an Aerosol Scientist with the Aerosol Physics Group, Air and Industrial Hygiene Laboratory, State of California Department of Health Services in Berkeley, CA since 1979. Also since 1979 he has been Research Scholar, Department of Chemical Engineering, University of California at Berkeley. He received his Ph.D. in Civil Engineering/Environmental Science from the University of California, Berkeley, and has been recipient of the Joseph E. Bonnheim Memorial Scholarship in Chemistry.

ALAN E. WALTER is co-founder of CFM technologies and has been involved in development of wet processing and wafer drying techniques for the past four years. Prior to joining CFM, he served as the project manager for high purity water systems at Crane Cochrane for seven years. He was awarded a BSChE degree from the University of Delaware. He is the holder of patents related to wet processing and drying of silicon wafers.

KENNETH J. WARD began working at Sandia National Laboratories in Albuquerque, NM in 1984 using FT-1R. He received his Ph.D. in Inorganic Chemistry from the University of Washington in Seattle in 1984, and during his graduate studies he was employed by the Lawrence Berkeley National Laboratory. Among his research interests, inorganic reaction kinetics continues to be a major research focus. In addition, he is developing combined instrumentation for analytical chemistry applications, specifically microscopy/FT-IR and gas chromatography/FT-IR, with particular emphasis in the use of quantitative spectroscopy with multivariate statistical analyses for the study of complex mixture characterization.

THOMAS M. WENTZEL is a graduate research assistant toward a Ph.D. in Physics at the University of Arizona where he received an M.S. in Physics in 1983. He has carried out research in adhesion forces between small particles since 1984 and worked in the field of solar Astrophysics for over two years before that.

CHARLES WESCHLER is presently a Distinguished Member of Professional Staff in the Environmental and Contaminants Research Group at Bellcore. He received his Ph.D. in Chemistry from the University of Chicago and joined Bell Laboratories in 1975. Since 1975 his research interests have centered on the chemical composition of airborne particles, as well as the chemical processes that are associated with these particles. He has been especially interested in the chemistry of the indoor environment, the chemistry of the outdoor environment as it impacts the indoor environment, and indoor-outdoor relationships.

CHARLES E. WILSON is Technical Specialist, Analytical Chemistry Laboratory, McDonnell Douglas Corp. in St. Louis. He received his B.S. degree in Chemistry in 1965 from Arkansas State University. He is currently working and directing activities in Fourier transform infrared spectroscopy. These activities include the development and application of analytical methods to solve material problems of a chemical nature associated with the development, production and operation of aerospace hardware.